*Economic Commission for Europe*
*Geneva*

# ECONOMIC SURVEY OF EUROPE

## 2001 No. 2

*Prepared by the*
SECRETARIAT OF THE
ECONOMIC COMMISSION FOR EUROPE
GENEVA

UNITED NATIONS
New York and Geneva, 2001

**NOTE**

The designations employed and the presentation of the material in this publication do not imply the expression of any opinion whatsoever on the part of the Secretariat of the United Nations concerning the legal status of any country, territory, city or area, or of its authorities, or concerning the delimitation of its frontiers or boundaries.

| UNITED NATIONS PUBLICATION |
|---|
| Sales No. E.01.II.E.26 |
| ISBN 92-1-116793-0 ISSN 0070-8712 |

Copyright © United Nations, 2001
All rights reserved
Printed at United Nations, Geneva (Switzerland)

# CONTENTS

| | Page |
|---|---|
| *Explanatory notes* .................................................................................................................................... | viii |
| *Abbreviations* ........................................................................................................................................... | ix |
| *About the main contributors to the Seminar* ............................................................................................ | xi |
| *Preface* ...................................................................................................................................................... | xii |

## Part One

### RECENT ECONOMIC DEVELOPMENTS AND THE SHORT-RUN OUTLOOK IN THE ECE REGION

| | | | Page |
|---|---|---|---|
| Chapter 1 | THE ECE ECONOMIES IN AUTUMN 2001 ..................................................................... | | 3 |
| | 1.1 | The global context and western Europe ............................................................... | 3 |
| | | (i) The global context ................................................................................ | 3 |
| | | (ii) Western Europe .................................................................................... | 10 |
| | 1.2 | The transition economies ...................................................................................... | 14 |
| | | (i) Introduction .......................................................................................... | 14 |
| | | (ii) Output and demand ............................................................................. | 16 |
| | | (iii) Costs and prices .................................................................................. | 24 |
| | | (iv) Labour markets .................................................................................. | 27 |
| | | (v) International trade .............................................................................. | 30 |
| | | (vi) Current accounts and external financing ........................................... | 36 |
| | 1.3 | The short-term oulook .......................................................................................... | 40 |
| | | (i) The western market economies ........................................................... | 40 |
| | | (ii) The transition economies .................................................................... | 43 |

## Part Two

### CREATING A SUPPORTIVE ENVIRONMENT FOR BUSINESS ENTERPRISE AND ECONOMIC GROWTH: INSTITUTIONAL REFORM AND GOVERNANCE
### Papers from the ECE Spring Seminar, May 2001

| | | | Page |
|---|---|---|---|
| INTRODUCTION AND SUMMARY OF DISCUSSION by the Economic Analysis Division, UNECE ....................... | | | 49 |
| Chapter 2 | PUBLIC AND CORPORATE GOVERNANCE: THE INSTITUTIONAL FOUNDATIONS OF THE MARKET ECONOMY | | |
| | *William Lazonick* ........................................................................................................................ | | 59 |
| | 2.1 | The triumph of the "market economy"? ............................................................... | 59 |
| | 2.2 | The theory of the market economy ...................................................................... | 61 |
| | 2.3 | Innovation and development in a market economy ............................................ | 63 |
| | 2.4 | The social conditions of innovative enterprise .................................................... | 65 |
| | 2.5 | Economic theory and economic development ..................................................... | 70 |
| | *Discussants' comments on Professor Lazonick's paper* | | |
| | 2.A | Victor Polterovich ................................................................................................ | 73 |
| | 2.B | Eugen Jurzyca ...................................................................................................... | 75 |

*Page*

| | | | | Page |
|---|---|---|---|---|
| *Chapter 3* | **INSTITUTIONAL CHANGE AND ECONOMIC PERFORMANCE IN THE TRANSITION ECONOMIES** | | | |
| | *Paul G. Hare* | | | 77 |
| | 3.1 | Introduction | | 77 |
| | | (i) | Political configuration | 78 |
| | | (ii) | State capacity | 78 |
| | | (iii) | State objectives | 78 |
| | | (iv) | Outline of paper | 79 |
| | 3.2 | Institutional change – concepts and definitions | | 79 |
| | | (i) | Definitions | 79 |
| | | (ii) | Typical institutions | 80 |
| | | (iii) | Economic behaviour with missing institutions | 80 |
| | | (iv) | Theories of institutional change | 81 |
| | 3.3 | Institutional change in transition economies | | 83 |
| | 3.4 | Links between institutional change and economic performance | | 86 |
| | | (i) | Macroeconomic studies | 86 |
| | | (ii) | Enterprise level studies | 88 |
| | 3.5 | Conclusions and policy implications | | 91 |
| | | (i) | Findings and conclusions | 91 |
| | | (ii) | Institutions | 91 |
| | | (iii) | Policy implications | 92 |
| | *Discussants' comments on Professor Hare's paper* | | | |
| | 3.A | Danica Popovic | | 95 |
| | 3.B | László Csaba | | 97 |
| *Chapter 4* | **CORRUPTION IN ECONOMIC DEVELOPMENT: GREASE OR SAND?** | | | |
| | *Shang-Jin Wei* | | | 101 |
| | 4.1 | Introduction | | 101 |
| | 4.2 | Corruption: which corruption? | | 101 |
| | | (i) | Corruption ratings based on "expert opinions" | 102 |
| | | (ii) | Corruption ratings based on surveys of firms or citizens | 102 |
| | | (iii) | Corruption ratings based on "a poll of polls" | 102 |
| | | (iv) | Corruption ratings based on more "objective" and "harder" data | 103 |
| | 4.3 | Corruption and economic growth | | 103 |
| | | (i) | Rate of economic growth | 103 |
| | | (ii) | Firm-level evidence | 104 |
| | | (iii) | Quality of growth | 106 |
| | 4.4 | Corruption in a globalizing world | | 107 |
| | | (i) | Corruption reduces the benefits of globalization: a specific example on transition economies | 107 |
| | | (ii) | Corruption increases the risks of globalization | 107 |
| | 4.5 | Corruption control: what can be done? | | 108 |
| | | (i) | Opportunities derived from government's role in the economy | 108 |
| | | (ii) | Civil servant recruitment and promotion system | 109 |
| | | (iii) | Compensation for civil servants | 109 |
| | | (iv) | Other elements in an anti-corruption strategy | 110 |
| | | (v) | International pressure | 111 |
| | | (vi) | Political economy considerations and "special governance zones" | 112 |
| | 4.6 | Concluding remarks | | 112 |
| | *Discussants' comments on Professor Wei's paper* | | | |
| | 4.A | Krassen Stanchev | | 113 |
| | 4.B | Ivo Bićanić | | 115 |

| | | | Page |
|---|---|---|---|
| Chapter 5 | GOVERNANCE IN TRANSITION: THE CHALLENGE OF SUBVERTING CORRUPTION | | |
| | *Antoni Z. Kamiński andBartlomiej Kamiński* | | 117 |
| | 5.1 Introduction | | 117 |
| | 5.2 Corruption, governance and transition | | 119 |
| | | (i) Corruption in post-communist societies and progress in transition | 119 |
| | | (ii) Strategic political choices during transition | 125 |
| | | (iii) Conclusion | 132 |
| | 5.3 Governance and combating corruption | | 133 |
| | | (i) Capture of policies by narrow interest groups: democracies versus electocracies | 135 |
| | | (ii) Combating administrative corruption | 137 |
| | | (iii) NGOs and independent media | 140 |
| | | (iv) Concluding comments | 141 |
| | 5.4 Corruption and development: policy convergence | | 141 |
| | 5.5 Conclusion | | 143 |
| | *Discussants' comments on Professors A. and B. Kamińskis' paper* | | |
| | 5.A Mark Pieth | | 144 |
| | 5.B Jorge Braga de Macedo | | 145 |

# Part Three

## *STATISTICAL APPENDIX*

| | |
|---|---|
| STATISTICAL APPENDIX | 149 |
| Corrections to *Economic Survey of Europe, 2001 No. 1* | 178 |

## LIST OF TABLES

| Table | | Page |
|---|---|---|
| 1.1.1 | Quarterly changes in real GDP, 2000 QI-2001 QII | 5 |
| 1.1.2 | Quarterly changes in real GDP and main expenditure items, 2001 QI-2001 QII | 7 |
| 1.2.1 | Basic economic indicators for the ECE transition economies, 1999-2002 | 15 |
| 1.2.2 | International trade and external balances of the ECE transition economies, 1999-2001 | 16 |
| 1.2.3 | GDP and industrial output in the ECE transition economies, 2000-2001 | 17 |
| 1.2.4 | Volume of retail trade in selected transition economies, 2000-2001 | 18 |
| 1.2.5 | Consumer prices, industrial producer prices and labour costs in industry in the ECE transition economies, 2000-2001 | 26 |
| 1.2.6 | Total employment and registered unemployment in the transition economies, 1998-2001 | 28 |
| 1.2.7 | Foreign trade of the ECE transition economies by direction, 1999-2001 | 31 |
| 1.2.8 | Changes in the volume of foreign trade in selected transition economies, 1997-2001 | 32 |
| 1.2.9 | CIS countries' trade with CIS and non-CIS countries, 1999-2001 | 36 |
| 1.2.10 | Current account balances of the ECE transition economies, 2000-2001 | 37 |
| 1.2.11 | Foreign direct investment in the ECE economies, 2000-2001 | 38 |
| 1.2.12 | Net capital flows into the ECE transition economies, 2000-2001 | 39 |
| 1.2.13 | Net capital flows into eastern Europe, the Baltic states and selected members of the CIS, by type of capital, 1999-2001 | 40 |
| 1.3.1 | Changes in the consensus forecasts of economic growth in 2001 and 2002 | 41 |
| 1.3.2 | Real GDP in the ECE market economies, 1999-2002 | 42 |
| 3.3.1 | Economic growth and inflation in selected transition economies, 1989-2000 | 84 |
| 3.3.2 | Progress with institutional reform in selected transition economies | 85 |
| 3.3.3 | Indicators of legal reform and effectiveness in selected transition economies | 86 |
| 4.2.1 | Measures of corruption in the transition economies of Europe and the CIS | 103 |
| 4.5.1 | How much should civil servants' legal pay be increased to lower corruption? | 110 |
| 5.2.1 | Measures of corruption and governance in the CEECs and the CIS | 122 |

| Table | | Page |
|---|---|---|
| A.1 | Real GDP in western Europe, North America and Japan, 1986-2000 | 151 |
| A.2 | Real private consumption expenditure in western Europe, North America and Japan, 1986-2000 | 152 |
| A.3 | Real general government consumption expenditure in western Europe, North America and Japan, 1986-2000 | 153 |
| A.4 | Real gross domestic fixed capital formation in western Europe, North America and Japan, 1986-2000 | 154 |
| A.5 | Real total domestic expenditures in western Europe, North America and Japan, 1986-2000 | 155 |
| A.6 | Real exports of goods and services in western Europe, North America and Japan, 1986-2000 | 156 |
| A.7 | Real imports of goods and services in western Europe, North America and Japan, 1986-2000 | 157 |
| A.8 | Industrial output in western Europe, North America and Japan, 1986-2000 | 158 |
| A.9 | Total employment in western Europe, North America and Japan, 1986-2000 | 159 |
| A.10 | Standardized unemployment rates in western Europe, North America and Japan, 1986-2000 | 160 |
| A.11 | Consumer prices in western Europe, North America and Japan, 1986-2000 | 161 |
| B.1 | Real GDP/NMP in eastern Europe, the Baltic states and the CIS, 1980, 1987-2000 | 162 |
| B.2 | Real total consumption expenditure in eastern Europe, the Baltic states and the CIS, 1980, 1987-2000 | 163 |
| B.3 | Real gross fixed capital formation in eastern Europe, the Baltic states and the CIS, 1980, 1987-2000 | 163 |
| B.4 | Real gross industrial output in eastern Europe, the Baltic states and the CIS, 1980, 1987-2000 | 164 |
| B.5 | Total employment in eastern Europe, the Baltic states and the CIS, 1980, 1987-2000 | 165 |
| B.6 | Employment in industry in eastern Europe, the Baltic states and the CIS, 1989-2000 | 166 |
| B.7 | Registered unemployment in eastern Europe, the Baltic states and the CIS, 1990-2000 | 167 |
| B.8 | Consumer prices in eastern Europe, the Baltic states and the CIS, 1990-2000 | 168 |
| B.9 | Producer price indices in eastern Europe, the Baltic states and the CIS, 1990-2000 | 169 |
| B.10 | Nominal gross wages in industry in eastern Europe, the Baltic states and the CIS, 1990-2000 | 170 |
| B.11 | Merchandise exports of eastern Europe, the Baltic states and the CIS, 1980, 1988-2000 | 171 |
| B.12 | Merchandise imports of eastern Europe, the Baltic states and the CIS, 1980, 1988-2000 | 172 |
| B.13 | Balance of merchandise trade of eastern Europe, the Baltic states and the CIS, 1980, 1988-2000 | 173 |
| B.14 | Merchandise trade of eastern Europe and the Russian Federation, by direction, 1980, 1988-2000 | 174 |
| B.15 | Exchange rates of eastern Europe, the Baltic states and the CIS, 1980, 1988-2000 | 175 |
| B.16 | Current account balances of eastern Europe, the Baltic states and the CIS, 1990-2000 | 176 |
| B.17 | Inflows of foreign direct investment in eastern Europe, the Baltic states and the CIS, 1990-2000 | 177 |

## LIST OF CHARTS

*Chart*                                                                               *Page*

| | | |
|---|---|---|
| 1.1.1 | Quarterly changes in real GDP in the developed market economies, 1999 QI-2001 QII | 5 |
| 1.1.2 | Monthly changes in industrial output, January 1995-August 2001 | 5 |
| 1.1.3 | International share prices, 4 January 1999 to 12 October 2001 | 6 |
| 1.1.4 | Business and consumer surveys in the United States, January 1995-September 2001 | 8 |
| 1.1.5 | United States labour market, January 2000-September 2001 | 8 |
| 1.1.6 | Consumer prices in the United States, January 2000-August 2001 | 9 |
| 1.1.7 | World commodity prices, January 1998-September 2001 | 9 |
| 1.1.8 | Business and consumer surveys in the European Union and the euro area, January 1995-September 2001 | 11 |
| 1.1.9 | Consumer prices in the euro area, January 2000-September 2001 | 12 |
| 1.1.10 | Standardized unemployment rates in the euro area and selected countries, January 1999-September 2001 | 12 |
| 1.1.11 | Short-term and long-term interest rates in the euro area, Japan and the United States, January 1999-September 2001 | 13 |
| 1.2.1 | Monthly dollar exports and imports in selected transition economies, 2000-2001 | 33 |
| 2.4.1 | Industrial, organizational and institutional conditions in the innovation process | 67 |
| 5.2.1 | Administrative corruption and state capture in transition economies | 123 |
| 5.2.2 | Progress in transition (governance) and corruption | 125 |
| 5.B.1 | G&G links to national economic performance | 145 |
| 5.B.2 | European institutional architecture | 147 |

## LIST OF BOXES

*Box*                                                                *Page*

| | | |
|---|---|---|
| 1.2.1 | Fiscal turmoil in Poland | 20 |
| 1.2.2 | The Elcoteq case | 22 |
| 5.2.1 | The 1997 Czech foreign exchange crisis and governance | 128 |
| 5.3.1 | Good economics and a corruption reducing measure: the case for a uniform tariff structure | 139 |

# EXPLANATORY NOTES

The following symbols have been used throughout this *Survey*:

- .. = not available or not pertinent
- – = nil or negligible
- \* = estimate by the secretariat of the Economic Commission for Europe
- | = break in series

In referring to a combination of years, the use of an oblique stroke (e.g. 1998/99) signifies a 12-month period (say, from 1 July 1998 to 30 June 1999). The use of a hyphen (e.g. 1998-2000) normally signifies either an average of, or a total for, the full period of calendar years covered (including the end-years indicated).

Unless the contrary is stated, the standard unit of weight used throughout is the metric ton. The definition of "billion" used throughout is a thousand million. The definition of "trillion" used throughout is a thousand billion. Minor discrepancies in totals and percentages are due to rounding.

References to dollars ($) are to United States dollars unless otherwise specified.

The membership of the United Nations Economic Commission for Europe (UNECE) consists of all the states of western Europe, eastern Europe and the territory of the former Soviet Union, North America and Israel.

The term *transition economies*, as used in the text and tables of this publication, refers to the formerly centrally planned economies of the ECE regions. *Eastern Europe* refers to the economies of Albania, Bosnia and Herzegovina, Bulgaria, Croatia, the Czech Republic, Hungary, Poland, Romania, Slovakia, Slovenia, The former Yugoslav Republic of Macedonia and Yugoslavia. The *Baltic states* refers to Estonia, Latvia and Lithuania and the *CIS countries* refers to Armenia, Azerbaijan, Belarus, Georgia, Kazakhstan, Kyrgyzstan, Republic of Moldova, Russian Federation, Tajikistan, Turkmenistan, Ukraine and Uzbekistan.

# ABBREVIATIONS

| | |
|---|---|
| BBC | British Broadcasting Corporation |
| BIS | Bank for International Settlements |
| CEE | central and eastern Europe |
| CEECs | central and east European countries |
| CEFTA | Central European Free Trade Agreement |
| CEO | Chief Executive Officer |
| CEPR | Centre for Economic Policy Research |
| CETE | central European transition economies |
| c.i.f. | cost, insurance and freight |
| CIS | Commonwealth of Independent States |
| CMEA | (former) Council for Mutual Economic Assistance |
| CPI | corruption perception index |
| EBRD | European Bank for Reconstruction and Development |
| ECB | European Central Bank |
| ECOSOC | Economic and Social Council (of the United Nations) |
| EMU | economic and monetary union |
| ESA | European System of Integrated Economic Accounts |
| EU | European Union |
| FCPA | Foreign Corrupt Practices Act |
| FDI | foreign direct investment |
| GATT | General Agreement on Tariffs and Trade |
| GCR | Global Competitiveness Report (corruption index) |
| GDP | gross domestic product |
| GDR | (former) German Democratic Republic |
| G&G | globalization and governance |
| GNP | gross national product |
| HS | Harmonized Commodity Description and Coding System |
| HWWA | Hamburg Institute for Economic Research |
| IATA | International Air Transport Association |
| ICRG | International Country Risk Guide (corruption index) |
| ICT | information and communications technology |
| IIF | Institute of International Finance, Inc. |
| ILO | International Labour Office |
| IMF | International Monetary Fund |
| INSEAD | The European Institute of Business Administration |
| IPO | initial public offering |
| LRS | license requiring state |
| M&A | mergers and acquisitions |

| | |
|---|---|
| NACE | Nomenclature générale des activités économiques dans les Communautés européennes (General Industrial Classification of Economic Activities within the European Communities) |
| NAPM | National Association of Purchasing Management |
| NATO | North Atlantic Treaty Organization |
| NBER | National Bureau of Economic Research, Inc. |
| NGO | non-governmental organization |
| NMP | net material product |
| OECD | Organisation for Economic Co-operation and Development |
| OPEC | Organization of the Petroleum Exporting Countries |
| OPT | outward processing trade |
| PHARE | Assistance programmes for the countries of central and eastern Europe (of the EU) |
| PMI | Purchasing Managers' Index |
| PPP | purchasing power parity |
| R&D | research and development |
| RFE/RL | Radio Free Europe/Radio Liberty |
| RPIX | retail price index excluding mortgage interest payments |
| SDR | special drawing right |
| SETE | south-east European transition economies |
| SGZ | special governance zone |
| SNA | System of National Accounts |
| TACIS | Technical Assistance for the Commonwealth of Independent States (of the EU) |
| TI | Transparency International |
| UNCTAD | United Nations Conference on Trade and Development |
| UNECE | United Nations Economic Commission for Europe |
| USSR | (former) Union of Soviet Socialist Republics |
| WDR | World Development Report (corruption index) |
| WTO | World Trade Organization |

## ABOUT THE MAIN CONTRIBUTORS TO THE SEMINAR

- *Ivo Bićanić,* University of Zagreb and Central European University, Budapest (iva.bicanic@zg.tel.hr)

- *Jorge Braga de Macedo,* OECD Development Centre, Paris (pamela.marqueyrol@oecd.org)

- *László Csaba,* Central European University and Budapest University of Economics and Public Administration, Budapest (csabal@ceu.hu)

- *Paul G. Hare,* Heriot-Watt University, Edinburgh (p.g.hare@hw.ac.uk)

- *Eugen Jurzyca,* Institute for Economic and Social Reforms and the Centre for Economic Development, Bratislava (jurzyca@ineko.sk)

- *Antoni Kamiński,* Institute of Political Science, Polish Academy of Sciences, Warsaw and Transparency International, Poland's Chapter (kaminski@omega.isppan.waw.pl)

- *Bartlomiej Kamiński,* University of Maryland, College Park, Maryland (bkaminski@worldbank.org)

- *William Lazonick,* University of Massachusetts Lowell and INSEAD (The European Institute of Business Administration), Fontainebleau (william.lazonick@insead.fr)

- *Mark Pieth,* University of Basle and Chairman of the OECD Working Group on Bribery in International Business Transactions (mark.pieth@unibas.ch)

- *Victor Polterovich,* Central Economics and Mathematics Institute, Russian Academy of Sciences and New Economic School, Moscow (victor@polter.mccme.rssi.ru)

- *Danica Popovic,* Univeristy of Belgrade, Belgrade (dpopovic@eunet.yu)

- *Krassen Stanchev,* Institute for Market Economics, Sofia (stanchev@ime.bg)

- *Shang-Jin Wei,* Harvard University and Brookings Institution, Washington (swei@brook.edu)

# PREFACE

The present *Survey* is the fifty-fourth in a series of annual reports prepared by the secretariat of the United Nations Economic Commission for Europe to serve the needs of the Commission and of the United Nations in reporting on and analysing world economic conditions.

Until 1997 the *Economic Survey of Europe* was issued once a year as was the *Economic Bulletin for Europe*, the secretariat's second publication, which focused on trade and payments issues. At its 52nd Session, in April 1997, the Commission decided to replace these two publications with an annual *Survey* of several issues. In 1998 and 1999 there were three issues each. There are now two issues a year published in April and November.

The Survey is published on the sole responsibility of the Executive Secretary of ECE and the views expressed in it should not be attributed to the Commission or to its participating governments.

The analysis in this issue is based on data and information available to the secretariat in late October 2001.

Economic Analysis Division
United Nations Economic Commission for Europe
Geneva

# PART ONE

# RECENT ECONOMIC DEVELOPMENTS AND THE SHORT-RUN OUTLOOK IN THE ECE REGION

# CHAPTER 1

# THE ECE ECONOMIES IN AUTUMN 2001

## 1.1 The global context and western Europe

### (i) The global context

#### (a) Increased downside risks and uncertainties

In early autumn 2001, the outlook for the global economy has become very worrisome. The terrorist attacks in New York and Washington on 11 September and the beginning of military reprisals in Afghanistan on 7 October 2001 have amplified the uncertainty that was already considerable in early September about the course of the United States economy and the world economy as a whole. At the time of the attacks, the United States economy and those in other major regions were on the edge of stagnation and short-term prospects appeared to be broadly balanced between the possibility of a moderate upturn and a slide into recession in the final months of 2001. The general effect of the attacks has been to worsen the economic outlook in the short term and possibly in the medium term as well.

It is still not possible to gauge with any reasonable degree of certainty all the short-term economic consequences of these shocks on economic activity let alone their possible longer-term implications. The immediate direct effects of the attacks on the United States economy – the loss of human lives, the destruction of physical assets (buildings, machinery and equipment) can be compared with those inflicted by a natural disaster such as an earthquake, i.e. a loss of productive capacity and a once-for-all reduction of effective demand.[1] To this should be added the adverse secondary consequences on domestic economic activity, especially on the air transport sector, tourism, hotels, restaurants and insurance companies, as well as disruptions to supply chains in manufacturing with subsequent negative multiplier effects throughout the economy. Rough estimates point to a possible total cost of $200 billion or 2 per cent of United States GDP. For comparison, the direct material damage of the Kobe earthquake in Japan in 1995 amounted to $120 billion or 2.5 per cent of Japan's GDP.[2] The upshot is that the magnitude of costs appears to be manageable both in absolute and relative terms given the enormous size of the United States economy.

But unlike an earthquake, the terrorist attacks and their aftermath may have a deeper and longer-lasting psychological effect to the extent that they have eroded people's sense of political and economic security.[3] A central question is therefore whether these attacks will have a sustained effect on consumer and business confidence, and thus on spending habits and on investors' perception of risk.[4] A related question is to what extent these factors can be offset by economic policy measures.

The secondary effects of the events of 11 September were not limited to the United States but spilled over to other countries. This particularly affected the global airline industry, which was faced with a sharp fall of demand and increased costs of security and insurance. It seems unlikely, however, that all the announced job losses[5] are directly related to the attacks: they also reflect long-standing structural problems that had to be addressed anyway.

Similarly, the perception of significantly increased risk by investors, and the associated preference for safe havens, has tended to accentuate a decline in net private capital flows to emerging markets and a rise in financing costs that were already underway in reaction to the crises in Argentina and Turkey and the weakening in global economic prospects.[6] More generally, the unusually large

---

[1] It has been estimated that the destruction and damage to property as a result of the attack on the World Trade Center together with the wider economic costs to the economy of New York City (including cleanup, security, business interruption, tax losses, etc.) could amount to $90-$105 billion. This corresponds to some 1 per cent of United States GDP in 2000. The City of New York, Office of the Comptroller, *The Impact of the September 11 WTC Attack on NYC's Economy and City Revenues,* 4 October 2001 (comptroller.nyc.ny.us).

[2] IMF, *World Economic Outlook* (Washington, D.C.), October 2001, p. 17, box 1.3 (www.imf.org).

[3] "In contrast to natural disasters, [the] ... events are of a far greater concern because they strike at the roots of our free society, one aspect of which is a market economy". Testimony of Chairman Alan Greenspan, *The Condition of the Financial Markets*, before the Committee on Banking, Housing and Urban Affairs, United States Senate, 20 September 2001 (www.federalreserve.gov).

[4] Peace and security are international public goods. "Where peace and security prevail, everyone can enjoy the fact that there is no war or threat of it, international travel and trade are unimpeded, people can go about their work without fear and the like". R. Mendez, "Peace as a global public good", in I. Kaul et al. (eds.), *Global Public Goods* (New York and Oxford, Oxford University Press, 1999), p. 388.

[5] According to IATA, some 200,000 jobs were shed in the global airline industry as a result of the events of 11 September. *Financial Times,* 10 October 2001.

[6] Institute of International Finance, Inc. (IIF), *Capital Flows to Emerging Market Economies* (Washington, D.C.), 20 September 2001 (www.iif.org).

degree of uncertainty, not least that pertaining to the length and scope of military reprisals and possible responses to them, could lead to an extended period of "wait-and-see" on the part of households and businesses outside the United States.

Another likely consequence of the attacks is an increase in transaction costs not only in the United States but also in the world economy at large due to stricter security measures and higher insurance premiums on cross-border trade and transport services. These could affect business productivity and the volume of international trade although orders of magnitude are difficult to gauge.[7] Also government spending on public security and military defence can be expected to increase. In a context of tight budget deficit targets this would have to be offset by cutbacks of other, mainly discretionary expenditure items, such as public investment, or entail the need to raise taxes.

### (b) A simultaneous cyclical downswing

A major feature of economic developments in 2001 (and already apparent before 11 September) has been the simultaneous weakening of economic activity in the major seven industrialized economies (chart 1.1.1), with concomitant adverse repercussions on other countries and regions of the global economy. In the United States and the euro area, real GDP rose by only 0.1 per cent between the first and second quarters of 2001. In western Europe, the increase was only slightly higher, and in Japan there was a sharp fall in economic activity over the same period (table 1.1.1). In the emerging markets of Asia and Latin America economic activity was already increasingly sluggish even before the attacks with several countries in or on the edge of recession. The transition economies of eastern Europe and the CIS remained surprisingly resilient to the general slowdown in the world economy in the first half of 2001, but by mid-year activity was starting to weaken, especially in central Europe.

Against this background, forecasts for world output growth in 2001 had already been lowered before the attacks of 11 September. In its September assessment of the global economic situation the IMF lowered its forecast for global growth to 2.6 per cent in 2001 from the 3.2 per cent expected in the spring.[8] Forecasts for world output growth in 2001 were lowered further in the aftermath of the attacks and now range from 1.5 to 2 per cent. This is very close to the previous lows of 1.4 per cent growth in 1991 and 1.1 per cent in 1982, the two preceding years of global recession. A consequence of the simultaneous cyclical weakening of economic activity across the globe has been a sharp deceleration in world trade. The volume of world trade in goods and services is now expected to increase, at best, by only 2 per cent in 2001, down from a growth rate of some 12.5 per cent in 2000.

With hindsight, it is clear that most economic forecasters underestimated the strength of the forces working towards a prolonged loss of cyclical momentum in the United States economy rather than a short V-shaped slowdown and recovery. This was largely due to a misreading of the implications of the nature of the economic upswing, which was fuelled by an investment boom in information and communications technology (ICT) goods and, related to that, an enormous stock market boom. Both booms ended in the first half of 2000. The ICT boom ended abruptly as companies (and investors) realized that the expansion of their supply capacities had by far outpaced effective demand. In other words, there was massive overinvestment. The stock market boom, long driven by irrational exuberance, collapsed under the pressure of weakening expectations of sales and profits. The economic boom led also to the accumulation of other imbalances such as the fall in personal savings to very low levels, high levels of private sector indebtedness, and a very large current account deficit. What is now underway is a process of capacity, inventory, and corporate and household balance sheet adjustments in the face of drastically changed asset valuations and economic prospects. This combination of circumstances helps to explain the lack of response of private sector spending to the significant easing of monetary policy by the Federal Reserve since the beginning of 2001.[9]

The impact of the abrupt weakening of economic growth in the United States on other parts of the world economy, notably western Europe, was also greatly underestimated. The United States cyclical weakening was not only transmitted via the traditional international trade channel but also via financial links between companies and countries, links which had been intensified by the recent surge of direct investment, much of it consisting of mergers and acquisitions, by west European companies in the United States. In a deteriorating economic situation, these financial links can have adverse spillover effects, via a deterioration of consolidated balance sheets, on investment both at home and abroad. Similarly, the sharp decline of activity in the global information and technology sectors illustrates how intricate supply linkages between companies can have powerful effects on economic activity in many regions of the world economy.[10] This holds especially for many Asian emerging markets and Japan, which have a strong specialization in information and communication technology goods. Furthermore, the increased perception of risk by investors, triggered by the financial crises in Argentina and Turkey and with concomitant reductions in capital flows to many emerging markets, aggravated the situation.

---

[7] S. Roach, "Back to borders", *Financial Times*, 28 September 2001.

[8] This latest forecast was made before but published after the September events and therefore does not take into account the possible adverse effects of 11 September. IMF, op. cit.

[9] UNECE, *Economic Survey of Europe, 2001 No. 1*, p. 5.

[10] BIS, *71st Annual Report* (Basle), 11 June 2001, p. 143.

## CHART 1.1.1
**Quarterly changes in real GDP in the developed market economies, 1999 QI-2001 QII**
(Percentage change over same period of previous year)

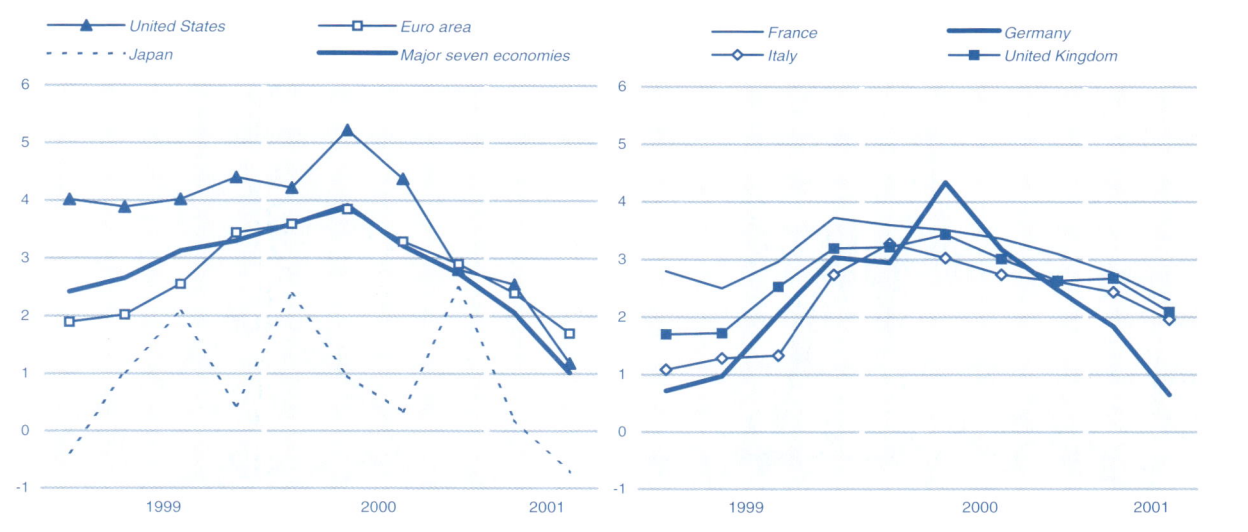

*Source:* National statistics; Eurostat, New Cronos Database.
*Note:* Based on seasonally adjusted data. Major seven economies: Canada, France, Germany, Italy, Japan, United Kingdom and United States.

## TABLE 1.1.1
**Quarterly changes in real GDP, 2000 QI-2001 QII**
(Percentage change over previous quarter)

|  | 2000 QI | 2000 QII | 2000 QIII | 2000 QIV | 2001 QI | 2001 QII |
|---|---|---|---|---|---|---|
| France | 0.7 | 0.7 | 0.8 | 0.8 | 0.4 | 0.3 |
| Germany | 1.0 | 1.2 | 0.1 | 0.2 | 0.4 | – |
| Italy | 0.9 | 0.5 | 0.4 | 0.8 | 0.8 | – |
| United Kingdom | 0.4 | 0.8 | 0.7 | 0.5 | 0.6 | 0.4 |
| Canada | 1.5 | 0.5 | 1.1 | 0.4 | 0.5 | 0.1 |
| United States | 0.6 | 1.4 | 0.3 | 0.5 | 0.3 | 0.1 |
| Japan | 2.4 | 0.1 | -0.7 | 0.6 | 0.1 | -0.8 |
| **7 major economies** | 1.0 | 0.9 | 0.2 | 0.5 | 0.4 | -0.1 |
| *Memorandum items:* | | | | | | |
| **Euro zone** | 0.9 | 0.8 | 0.5 | 0.6 | 0.5 | 0.1 |
| **European Union** | 0.8 | 0.8 | 0.5 | 0.6 | 0.5 | 0.1 |
| **Western Europe** [a] | 0.8 | 0.8 | 0.5 | 0.6 | 0.5 | 0.2 |
| Western Europe, North America and Japan | 1.0 | 0.9 | 0.3 | 0.5 | 0.4 | – |

*Source:* Eurostat, New Cronos Database; national statistics.
*Note:* Data are seasonally adjusted.
[a] Western Europe: European Union plus Norway and Switzerland.

## CHART 1.1.2
**Monthly changes in industrial output, January 1995-August 2001**
(Indices, 1995=100)

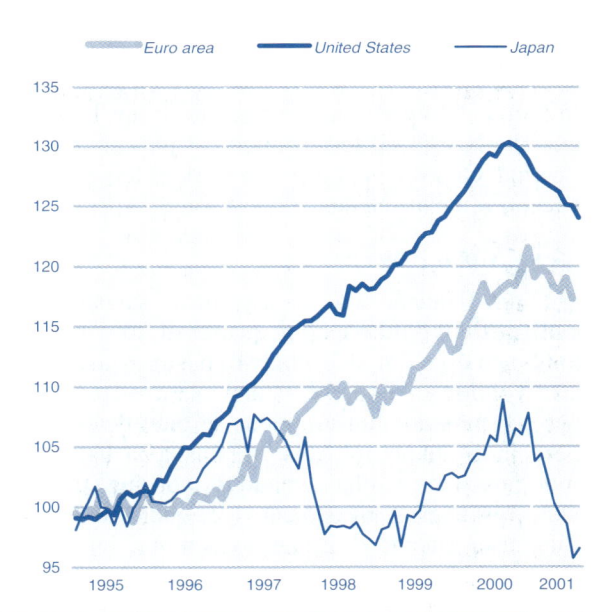

*Source:* OECD, *Main Economic Indicators* (Paris), various issues; Eurostat, New Cronos Database.

The weakening of cyclical economic growth in the rest of the world, in turn, has had negative feedback effects not only on the United States economy, but also across the major regions themselves, leading to a mutually reinforcing and synchronized downswing. This is most visible in a sharp industrial recession in the developed market economies, which has spilled over to the rest of the world economy (chart 1.1.2). This weakening of industrial activity, largely reflecting the fall in demand for information and communication technology products, was increasingly felt in the services sector and thereby accentuated the cyclical downswing.

Moreover, the sharp fall in international equity prices, already advanced before September 2001 (chart 1.1.3) entailed a sizeable loss of net financial wealth,

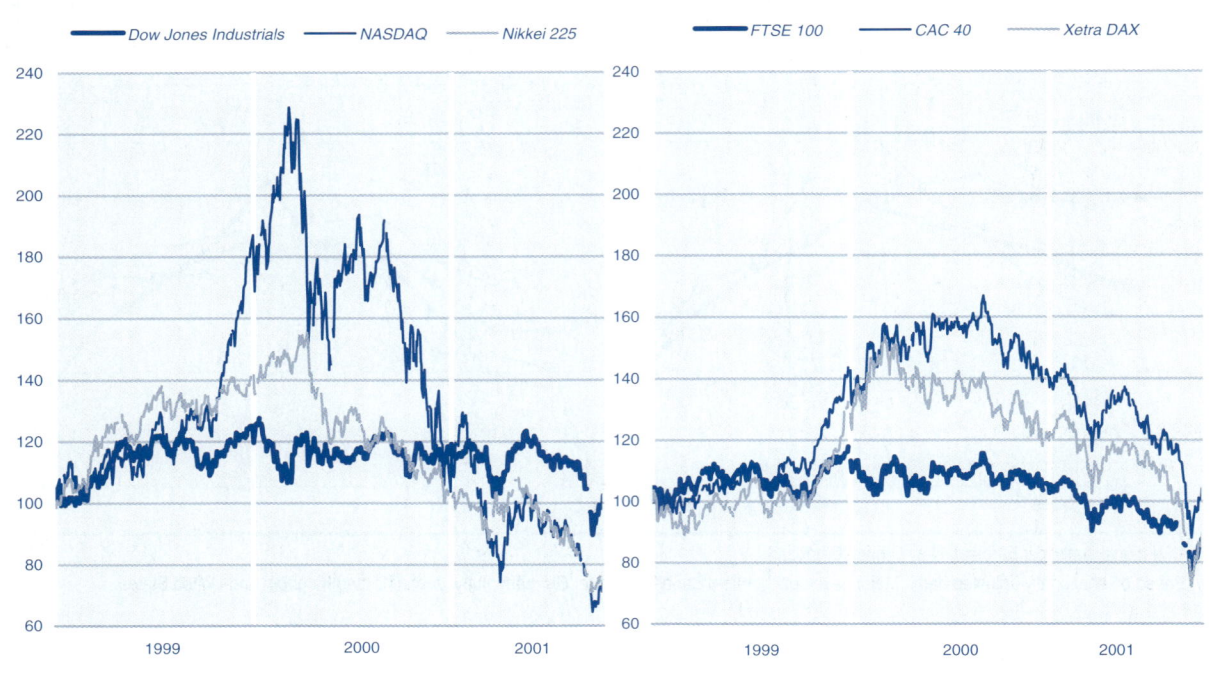

CHART 1.1.3

**International share prices, 4 January 1999 to 12 October 2001**
(Indices, 4 January=100)

Source: Bloomberg L.P. (www.bloomberg.com).
Note: Daily values excluding national holidays.

which dampened private consumption expenditures. By mid-October 2001 the Dow Jones Industrial Average index had lost around 14 per cent compared to its level one year earlier. The NASDAQ fell by about 54 per cent over the same period. On average, European markets fell by around 25 per cent, while the Nikkei 225 lost some 34 per cent.

### (c) North America

In the United States, the near stagnation of economic activity in the second quarter of 2001 reflected the combined effect of sharp falls in business investment and the volume of exports, which were offset by the strength of private consumption despite deteriorating prospects in the labour markets. Against a background of low mortgage rates, strong demand for housing continued to buoy residential investment. Higher government spending also supported domestic activity, but all the other major components of domestic demand contracted. Business fixed investment fell sharply in the face of falling corporate profits, large margins of excess capacity in industry and weakening sales prospects. The decline of spending on information processing equipment and software accelerated, but there was also a large cut of investment in non-residential buildings and the process of reducing excess inventories continued to be a drag on economic growth. The weakening of global economic activity was reflected in a marked decline in exports in the second quarter with a further negative impact on manufacturing industry. At the same time, stagnating domestic demand in the United States depressed imports from the rest of the world with the result that changes in real net exports were neutral in their effect on economic growth (table 1.1.2).

Short-term economic indicators, reflecting activity prior to the September attacks, point to continued weakness of economic activity in the third quarter of 2001.[11] Industrial output continued to fall in September for the twelfth consecutive month and capacity utilization rates fell to their lowest level since the end of the 1960s. The Purchasing Managers' Index continued to signal sluggish activity in the manufacturing sector and the economy as a whole. Personal spending turned increasingly sluggish suggesting that a large proportion of private households were unwilling to spend all of their tax rebates. Consumer confidence dropped sharply in September 2001 to its lowest level since September 1995 (chart 1.1.4). In fact, this was the largest monthly decline since October 1990, the time of the Gulf crisis. While the fall was likely to have been influenced by the attacks on 11 September, the survey results were already pointing to a sharp downward trend before then.[12] But the

---

[11] Federal Reserve Board, *Beige Book* (Washington, D.C.), 19 September 2001 (www.federalreserve.gov).

[12] The Conference Board, *Consumer Confidence Index Drops Sharply*, Press Release, 25 September 2001 (www.conference-board.org).

### TABLE 1.1.2
**Quarterly changes in real GDP and main expenditure items, 2001 QI-2001 QII**
*(Percentage change over preceding quarter)*

|  | Western Europe Total[a] QI | QII | Euro area QI | QII | United States QI | QII | Japan QI | QII | Western Europe, North America and Japan QI | QII |
|---|---|---|---|---|---|---|---|---|---|---|
| Private consumption | 0.8 | 0.5 | 0.8 | 0.4 | 0.7 | 0.6 | 0.6 | 0.5 | 0.8 | 0.5 |
| Government consumption | 0.5 | 0.2 | 0.6 | 0.2 | 1.3 | 1.2 | – | 0.8 | 0.8 | 0.7 |
| Gross fixed investment | -0.3 | -0.2 | 0.1 | -0.4 | 0.5 | -2.5 | – | -4.0 | 0.1 | -1.7 |
| Final domestic demand | 0.5 | 0.3 | 0.6 | 0.2 | 0.8 | 0.2 | 0.3 | -0.7 | 0.6 | 0.1 |
| Stockbuilding[b] | -0.3 | – | -0.6 | 0.2 | -0.8 | -0.1 | – | – | -0.4 | – |
| Total domestic demand | 0.2 | 0.3 | – | 0.3 | 0.1 | – | 0.3 | -0.7 | 0.2 | 0.1 |
| Net exports[b] | 0.2 | -0.2 | 0.5 | -0.2 | 0.2 | – | -0.2 | -0.1 | 0.2 | -0.1 |
| Exports | 0.4 | -1.0 | 0.2 | -0.8 | -0.3 | -3.1 | -3.6 | -2.9 | -0.6 | -2.1 |
| Imports | -0.5 | -0.5 | -1.2 | -0.2 | -1.3 | -2.2 | -2.1 | -2.5 | -1.1 | -1.5 |
| GDP | 0.5 | 0.2 | 0.5 | 0.1 | 0.3 | 0.1 | 0.1 | -0.8 | 0.4 | – |

*Source:* Eurostat, New Cronos Database; national statistics.

[a] Seventeen countries: EU15 plus Norway and Switzerland.

[b] Growth contribution to change in real GDP (percentage points).

heightened uncertainty and associated psychological stress of households contributed to a slump in retail sales in September.

The number of workers on non-farm payrolls fell by nearly 200,000 between August and September – the largest monthly decline since the recession of 1990-1991. The unemployment rate was 4.9 per cent, the same as in August but up from 4.2 per cent at the beginning of the year (chart 1.1.5). The announcements of massive layoffs in the airline, financial and ICT sectors, together with the worsening economic outlook, are bound to lead to a further rise in unemployment. This will tend to further erode consumer confidence, already weakened by deteriorating economic conditions, including the fall in financial wealth, which will have adverse effects on households' balance sheets and on their ability to service their debts.

Against the backdrop of weak economic activity, easing labour market pressures and falling energy prices, inflation remained weak in the third quarter. The consumer price index in September was 2.7 per cent higher than 12 months earlier, down from a recent peak of 3.6 per cent in May (chart 1.1.6).

In Canada, the economy was also increasingly affected by the global downturn, and especially by the weakness of the United States economy, its main trading partner. Real GDP rose by only 0.1 per cent between the first and second quarters of 2001 reflecting a weaker growth of private household expenditures and a fall in exports. The sharp deceleration in the rate of growth has started to push up the unemployment rate, which rose to 7.2 per cent in August 2001. Inflation peaked at 3.9 per cent in May 2001 but fell back to 2.8 per cent (i.e. below the 3 per cent ceiling of the central bank's target range) in August. This was largely attributed to falls in gasoline and natural gas prices, but it also partly reflects the overall weakening of domestic demand.

### (d) Japan

Real GDP fell sharply in the second quarter of 2001 and the economy probably moved into recession in the third quarter. This would be the fourth recession in a decade. This outcome reflects large falls in business investment and exports, which in turn, reflect the overall deterioration of the global economy (table 1.1.2). Exports were particularly hard hit by the global slump in demand for information and communications technology products. Changes in real net exports have been subtracting from overall economic growth since the third quarter of 2000, while consumer spending has remained lacklustre in the presence of shrinking employment and stagnating real income. Corporate restructuring and an increasing number of bankruptcies raised the unemployment rate to a record 5 per cent in July 2001 (and it remained unchanged in August). Industrial output in August 2001 was nearly 12 per cent lower than a year earlier. Deflationary pressures continued to prevail, with both consumer and producer prices falling. Real estate prices have been declining for 10 years. Share prices (Nikkei 225) in mid-September 2001 reached their lowest level in 18 years.

Monetary policy was eased again in August and September, but as interest rates were already close to zero there were no indications of any expansionary effects (the "liquidity trap"). The Bank of Japan also conducted non-sterilized interventions in the foreign exchange market to stem an appreciation of the yen against the dollar. Faced with the threat of a new recession the government introduced a new fiscal spending package in early October. The worsening of the economic outlook could lead to the postponement of previously announced plans

### CHART 1.1.4
**Business and consumer surveys in the United States, January 1995-September 2001**
*(Indices)*

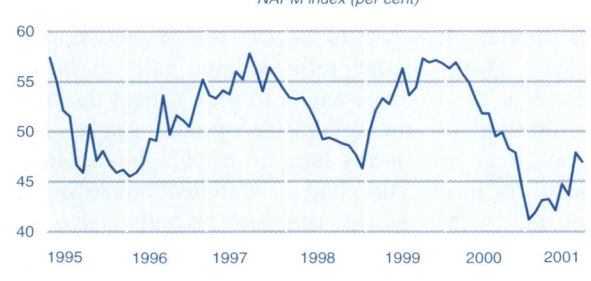

*Source:* The Conference Board (www.conference-board.org) and National Association of Purchasing Management (NAPM) (www.napm.org).

*Note:* The Purchasing Managers' Index (PMI) is a composite index pertaining to the business situation in manufacturing industry. An index value above (below) 50 per cent indicates that manufacturing industry is generally expanding (contracting). A PMI above (below) 44.5 per cent, over a period of time, indicates that overall economic activity, as measured by real GDP, is generally expanding (contracting).

### CHART 1.1.5
**United States labour market, January 2000-September 2001**
*(Per cent of civilian labour force, index)*

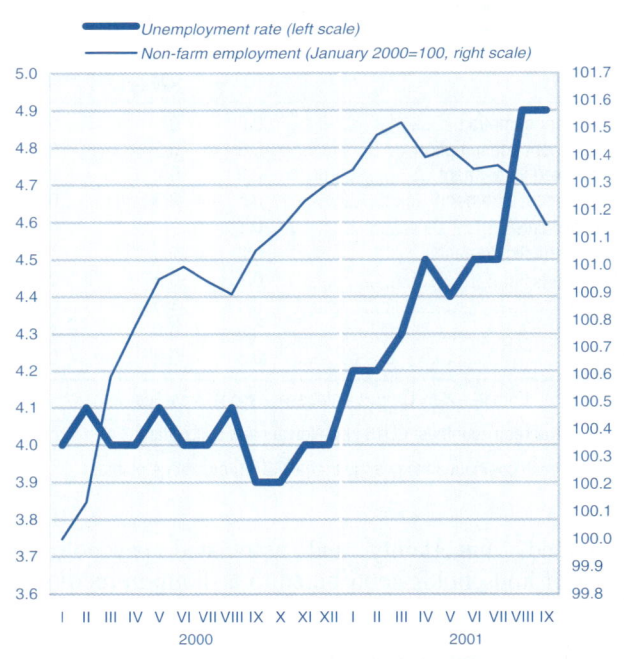

*Source:* United States Bureau of Labor Statistics (www.bls.gov).

for the fiscal year starting in April 2002 to reverse the rise in government debt which amounted to some 130 per cent of GDP in 2000. The government has also announced a programme of structural reforms designed, *inter alia*, to solve the long-standing problem of non-performing loans accumulated in the banking sector.

### (e) Emerging markets of Asia and Latin America

The weakening of world economic activity in 2001 had a negative impact on emerging market economies, mainly through the channels of foreign trade and international capital flows. The economic slowdown accentuated the risk aversion of international investors, which had already increased after the crises in Argentina and Turkey in the first half of the year. This was reflected in efforts to limit exposure to emerging markets and the repatriation of funds in a search for quality and safety. Both wider yield spreads and reduced private capital inflows signalled more difficult external financing conditions. This was particularly the case for countries more dependent on external financing, such as Argentina, Brazil, the Philippines and Turkey. All these tendencies strengthened considerably in the aftermath of the terrorist attacks of September, further accentuating risk aversion.

As a result of all these factors, total private capital flows to emerging market economies are forecast to fall from $167 billion in 2000 to $106 billion in 2001. While foreign direct investment (FDI) flows to developing countries are projected to decline only slightly in 2001, private credit is expected to turn from an inflow of about $20 billion to an outflow of a similar magnitude.[13] Most of the shortfall should be offset by public institutions, chiefly the IMF.

The east Asian emerging markets were strongly affected by the global crisis in the information and telecommunication sectors. Falling exports of electronic goods, which account for more than half of total exports from many of these countries, contributed to a sharp slowdown in the rate of economic growth, which in some countries (Singapore, Taiwan) turned into a full-fledged

---

[13] IIF, op. cit.

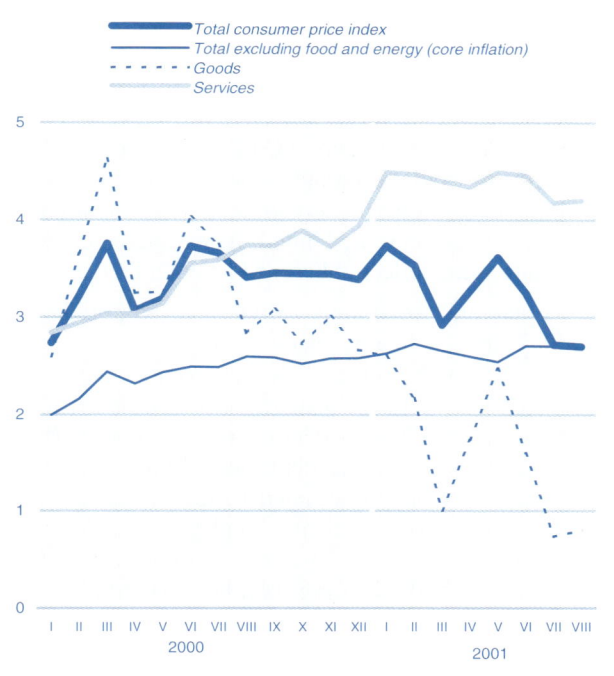

**CHART 1.1.6**
**Consumer prices in the United States, January 2000-August 2001**
*(Percentage change over same month of previous year)*

Source: United States Bureau of Labor Statistics (www.bls.gov).

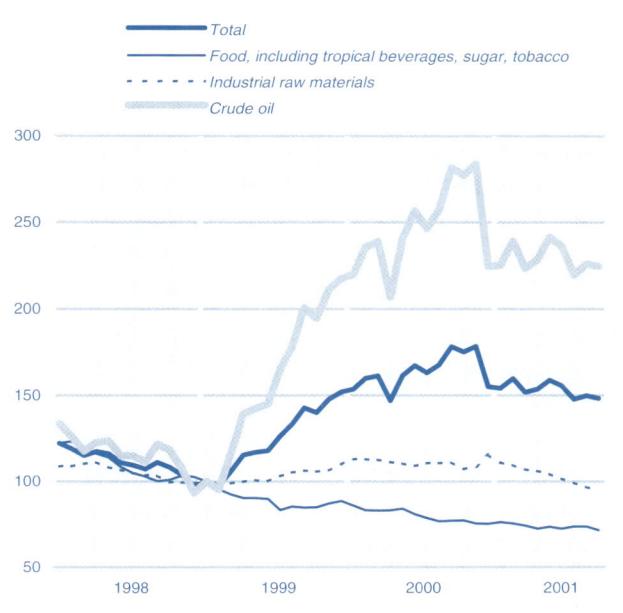

**CHART 1.1.7**
**World commodity prices, January 1998-September 2001**
*(Indices, 1999=100)*

Source: Hamburg Institute for Economic Research (HWWA).
Notes: Indexes calculated on the basis of current dollar prices.

recession that the authorities tried to counter with expansionary monetary and fiscal policies. The deteriorating economic outlook has raised concerns about the resilience of financial systems in the region in view of the large increases in public debt and non-performing loans. In contrast, the two very large and relatively closed economies of China and India appear so far to have been less affected by the downturn in international trade.

Economic growth also slowed sharply in Latin America. A major concern remains the economic and financial situation in Argentina, which is experiencing the third consecutive year of recession brought about largely by depressed domestic demand and an overvalued peso, which has been pegged to the dollar in the framework of a currency board since 1991. Fears of a public debt default and/or exit from the currency board arrangement led to capital flight and the withdrawal of funds by international private creditors. The yield spread over United States treasury bonds has surged to very high levels (nearly 19 percentage points in mid-October 2001), which effectively bars access to international capital markets. Against this backdrop, the IMF has arranged for a new support programme involving stringent fiscal austerity. The principal motive for this is that a debt default could have significant spillover effects in the region and lead to contagion of other emerging markets. The future solvency of the government depends on its capacity to reschedule its existing debt, backed by additional loans from official foreign institutions. The instability in Argentina spilled over to Brazil, given the two countries' strong economic links. A large current account deficit in the face of declining FDI led to tension in the foreign exchange market and a sizeable cumulative depreciation of the national currency. A tight monetary policy, designed to defend the currency, has added to the contractionary effects of a domestic energy crisis and of a restrictive fiscal policy adopted in the context of an IMF-supported programme. The IMF agreed to a new stand-by loan to Brazil in August 2001, designed to limit the adverse financial market spillovers from the crisis in Argentina.

### (f) Commodity prices

Primary commodity prices during the first nine months of 2001 came under downward pressure from the strong downturn in world economic activity. The lowering of growth expectations for the world economy after 11 September accentuated the weakness of demand. In the event, average commodity prices in September 2001 were 17 per cent below their recent cyclical peak of one year earlier (chart 1.1.7).

The price of crude oil (OPEC basket) fluctuated within OPEC's target range of $22-$28 between January and August 2001. In order to offset the downward pressure on prices, OPEC reduced its target output three times between February and September, resulting in a cumulative 11 per cent cut in OPEC's oil supply. The effects of the terrorist attacks in September on the global airline sector and the deteriorating global economic outlook led to a sharp fall in oil prices, the price of Brent crude falling from $27 per barrel in August to $21 per barrel in mid-September 2001. OPEC has stated its intention to supply enough oil to prevent a surge in prices, but it is now considering cuts in order to sustain prices. However, if military tensions were to spread to other countries, oil prices might eventually rise steeply.

Prices for industrial raw materials (essentially base metals and agricultural raw materials) fell by 15 per cent in the 12 months to September 2001. Even before 11 September, it was projected that world consumption of base metals in 2001 would experience its largest fall since 1982[14] with prices dropping to levels not seen in many years. Thus, the price of rubber in September reached its lowest level in over 15 years. Although the markets for temperate-zone food commodities are more balanced, prices for tropical products (particularly coffee and sugar) were squeezed by oversupply and high levels of stocks.

### (ii) Western Europe

In the euro area, economic growth virtually came to a halt in the second quarter of 2001. The slight rise in GDP conceals falling activity in industry, which was only just offset by continued growth in the service sector. Viewed from the demand side, the slowdown mainly reflected a pronounced weakening of the growth of private consumption and the effects of the deteriorating external environment on exports, which declined markedly between the first and second quarters (table 1.1.2). Fixed investment also fell against a background of weakening industrial confidence, increased margins of idle capacity and a bleaker outlook for sales and profits. The general deterioration in economic performance also depressed consumer confidence (chart 1.1.8).

Although there continued to be considerable differences between individual countries, there was a general weakening in the second quarter of 2001. Quarterly growth rates ranged from 0.5 per cent in Spain to a sharp fall by 1.7 per cent in Finland. (This was the second consecutive fall in quarterly GDP in Finland, the technical definition of recession.) Among the three largest economies, economic activity stagnated in Germany and Italy, and rose only slightly in France (table 1.1.1 above). In Germany and Italy, income tax cuts continued to support private consumption, but fixed investment fell and changes in real net exports also depressed domestic activity. Growth in Germany continued to be restrained by the persistent crisis in the construction sector, which is now in the process of reducing capacity after the post-unification boom.[15] In France, the growth of private consumption slowed markedly between the first and second quarters of 2001, while fixed investment stagnated. Government consumption and changes in stockbuilding also supported domestic demand but, as in Germany and Italy, the change in real net exports was a drag on overall economic growth.

The available short-term economic indicators suggest that economic activity in the euro area remained weak in the third quarter of 2001. The service sector appears to be increasingly affected by the recession in industry. Consumer and business confidence have continued to fall, although the latest figures for September do not provide any indication of the impact of the terrorist attacks on the United States since most of the survey data were collected before 11 September.

Weakening economic activity and the fading of temporary special factors (such as the effects of animal diseases on food prices and the temporary surge in oil prices) led to a fall in the headline inflation rate from 3.4 per cent in May to 2.5 per cent in August. Core inflation has levelled off at 2.1 per cent since May (chart 1.1.9) and labour costs have continued to rise moderately. In view of the weak cyclical conditions, inflation is expected to approach the 2 per cent ceiling of the ECB's inflation target in the coming months. Conditions in the labour markets – a lagging cyclical indicator – were also increasingly influenced by the sharp deceleration of economic growth. The fall in the unemployment rate appears to have bottomed out in the summer at 8.3 per cent, its lowest level in almost a decade. The preceding cyclical upswing led to a marked fall in the unemployment rate in France to 8.5 per cent in August, down from 11.7 per cent in January 1999 (chart 1.1.10). In Spain, the rate fell from 17.2 per cent to 13 per cent over the same period.

West European economies outside the euro area could not entirely escape from the dampening effects of the global economic slowdown. Nevertheless, the growth of real GDP held up better in the United Kingdom than in the euro area. The growth of real GDP slowed to 0.4 per cent in the second quarter (table 1.1.1) to attain a level 2.3 per cent higher than in the same period of 2000. The private consumption boom, underway since 1998, remained the mainstay of economic growth. Household spending was supported by further gains in real disposable incomes associated with higher levels of employment. In addition, consumer demand was stimulated by a conducive environment for personal borrowing created by soaring house prices and low interest rates. Business investment picked up after a decline in the first quarter. Economic activity was also supported by strong growth in government spending. The strength of domestic demand contrasts with the sluggishness of the tradeables sector, which was hit by the strong exchange rate of sterling and the global cyclical downturn (the "two-speed economy"). Exports fell sharply, but the effect on aggregate domestic output was offset by a contraction of imports. Inflation (RPIX) has remained moderate but, due to special factors, edged slightly above the government's target of 2.5 per cent in August 2001. Inflation has been below the central

---

[14] E. Fam, "Les marchés mondiaux de matières premières au deuxième trimestre 2001", *Bulletin de la Banque de France*, No. 92, (Paris), August 2001.

[15] Construction investment fell by 55 per cent in the first half of 2001 compared with the same period of 1995.

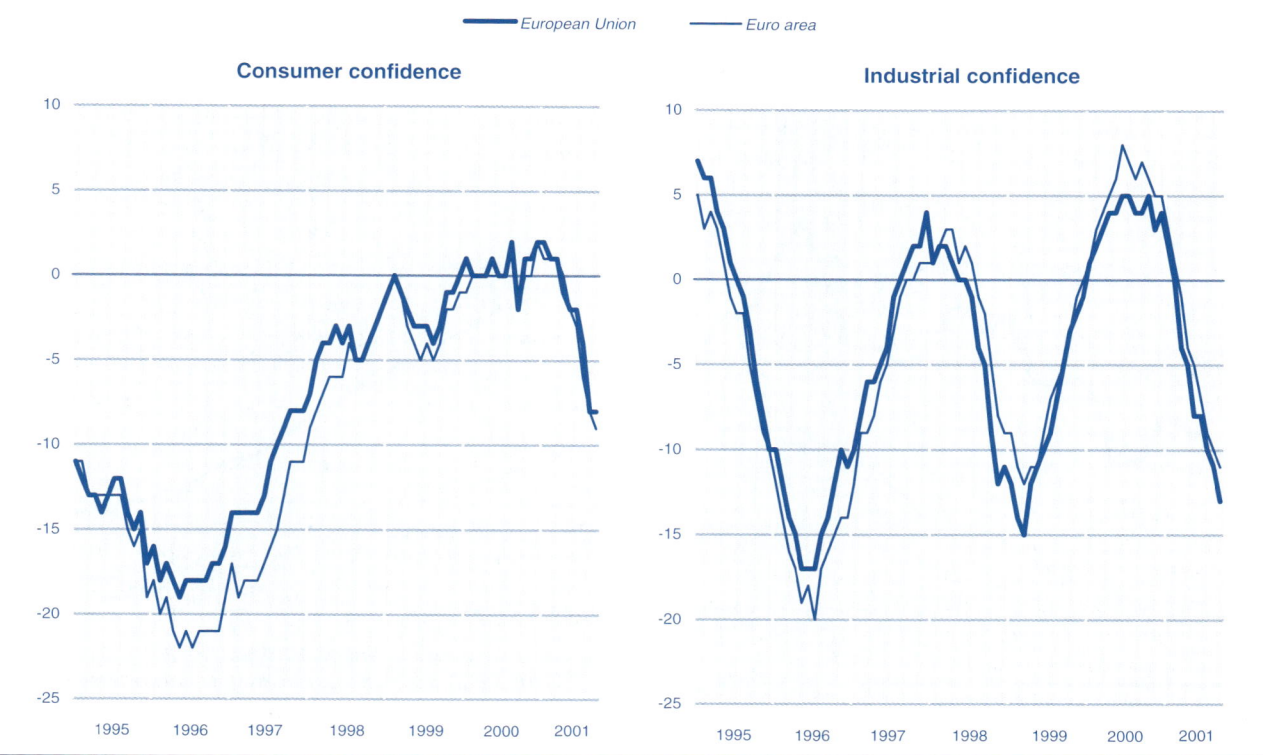

CHART 1.1.8

Business and consumer surveys in the European Union and the euro area, January 1995-September 2001

Source: Commission of the European Communities, *European Economy*, Supplement B (Luxembourg), various issues and direct communications.

Note: European Union data show net balances between the percentages of respondents giving positive and negative answers to specific questions. For details see any edition of the source.

bank's target of 2.5 per cent for more than two years. Developments in the labour markets remained favourable, with further gains in employment and unemployment falling to its lowest level for over a quarter of a century.

In Turkey, the economy is currently in deep recession. The financial-cum-exchange rate crisis of February 2001 and the subsequent free floating of the currency have led to a sharp depreciation, which amounted to some 55 per cent in early October. The resulting stimulus to net exports, however, was more than offset by the erosion of consumer and business confidence, which not only depressed domestic demand but also limited the ability of the government to borrow at the very high interest rates resulting from the increased risk aversion of domestic and international investors. The government is therefore dependent on financial support from the international financial institutions, notably the IMF. Real GDP is forecast to decline by 6 per cent in 2001, but the outcome could very well be worse.

### *Macroeconomic policy stances*

The drastic change in the economic outlook, especially since 11 September, has led in general to a more expansionary stance of monetary and fiscal policies. In the United States, monetary policy had already been eased significantly before 11 September 2001.[16] Also the federal income tax rebate was expected to support private consumption in the final months of 2001.[17] More or less in tandem with the Federal Reserve, there was also a progressive easing of monetary policy in Canada.

In the euro area, the ECB reacted more hesitantly to the increasing signs of cyclical weakening having assumed for too long that the euro area would be largely immune to developments in the United States. The ECB also felt constrained by the fact that headline inflation was significantly above target, although this was largely acknowledged to reflect transitory factors.[18] But in the face of increasingly sluggish economic activity in the euro area and forecasts of declining inflation, the ECB decided to lower the minimum bid rate on its main refinancing operations in two steps by half a percentage point to 4.25 per cent between early May and late August 2001.

---

[16] The Federal Reserve lowered the target for the federal funds rate in seven steps from 6.5 per cent at the beginning of January 2001 to 3.5 per cent in August 2001.

[17] Households received cheques ranging from $300 to $600 between late July and late September 2001, totalling some $40 billion or slightly less than 0.5 per cent of GDP in 2000.

[18] UNECE, *Economic Survey of Europe, 2001 No. 1*, chap. 1.

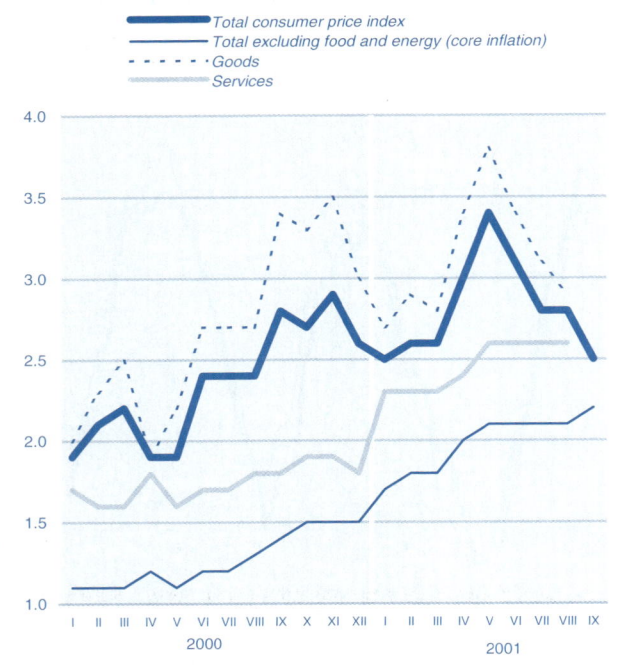

**CHART 1.1.9**

**Consumer prices in the euro area, January 2000-September 2001**
*(Percentage change over same month of previous year)*

Source: Eurostat, New Cronos Database.
Note: September data for goods and services not yet available.

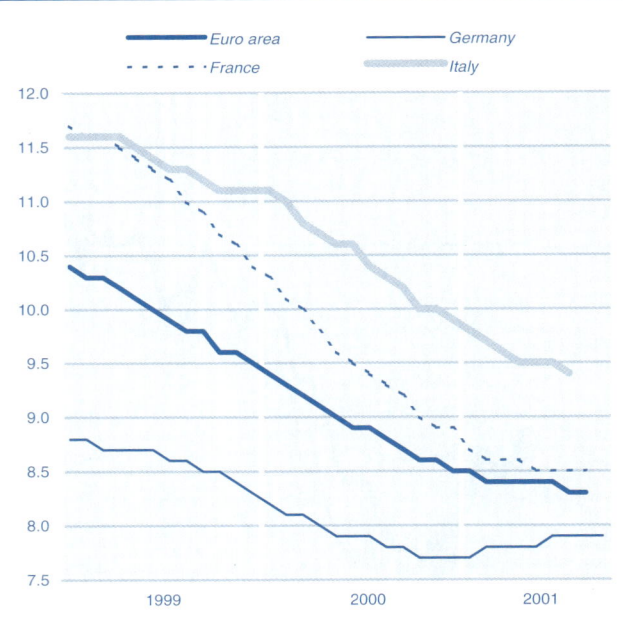

**CHART 1.1.10**

**Standardized unemployment rates in the euro area and selected countries, January 1999-September 2001**
*(Per cent of civilian labour force, seasonally adjusted)*

Source: OECD, *Main Economic Indicators* (Paris), various issues.
Note: Data for the euro area and France only available until August 2001; for Italy only until July 2001.

In the spring of 2001, the average fiscal policy stance in the euro area was projected to be mildly expansionary in 2001. Several countries enacted tax cuts (France, Germany, Italy, the Netherlands), but sizeable fiscal easing was projected only in Finland, Germany, Ireland and the Netherlands.[19] Weaker than expected economic growth and the working of the automatic stabilizers has undermined the ability of countries to achieve their budgetary targets for this year which were set in line with the requirements of the Stability and Growth Pact. This has created problems for France, Germany, Italy and Portugal where even under the optimistic growth assumptions of earlier in the year government budgets were not expected to be "close to balance or in surplus" over the medium term as required by the rules of the Stability and Growth Pact. In principle, therefore, these countries were expected to limit the deviations of the actual deficit from the target fixed in their stability programmes or, in other words, to prevent the automatic stabilizers from operating freely in view of the risk that the deficits could approach the 3 per cent ceiling.[20] But this has created a dilemma for policy since restricting the automatic fiscal stabilizers would have a procyclical effect and therefore risk accentuating the downturn.[21]

In the United Kingdom, against a background of concern about the effects of deteriorating economic conditions in the United States and continental Europe on the domestic economy, the Bank of England lowered its key interest rate in four steps from 6 per cent in early February 2001 to 5 per cent in early August. Economic activity was also supported by an expansionary fiscal policy.

In the immediate aftermath of the attacks on 11 September, the Federal Reserve together with the ECB and other major central banks ensured the normal functioning of the banking system by injecting additional liquidity into the money markets and by conducting foreign exchange swaps. On 17 September, when United States stock markets reopened after a closure of four days, and against lingering concerns about the stability of international financial markets and the increased cyclical downside risks, the Federal Reserve cut the target for the federal funds rate by a further half a percentage point to 3 per cent. The move was rapidly followed, possibly as a sign of solidarity, by the ECB, which lowered its key

---

[19] Commission of the European Communities, "Public finances in EMU – 2001", *European Economy*, Reports and Studies, No. 3 (Luxembourg), 2001, pp. 16-19.

[20] European Central Bank, *Monthly Bulletin* (Frankfurt am Main), September 2001, p. 6.

[21] In Germany, the government decided to raise taxes on tobacco and insurance premiums to finance anti-terrorist spending, arguing that the increases were the only option given the commitment to reduce net borrowing in 2002 and to balance the budget by 2006. *Financial Times*, 21 September 2001.

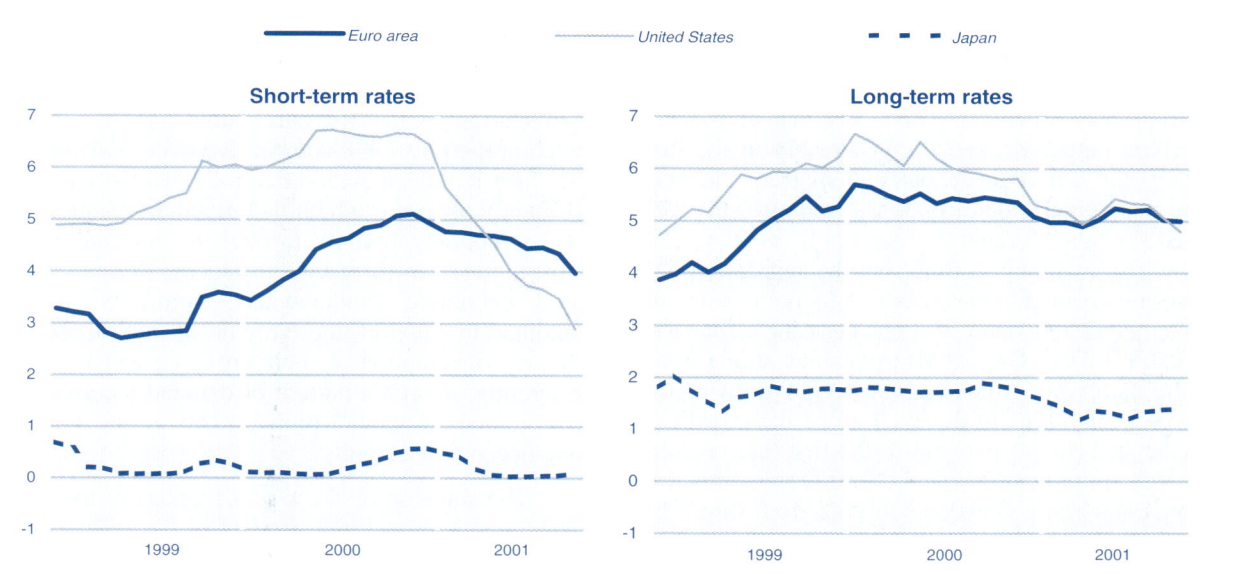

CHART 1.1.11

**Short-term and long-term interest rates in the euro area, Japan and the United States, January 1999-September 2001**
(Per cent per annum)

Source: OECD, Main Economic Indicators (Paris), various issues; European Central Bank (www.ecb.int); United States Federal Reserve (www.federalreserve.gov).
Note: Short-term interest rates: three-month money market rates. Long-term interest rates: yields on 10-year government bonds.

interest rate by the same amount to 3.75 per cent. But the ECB was also concerned about the rapid loss of cyclical momentum in the euro area. The central banks of Canada, Japan, Sweden, Switzerland and the United Kingdom also reduced their official interest rates in the wake of the decision by the Federal Reserve.[22] In the United Kingdom, the main refinancing rate was reduced to its lowest level since 1964. In Switzerland, the cut in interest rates was also intended to arrest the rapid appreciation of the Swiss franc, which had been triggered by the strong inflow of capital in search of a safe haven after 11 September. Concerns about the dampening effect of increased uncertainty on business and consumer spending led the Federal Reserve to lower the target for the federal funds rate by a further half a percentage point to 2.5 per cent on 2 October 2001. At the time of writing the ECB's main refinancing rate was unchanged at 3.75 per cent.

As a result of these monetary policy measures, short-term interest rates fell sharply in the United States; they fell also in the euro area, but given the differential degree of monetary easing the decline has been less pronounced (chart 1.1.11). Real short-term interest rates in the United States were negative (about -0.5 per cent) and around 1 per cent in the euro area in early October 2001. United States long-term interest rates (yields on 10-year government bonds) fell below those in the euro area in September. By mid-October the differential amounted to about half a percentage point. The combined effect of the differential changes in short- and long-term interest rates was a significant steepening of the yield curve in the United States. This reflected not only the sizeable monetary stimulus in the pipeline but also some upward pressure on long-term yields resulting from expectations of future fiscal imbalances. Longer-term inflationary expectations, however, have remained moderate.

In the United States, the response of fiscal policy to the disruptions caused by the September attacks has also been swift. The federal government quickly announced a $40 billion emergency relief package designed, *inter alia*, to assist in the rescue of victims and rebuilding. This package was supplemented by $15 billion of support for the airline industry. The combined package of $55 billion corresponds to 0.5 per cent of GDP in 2000. In early October, the administration proposed a further package of tax cuts and spending increases totalling some $60-$75 billion or 0.6-0.7 per cent of GDP to strengthen domestic demand. This package, which has yet to be adopted by Congress, would put the total post-attack fiscal stimulus at some 1¼ per cent of GDP. This comes on top of the fiscal stimulus from the tax rebate in the second half of 2001, which was already equivalent to some 0.5 per cent of GDP, and the income tax cuts to be enacted in the context of the tax reforms in 2002. The total fiscal stimulus could therefore amount to some 2 per cent of GDP. The deteriorating economic performance and weaker growth prospects are reflected in projections of a significantly reduced budget surplus (if any) in 2002 and the medium term.

---

[22] Central banks in some of the emerging markets of Asia (Hong Kong, Taiwan) also lowered their official interest rates.

## 1.2 The transition economies

### (i) Introduction

In the first half of 2001, the ECE transition economies generally showed some resilience in the face of a deteriorating international environment. The immediate impact of the global slowdown on eastern Europe and the CIS has so far been much less pronounced than on western Europe where output weakened considerably in the first half of the year. In contrast, output remained relatively robust in many transition economies and growth even accelerated in some countries, especially in the CIS. During the first six months of the year the aggregate GDP of the transition economies increased by 5 per cent (year-on-year) largely a reflection of a 6.1 per cent increase in the CIS (table 1.2.1). Growth also remained strong in all three of the Baltic states (their aggregate GDP growing 6.3 per cent, year-on-year). In eastern Europe the growth of GDP decelerated (to 3.1 per cent in the first half of 2001 from 4.5 per cent in the same period of 2000) but this was mostly due to the sharp slowdown in Poland. It should be added that there are marked differences in the current sources of growth – and hence in the short-run growth prospects – between the different subgroups of transition economies, notably, between the countries of eastern Europe (including the Baltic area) and the CIS (sections (1.2(ii) and 1.3).

Although these figures look impressive against the background of a general slowdown in the world economy, their significance should not be overstated. All the ECE transition economies are highly dependent on their international markets and the sharp deterioration in world market conditions will, sooner or later, start to take its toll – the varying lags depending on the specific transmission channels. In fact, all the indications are that by mid-year economic activity was rapidly weakening in most transition economies (with the possible exception of a few CIS countries).

The Russian economy has been the main support of the continuing recovery in the CIS. Much of Russia's present economic strength is due to the effects of the ongoing restructuring of the economy after the 1998 financial crisis and, especially, the comprehensive reforms initiated by the current administration. The presently prevailing sense of political stability – after a decade of volatility and frequent changes of government – has also contributed to an improving business climate in the country. In addition, the Russian authorities have adopted a more forward-looking approach to policy, which has had a positive effect on business and consumer confidence and expectations in general.

However, it should also be borne in mind that the main source of growth in Russia – and, indirectly, in much of the CIS as a whole – is the continuing strong recovery in Russian final domestic demand, in particular, private consumption. In turn, this has been largely driven by the windfall revenue gains due to the relatively high world oil prices in 2000 and part of 2001. Thus, in assessing Russia's short-term economic prospects, the main issue is whether and to what extent this domestic demand-led growth is sustainable. If world oil prices were to follow the downward trend in the global economy (as suggested by their direction in early autumn), a key stimulus to growth in Russia and the CIS could soon peter out. In addition, the continuing real appreciation of the Russian rouble has already eroded much of the post-1998 competitive gains – acquired from the sharp real depreciation that followed the August 1998 financial collapse – which had given an additional boost to the economy (mainly through its stimulus to import substitution). In these circumstances, Russia's currently good economic fundamentals would be unable to continue to support the growth of real incomes and private consumption at their present pace; or, to put it differently, if such a pattern of demand-led growth were to continue, it would probably be at the expense of macroeconomic stability.

The situation on the labour markets of the transition economies is mixed and in view of the relatively strong output growth in the first half of the year, the outcomes are less favourable than could have been expected: in June 2001 the average unemployment rates in eastern Europe and the Baltic states (14.9 per cent and 9.8 per cent, respectively) were higher than a year earlier and only a handful of countries (Hungary, Slovenia and partly the Czech Republic) have reported more lasting reductions in joblessness. Within the CIS, there was a notable easing in Russia's labour market and some improvement in Ukraine, but the situation in the other countries is difficult to assess due to the unreliability of their statistics.

Most transition economies made further progress in reducing inflation in the period to August 2001. Both internal (prudent macroeconomic policies but also, in some cases, the exchange rate regime and lower food prices due to better harvests) and external factors (especially the easing of oil prices) contributed to this development. Countries with double-digit inflation rates are now a minority among the transition economies and their number is falling steadily (table 1.2.1). Nonetheless, as indicated by several crises in the past, it cannot be taken for granted that this process is irreversible: sustaining macroeconomic stability requires constant effort by policy makers.

The merchandise trade of the transition economies weakened somewhat but still remained quite buoyant in the first half of the year: in dollar terms, both their aggregate exports and imports were increasing at double-digit rates (table 1.2.2). However, in eastern Europe and in the Baltic region, their growth has been rapidly decelerating and the outcome for the second half of the year is likely to be considerably worse than in the first. The continuing recovery in Russia has given a strong boost to intra-CIS trade: this, together with the strong real appreciation of the rouble, has benefited exporters in neighbouring CIS countries.

## TABLE 1.2.1

**Basic economic indicators for the ECE transition economies, 1999-2002**

*(Rates of change and shares, per cent)*

| | | | GDP (growth rates) | | | | Industrial output (growth rates) | | | Inflation (per cent change, Dec./Dec.) | | | Unemployment rate (end of period, per cent) | | |
|---|---|---|---|---|---|---|---|---|---|---|---|---|---|---|---|
| | | | | 2001 | | | | | | | | | | | |
| | 1999 | 2000 | Apr. official forecast | Jan.-Jun. actual[a] | Oct. official forecast | 2002 official forecast | 1999 | 2000 | Jan.-Jun. 2001[a] | 1999 | 2000 | 2001[b] | Jun. 1999 | Jun. 2000 | Jun. 2001 |
| **Eastern Europe** | 1.5 | 3.7 | 4.2 | 3.1 | 3.1 | 3½ | -0.4 | 8.3 | 4.8 | .. | .. | .. | 13.6 | 14.7 | 14.9* |
| Albania[c] | 7.3 | 7.8 | 5-7 | .. | 7 | 7 | 16.0 | 12 | -20* | -1.0 | 4.2 | 4.0 | 18.0 | 17.6 | 15.1 |
| Bosnia and Herzegovina[d] | .. | 9.1 | 7-9 | .. | 7-9 | .. | 10.6 | 8.8 | 14.6 | -0.4 | 3.4 | 3.3 | 39.1 | 39.1 | 39.3 |
| Bulgaria | 2.4 | 5.8 | 5 | 4.8 | 4.5 | 4 | -9.3 | 5.8 | 1.7 | 6.9 | 11.2 | 9.3 | 12.8 | 18.2 | 17.1 |
| Croatia | -0.4 | 3.7 | 3-4 | 4.5 | 4 | 3-4.2 | -1.4 | 1.7 | 5.9 | 4.6 | 7.5 | 5.0 | 18.9 | 20.5 | 21.5 |
| Czech Republic | -0.4 | 2.9 | 3 | 4.0 | 3.6-3.8 | 3.8 | -3.1 | 5.4 | 8.6 | 2.5 | 4.1 | 5.5 | 8.4 | 8.7 | 8.1 |
| Hungary | 4.2 | 5.2 | 4.5-5 | 4.2 | 4.3 | 3.7-4.2 | 10.4 | 18.3 | 7.4 | 11.3 | 10.1 | 10.6 | 9.4 | 8.9 | 8.4 |
| Poland | 4.1 | 4.0 | 4.5 | 1.6 | 1.5 | 2.0-2.5 | 3.6 | 6.8 | 1.6 | 9.9 | 8.6 | 6.1 | 11.6 | 13.6 | 15.9 |
| Romania | -2.3 | 1.6 | 4.1 | 4.9 | 4.5 | 5.1 | -7.9 | 8.2 | 10.4 | 54.9 | 40.7 | 35.8 | 11.4 | 10.8 | 8.8 |
| Slovakia | 1.9 | 2.2 | 3.2 | 2.9 | 2.8-3.0 | 3.6 | -3.1 | 9.0 | 6.2 | 14.2 | 8.3 | 7.9 | 17.7 | 19.1 | 17.8 |
| Slovenia | 5.2 | 4.6 | 4.5 | 2.9 | 3.7 | 3.6 | -0.5 | 6.2 | 3.2 | 8.1 | 9.0 | 9.6 | 13.4 | 11.8 | 11.1 |
| The former Yugoslav Republic of Macedonia | 4.3 | 4.3 | 6 | -5.0 | -5 | 4 | -2.6 | 3.5 | -8.8 | 2.3 | 10.8 | 6.3 | 42.3 | 43.6 | .. |
| Yugoslavia[e] | -17.7 | 7.0 | 5 | .. | 5 | .. | -23.1 | 10.9 | -2.4 | 54.0 | 115.1 | 125.1 | 26.7 | 26.5 | 27.1 |
| **Baltic states** | -1.7 | 5.4 | 4.7 | 6.3 | 5.7 | 4¾ | -8.0 | 7.6 | 11.7 | .. | .. | .. | 8.1 | 9.3 | 9.8 |
| Estonia | -0.7 | 6.9 | 6 | 5.4 | 4.8 | 4 | -3.4 | 12.8 | 5.9 | 3.9 | 4.9 | 6.9 | 6.4 | 6.2 | 7.5 |
| Latvia | 1.1 | 6.6 | 5-6 | 8.8 | 8 | 5-5.5 | -5.4 | 3.2 | 7.9 | 3.3 | 1.9 | 3.1 | 10.0 | 8.4 | 7.8 |
| Lithuania | -3.9 | 3.9 | 3.7 | 5.1 | 4.5-5.1 | 4.7 | -11.2 | 7.0 | 16.4 | 0.3 | 1.5 | 1.7 | 7.5 | 11.1 | 12.1 |
| **CIS** | 4.5 | 7.8 | 4.2 | 6.1 | 6.1 | 5 | 9.2 | 11.6 | 7.6 | .. | .. | .. | 8.1 | 7.2 | 6.4 |
| Armenia | 3.3 | 6.0 | 6.5 | 6.6 | 6.5 | 6 | 5.2 | 6.4 | 2.7 | 2.1 | 0.4 | 4.1 | 10.4 | 11.9 | 10.4 |
| Azerbaijan | 7.4 | 11.1 | 8.5 | 8.4 | 8.5 | 8.5 | 3.6 | 6.9 | 5.1 | -0.5 | 2.1 | 2.1 | 1.2 | 1.1 | 1.3 |
| Belarus | 3.4 | 5.8 | 3-4 | 3.7 | 3-4 | 4-5 | 10.3 | 7.8 | 4.1 | 251.3 | 108.0 | 65.3 | 2.1 | 2.0 | 2.2 |
| Georgia | 3.0 | 2.0 | 3-4 | 5.5 | 3-4 | 3.5 | 7.4 | 6.1 | -3.1 | 11.1 | 4.6 | 6.0 | 5.5 | .. | .. |
| Kazakhstan | 2.7 | 9.8 | 4 | 14.0 | 10 | 7 | 2.7 | 15.5 | 13.6 | 18.1 | 10.0 | 9.1 | 3.5 | 4.2 | 3.3 |
| Kyrgyzstan | 3.7 | 5.0 | 5 | 6.7 | 5.6 | 4 | -4.3 | 6.0 | 6.0 | 39.8 | 9.5 | 8.4 | 3.1 | 3.2 | 3.2 |
| Republic of Moldova[f] | -3.4 | 1.9 | 5 | 3.1 | 5 | 6 | -11.6 | 2.3 | 12.1 | 43.8 | 18.5 | 9.1 | 2.3 | 2.0 | 2.0 |
| Russian Federation | 5.4 | 8.3 | 4 | 5.1 | 5.5-6.0 | 4.3 | 11.0 | 11.9 | 5.5 | 36.6 | 20.1 | 23.7 | 12.1 | 10.1 | 8.4 |
| Tajikistan | 3.7 | 8.3 | 6.7 | 10.3 | 6.7 | 8 | 5.6 | 10.3 | 13.0 | 30.1 | 60.6 | 49.3 | 3.2 | 3.1 | 2.5 |
| Turkmenistan[g] | 16.0 | 17.6 | 16 | 15.0 | 16 | 18 | 15.0 | 28.6 | 8.5 | .. | .. | .. | .. | .. | .. |
| Ukraine | -0.4 | 5.8 | 3-4 | 9.1 | 7.3 | 6 | 4.3 | 12.9 | 18.5 | 19.2 | 25.8 | 11.6 | 4.0 | 4.3 | 3.8 |
| Uzbekistan | 4.4 | 4.0 | 4.4 | 4.2 | 4.4 | 4.5 | 6.1 | 6.4 | 7.5 | 26.0 | 28.2 | 28.0 | 0.6 | 0.7 | .. |
| **Total above** | 3.2 | 6.2 | 4.2 | 5.0 | 5.1 | 4½ | 4.6 | 10.2 | 6.5 | .. | .. | .. | .. | .. | .. |
| *Memorandum items:* | | | | | | | | | | | | | | | |
| **CETE-5** | 3.0 | 3.8 | 4.1 | 2.7 | 2.6 | 3 | 2.4 | 8.7 | 4.7 | .. | .. | .. | 11.4 | 12.6 | 13.8 |
| **SETE-7** | -2.3 | 3.4 | 4.5 | 4.4* | 4.2 | 4½ | -9.4 | 7.2 | 5.2 | .. | .. | .. | 16.2 | 17.1 | 16.6* |

*Source:* National statistics; CIS Statistical Committee; direct communications from national statistical offices to UNECE secretariat.

*Note:* Aggregates are UNECE secretariat calculations, using PPPs obtained from the 1996 European Comparison Programme. Output measures are in real terms (constant prices). Forecasts are those of national conjunctural institutes or government forecasts associated with the central budget formulation. Industrial output refers to gross output, not the contribution of industry to GDP. Inflation refers to changes in the consumer price index. Unemployment generally refers to registered unemployment at the end of the period (with the exceptions of the Russian Federation, where it is the Goskomstat estimate according to the ILO definition, and Estonia where until October 2000 it refers to job seekers). Aggregates shown are: *Eastern Europe* (the 12 countries below that line), with sub-aggregates *CETE-5* (central European transition economies: Czech Republic, Hungary, Poland, Slovakia, Slovenia) and *SETE-7* (south-east European transition economies: Albania, Bosnia and Herzegovina, Bulgaria, Croatia, Romania, The former Yugoslav Republic of Macedonia and Yugoslavia); *Baltic states* (Estonia, Latvia, Lithuania); and *CIS* (12 member countries of the Commonwealth of Independent States).

[a] January-June 2001 over January-June 2000.

[b] June 2001 over June 2000.

[c] Industrial output covers state sector only.

[d] Data reported by the Statistical Office of the Federation; these exclude the area of Republika Srpska.

[e] Gross material product instead of GDP. Data for 1999, 2000 and 2001 exclude Kosovo and Metohia.

[f] Excluding Transdniestria.

[g] Officially reported growth rates are of dubious reliability.

### TABLE 1.2.2

**International trade and external balances of the ECE transition economies, 1999-2001**

*(Rates of change and shares, per cent)*

|  | Merchandise exports in dollars (growth rates) ||| Merchandise imports in dollars (growth rates) ||| Trade balances (per cent of GDP) ||| Current account (per cent of GDP) |||
|---|---|---|---|---|---|---|---|---|---|---|---|---|
|  | 1999 | 2000 | 2001[a] | 1999 | 2000 | 2001[a] | 1999 | 2000 | 2001[a] | 1999 | 2000 | 2001[a] |
| **Eastern Europe** | -1.2 | 12.9 | 12.8 | -2.5 | 11.0 | 11.2 | -9.7 | -10.5 | -9.9 | -5.8 | -5.0 | -5.7 |
| Albania | 28.3 | -10.0 | 26.2 | 11.3 | 14.0 | 25.8 | -16.9 | -20.6 | -22.0 | -3.6 | -4.4 | -7.2 |
| Bosnia and Herzegovina | 47.3 | 30.2 | 20.1 | 14.7 | -5.8 | -4.9 | -42.1 | -38.3 | -27.1 | -21.4 | -21.6 | -17.1 |
| Bulgaria | -4.5 | 20.0 | 8.9 | 11.3 | 17.6 | 12.6 | -12.2 | -14.0 | -17.1 | -5.3 | -5.9 | -4.8 |
| Croatia | -5.3 | 3.0 | 0.7 | -7.0 | 1.6 | 22.9 | -17.4 | -18.3 | -23.7 | -6.9 | -2.1 | -14.7 |
| Czech Republic | -0.4 | 10.4 | 16.2 | -2.5 | 14.9 | 16.9 | -3.4 | -6.4 | -5.3 | -2.9 | -4.5 | -4.3 |
| Hungary | 8.7 | 12.3 | 11.8 | 9.0 | 14.5 | 11.4 | -6.2 | -8.7 | -7.7 | -4.3 | -3.3 | -3.0 |
| Poland | -2.9 | 15.5 | 16.7 | -2.4 | 6.6 | 2.9 | -11.9 | -11.0 | -8.4 | -7.4 | -6.3 | -5.2 |
| Romania | 2.4 | 21.9 | 16.0 | -12.2 | 25.6 | 32.4 | -5.4 | -7.3 | -12.4 | -4.2 | -3.7 | -8.1 |
| Slovakia | -4.6 | 15.8 | 8.6 | -13.4 | 12.5 | 18.5 | -5.0 | -4.0 | -8.3 | -5.0 | -3.7 | -8.1 |
| Slovenia | -5.6 | 2.2 | 6.5 | -0.2 | 0.3 | 0.6 | -7.7 | -7.6 | -5.1 | -3.9 | -3.4 | -0.6 |
| The former Yugoslav Republic of Macedonia | -9.1 | 11.3 | -10.9 | -7.2 | 16.3 | -25.5 | -15.9 | -20.7 | -13.2 | -3.1 | -3.2 | -11.0 |
| Yugoslavia | -47.6 | 15.1 | 3.0 | -31.8 | 12.6 | 9.0 | -17.7 | -24.6 | -26.1 | -7.5 | -4.2 | -9.0 |
| **Baltic states** | -12.5 | 24.9 | 18.9 | -13.7 | 15.0 | 13.1 | -18.8 | -17.2 | -15.0 | -9.3 | -6.3 | -5.2 |
| Estonia | -9.2 | 33.2 | 19.7 | -14.2 | 23.7 | 13.4 | -22.5 | -21.9 | -17.7 | -4.7 | -6.3 | -5.6 |
| Latvia | -4.9 | 8.1 | 7.9 | -7.6 | 8.1 | 9.3 | -18.4 | -18.5 | -17.3 | -9.8 | -6.9 | -5.3 |
| Lithuania | -19.0 | 28.1 | 24.0 | -16.6 | 13.0 | 15.2 | -17.2 | -14.3 | -12.4 | -11.2 | -6.0 | -5.0 |
| **CIS**[b] | -0.5 | 38.0 | 7.5 | -24.1 | 14.0 | 15.9 | 14.8 | 21.3 | 18.1 | 8.3 | 13.9 | 11.4 |
| Armenia | 5.1 | 29.7 | 9.3 | -10.1 | 9.0 | -9.2 | -31.4 | -30.5 | -34.2 | -16.6 | -14.5 | -14.2 |
| Azerbaijan | 53.3 | 87.8 | 57.1 | -3.8 | 13.1 | -3.9 | -2.3 | 10.9 | 27.6 | -13.0 | -3.2 | 4.3 |
| Belarus | -16.4 | 24.9 | 4.2 | -21.9 | 27.0 | -12.5 | -6.9 | -10.6 | -0.1 | -1.8 | -2.1 | 4.6 |
| Georgia | 23.7 | 38.5 | -2.5 | -31.9 | 20.7 | 24.6 | -12.9 | -13.2 | -15.0 | -6.9 | -8.7 | -12.4 |
| Kazakhstan | 2.9 | 63.4 | 5.8 | -15.2 | 37.0 | 45.6 | 11.3 | 22.3 | 12.7 | -1.4 | 4.1 | -5.5 |
| Kyrgyzstan | -11.6 | 11.2 | -4.0 | -28.7 | -7.6 | -16.9 | -11.9 | -3.8 | 2.6 | -14.7 | -5.9 | – |
| Republic of Moldova | -26.6 | 1.6 | 30.4 | -42.7 | 32.5 | 16.7 | -10.5 | -23.8 | -21.4 | -4.0 | -9.4 | -6.4 |
| Russian Federation | 1.0 | 39.5 | 5.4 | -30.5 | 8.9 | 24.2 | 22.1 | 27.5 | 22.5 | 12.8 | 18.4 | 15.2 |
| Tajikistan | 15.4 | 13.9 | -14.4 | -6.7 | 1.8 | 11.6 | 2.4 | 11.1 | -10.2 | -3.3 | -6.3 | -21.2 |
| Turkmenistan | 100.4 | 110.1 | 12.2 | 48.9 | 18.7 | 38.4 | -9.5 | 16.3 | 0.8 | -23.0 | 2.3 | -9.6 |
| Ukraine | -8.4 | 25.8 | 21.7 | -19.3 | 17.8 | 7.4 | -0.8 | 1.9 | 3.5 | 5.3 | 4.7 | 5.0 |
| Uzbekistan | -0.6 | 0.9 | 0.7 | -4.0 | -5.0 | 8.0 | 1.2 | 2.8 | 1.1 | -1.0 | 1.4 | -0.6 |
| **Total above** | -1.3 | 24.7 | 10.4 | -10.1 | 12.0 | 12.5 | – | 3.9 | 3.0 | -0.1 | 3.6 | 2.3 |
| *Memorandum items:* |  |  |  |  |  |  |  |  |  |  |  |  |
| **CETE-5** | 0.1 | 12.2 | 13.6 | -1.0 | 10.3 | 9.4 | -8.7 | -9.2 | -7.6 | -5.7 | -5.2 | -4.6 |
| **SETE-7** | -6.8 | 16.2 | 9.3 | -7.9 | 13.8 | 18.0 | -13.2 | -14.9 | -18.5 | -6.1 | -4.6 | -9.8 |

*Source:* National statistics; CIS Statistical Committee; direct communications from national statistical offices to UNECE secretariat; UNECE secretariat calculations.

*Note:* Foreign trade growth is measured in current dollar values. Trade and current account balances are related to GDP at current prices, converted from national currencies at current dollar exchange rates. Current price GDP values are in some cases estimated from reported real growth rates and consumer price indices. On regional aggregates, see the note to table 1.2.1.

[a] January-June. Change over the same period of previous year.

[b] Including intra-CIS trade.

Most of the east European and Baltic economies have been able to easily finance their current account deficits – even when they have been growing – mainly with the continuing inflow of FDI. There has been, however, a worrying escalation in the external imbalances of some countries (Croatia, Romania, Slovakia), which may soon be faced with the need to take policy measures to arrest this process. The external financial positions of the CIS countries varied considerably in the first half of 2001. Russia's current account remained in large surplus (although less than in 2000) and the balances of about one half of the other CIS members benefited from buoyant Russian demand. However, a majority of energy exporters saw their current account balances deteriorate due to rapid import growth (for example Kazakhstan, Turkmenistan and Uzbekistan). Many CIS countries face various degrees of financing constraints because of their inability to attract private capital (virtually all the small CIS countries and Ukraine) and, in some cases, multilateral funds as well (Belarus).

### (ii) Output and demand

During the first six months of 2001 output remained relatively strong in the ECE transition economies but, compared with the previous year, activity weakened somewhat in some countries, especially in eastern Europe. The main reason for the deceleration of growth

## TABLE 1.2.3

**GDP and industrial output in the ECE transition economies, 2000-2001**
*(Percentage change over the same period of the preceding year)*

|  | GDP 2000 Jan.-Mar. | GDP 2000 Jan.-Jun. | GDP 2000 Jan.-Sept. | GDP 2000 Jan.-Dec. | GDP 2001 Jan.-Mar. | GDP 2001 Jan.-Jun. | GDP 2001 Jan.-Sept. | Industrial output 2000 Jan.-Mar. | Industrial output 2000 Jan.-Jun. | Industrial output 2000 Jan.-Sept. | Industrial output 2000 Jan.-Dec. | Industrial output 2001 Jan.-Mar. | Industrial output 2001 Jan.-Jun. | Industrial output 2001 Jan.-Sept. |
|---|---|---|---|---|---|---|---|---|---|---|---|---|---|---|
| **Eastern Europe** | 4.7 | 4.5 | 4.1 | 3.7 | 3.4 | 3.1 | .. | 8.6 | 9.8 | 9.7 | 8.5 | 6.7 | 4.8 | .. |
| Albania | .. | .. | .. | 7.8 | .. | .. | .. | 8.8 | 7.6 | 18.4 | 12.0 | -15* | -20* | .. |
| Bosnia and Herzegovina | .. | .. | .. | 9.1 | .. | .. | .. | 17.8 | 15.3 | 10.1 | 8.8 | 14.4 | 14.6 | 14.3 |
| Bulgaria | 4.5 | 5.1 | 5.5 | 5.8 | 4.5 | 4.8 | .. | 5.2 | 3.2 | 3.3 | 2.3 | 5.6 | 1.7 | 2.6[a] |
| Croatia | 3.7 | 4.1 | 4.1 | 3.7 | 4.2 | 4.5 | .. | 3.7 | 2.8 | 2.7 | 1.7 | 5.6 | 5.9 | 5.9 |
| Czech Republic | 3.2 | 2.8 | 2.7 | 2.9 | 4.1 | 4.0 | .. | 4.8 | 5.0 | 5.5 | 5.4 | 10.0 | 8.6 | 7.9[a] |
| Hungary | 6.5 | 6.0 | 5.5 | 5.2 | 4.4 | 4.2 | .. | 20.9 | 20.7 | 20.4 | 18.7 | 10.2 | 7.4 | 6.2[a] |
| Poland | 5.9 | 5.4 | 4.6 | 4.0 | 2.3 | 1.6 | .. | 10.1 | 10.2 | 9.2 | 7.5 | 4.2 | 1.6 | 0.7 |
| Romania | 1.2 | .. | .. | 1.6 | 4.8 | 4.9 | .. | 0.5 | 5.0 | 7.7 | 8.2 | 10.8 | 10.4 | 9.0[a] |
| Slovakia | 1.5 | 1.7 | 2.0 | 2.2 | 3.0 | 2.9 | .. | 7.0 | 7.6 | 8.2 | 9.0 | 6.2 | 6.2 | 6.0[a] |
| Slovenia | 6.2 | 4.8 | 5.0 | 4.6 | 3.2 | 2.9 | .. | 7.2 | 8.4 | 7.7 | 6.2 | 4.7 | 3.2 | 3.6[a] |
| The former Yugoslav Republic of Macedonia | 13.3 | 10.4 | 7.0 | 4.3 | -5.1 | -5.0 | .. | 10.3 | 10.7 | 5.3 | 3.4 | -8.7 | -8.8 | -9.6[a] |
| Yugoslavia | .. | .. | .. | 7.0 | .. | .. | .. | -5.0 | 19.9 | 18.8 | 10.9 | -0.7 | -2.4 | -3.5 |
| **Baltic states** | 5.5 | 4.5 | 5.3 | 5.4 | 5.9 | 6.3 | .. | 9.5 | 7.1 | 7.7 | 7.6 | 9.6 | 11.7 | 11.9[a] |
| Estonia | 6.5 | 7.2 | 7.2 | 6.9 | 5.8 | 5.4 | .. | 13.2 | 14.3 | 13.8 | 12.9 | 6.4 | 5.9 | 6.5 |
| Latvia | 6.1 | 5.5 | 5.9 | 6.6 | 8.3 | 8.8 | .. | 4.4 | 4.2 | 3.2 | 3.2 | 7.5 | 7.9 | 8.1[a] |
| Lithuania | 4.6 | 2.6 | 3.9 | 3.9 | 4.4 | 5.1 | 5.1 | 10.1 | 4.8 | 6.8 | 7.0 | 12.2 | 16.4 | 15.1 |
| **CIS** | 8.1 | 7.9 | 8.1 | 7.8 | 5.4 | 6.1 | .. | 12.6 | 12.0 | 12.0 | 11.6 | 7.0 | 7.6 | 7.2 |
| Armenia | 0.3 | 2.4 | 2.7 | 6.0 | 12.3 | 6.6 | 9.9 | 0.3 | 2.9 | 3.9 | 6.4 | 9.2 | 2.7 | 3.8 |
| Azerbaijan | 6.7 | 8.5 | 9.9 | 11.1 | 8.0 | 8.4 | 9.3 | 3.5 | 4.7 | 5.8 | 6.9 | 6.1 | 5.1 | 5.4 |
| Belarus | 6.5 | 4.3 | 5.1 | 5.8 | 2.2 | 3.7 | 2.9 | 7.5 | 5.5 | 8.1 | 7.8 | 2.2 | 4.1 | 4.6 |
| Georgia | 8.1 | 0.8 | 0.1 | 2.0 | 1.7 | 5.5 | 4.8 | 14.1 | 9.1 | 7.2 | 6.1 | -10.6 | -3.1 | -0.7 |
| Kazakhstan | 9.4 | 10.6 | 10.8 | 9.8 | 11.2 | 14.0 | 12.3 | 12.2 | 16.0 | 15.4 | 15.5 | 11.1 | 13.6 | 13.8 |
| Kyrgyzstan | 0.6 | 6.5 | 5.7 | 5.0 | 5.6 | 6.7 | 6.7 | -4.8 | 3.3 | 6.7 | 6.0 | 11.0 | 6.0 | 7.1 |
| Republic of Moldova | 1.0 | 1.7 | 1.0 | 1.9 | 2.6 | 3.1 | .. | 3.3 | 3.6 | 2.0 | 2.3 | 3.1 | 12.1 | 12.2 |
| Russian Federation | 9.0 | 8.8 | 8.8 | 8.3 | 4.9 | 5.1 | .. | 14.3 | 13.1 | 12.7 | 11.9 | 5.2 | 5.5 | 5.2 |
| Tajikistan | 3.8 | 6.5 | 10.8 | 8.3 | 7.6 | 10.3 | 12.1 | 8.7 | 9.0 | 10.1 | 10.3 | 13.9 | 13.0 | 16.0 |
| Turkmenistan | 12.0 | 14.0 | 17.0 | 17.6 | .. | 15.0 | 18.1 | 14.0 | 14.0 | 21.0 | 28.6 | 4.5 | 8.5 | 9.3 |
| Ukraine | 5.5 | 5.0 | 5.2 | 5.8 | 7.7 | 9.1 | 9.3 | 9.7 | 10.8 | 11.6 | 12.9 | 17.4 | 18.5 | 16.6 |
| Uzbekistan | 3.0 | 3.8 | 4.2 | 4.0 | 2.8 | 4.2 | 4.5 | 5.1 | 6.2 | 6.4 | 6.4 | 7.0 | 7.5 | 7.6 |
| **Total above** | 6.8 | 6.6 | 6.5 | 6.2 | 4.7 | 5.0 | .. | 10.8 | 10.9 | 10.9 | 10.2 | 6.9 | 6.5 | .. |
| *Memorandum items:* | | | | | | | | | | | | | | |
| **CETE-5** | 5.1 | 4.6 | 4.2 | 3.8 | 3.1 | 2.7 | .. | 10.5 | 10.6 | 10.2 | 9.1 | 6.8 | 4.7 | .. |
| **SETE-7** | 3.6* | 4.1* | 4.0* | 3.4 | 4.1* | 4.4* | .. | 1.5 | 6.7 | 7.9 | 6.5 | 6.5 | 5.2 | .. |

*Source:* National statistics; CIS Statistical Committee; direct communications from national statistical offices to UNECE secretariat.

*Note:* Industrial output figures are based on monthly statistical reports. Because of differences in coverage, monthly cumulative figures for 2000 as a whole differ slightly from reported annual figures for some countries. For these countries reported annual figures have been used. On coverage of indicators and regional aggregates see the note to table 1.2.1.

[a] January-August.

---

seems to be the weakening of external demand, especially in western Europe, but real exchange rate appreciation may have also affected exports and output in some east European and Baltic economies.[23] As a result of these developments, a shift from external to domestic sources of growth has been taking place in 2001 in a number of countries in the region. This can be traced in the preliminary reports of their quarterly national accounts and also in the strengthening of retail sales volumes (table 1.2.4). Industrial output showed signs of weakening in 2001 in some transition economies but the picture varied considerably across countries (tables 1.2.1 and 1.2.3). While the growth of industrial production in the CIS as a whole also decelerated from 2000, the average rate of growth remained remarkably high, at 7.6 per cent (year-on-year) during the first half of 2001.

### (a) Eastern Europe and the Baltic states

During the first half of 2001, central Europe was the slowest growing subregion among the ECE transition economies, the weak performance of the Polish economy

---

[23] Most east European and Baltic currencies have appreciated considerably in real terms since early 1999; while a broad-based real appreciation against the euro has also been prevalent among CIS currencies, these economies are far less integrated with the euro area for this to matter much. In contrast, the significant depreciation of CIS currencies against the Russian rouble during the first half of this year has had a strong impact on intra-CIS trade (see section 1.2(v)).

weighing heavily on the aggregate figures. The deceleration of growth in Hungary has been perceptible since mid-2000 and GDP growth in the first half of 2001 was 4.2 per cent. So far, the slowdown has not given rise to any significant macroeconomic imbalance.[24] However, during the first half of 2001, the growth of exports slowed down and domestic demand, while rising steadily, has been unable to fully offset it. In this period growth was boosted by a 7.7 per cent, year-on-year, increase in construction output (real value added), with industry growing at a similar rate to GDP; agriculture and services made only small contributions to growth.[25] In Slovenia, weakening domestic demand (fixed capital formation fell by 3.8 per cent, year-on-year, in the first six months and consumption growth weakened due to modest increases in real household incomes) contributed to the deceleration of GDP growth. The curbing of domestic demand was partly policy-driven (fiscal restraint helped curb investment and imports while the central bank intervened on the foreign exchange market to prevent real exchange rate appreciation); through the course of the year, weakening external demand also added to the slowdown in domestic output. However, as import restraint was set in before the weakening of exports, the slowdown in Slovenian growth may not lead to the emergence of serious imbalances.

Slovakia's quarterly growth rate has been positive since the end of 1993 and this continued in 2001 as well. Unlike Hungary, however, the country has repeatedly gone through cycles of domestic versus external demand-led developments: after two years of export-driven recovery after the Russian crisis, preliminary data for the first half of 2001 suggest a strengthening of domestic demand (almost exclusively driven by a surge in investment) combined with an increasingly negative net trade contribution to GDP growth. During the first six months, the growth of industrial output decelerated to 6.2 per cent, while the volume of retail sales picked up only moderately (tables 1.2.3 and 1.2.4). Following nine consecutive quarters of uninterrupted growth, the Czech Republic was one of the few countries in the region to have raised its growth forecast for this year. The 4 per cent rate of GDP growth in the first half of 2001 was underpinned by continuing recovery in industry and in the trade and transport sectors. Stronger than expected domestic demand – contributing 6.2 percentage points to GDP growth in the first half of the year – largely offset a weakening, negative net trade position, in spite of a 9.8 per cent increase in the heavily export-oriented manufacturing output. The surge in domestic demand was partly due to rising real wages (giving a lift to household consumption) and large inflows of FDI (driving investment demand), but there was also some loosening of monetary and fiscal policies. However, given the current external environment, this policy mix is hardly sustainable: a further acceleration of domestic demand at a time when external demand is slowing down may give rise to macroeconomic imbalances.

In contrast to its neighbours, GDP growth in Poland has been slowing since the fourth quarter of 1999 when the quarterly rate peaked at 6.2 per cent, year-on-year. Restrictive monetary policy, aimed at meeting a medium-term inflation objective of 4 per cent by the end of 2003, has since contributed to curbing domestic demand. In the first half of 2001 aggregate domestic absorption fell by 1.3 per cent, year-on-year, largely due to sluggish

---

TABLE 1.2.4

**Volume of retail trade in selected transition economies, 2000-2001**
*(Percentage change over same period of preceding year)*

|  | 2000 |  |  |  | 2001 |  |  |
|---|---|---|---|---|---|---|---|
|  | Jan.-Mar. | Jan.-Jun. | Jan.-Sept. | Jan.-Dec. | Jan.-Mar. | Jan.-Jun. | Jan.-Sept. |
| Albania | .. | .. | .. | .. | 3.1 | .. | .. |
| Bosnia and Herzegovina | 23.4 | 36.1 | 35.5 | 59.4 | 40.7 | 27.6 | .. |
| Bulgaria | 6.7 | 3.7 | 3.1 | 0.6 | 1.1 | 3.3 | 4.6[a] |
| Croatia | 10.9 | 10.7 | 10.7 | 9.9 | 11.0 | 11.6 | 10.8[a] |
| Czech Republic | 6.6 | 5.8 | 4.8 | 4.3 | 3.6 | 3.8 | 4.0[a] |
| Hungary | 4.2 | 3.4 | 2.2 | 1.9 | 6.7 | 5.3 | 4.7[a] |
| Poland | 10.6 | 8.1 | 5.2 | 3.0 | -2.2 | -1.8 | -1.2 |
| Romania | -11.2 | -7.8 | -5.9 | -4.0 | -0.5 | -2.5 | -1.5[b] |
| Slovakia | -1.7 | -1.4 | 0.4 | 2.3 | 3.7 | 3.0 | 3.3[a] |
| Slovenia | 17.5 | 5.5 | 7.0 | 7.4 | 5.5 | 6.2 | 7.2[b] |
| The former Yugoslav Republic of Macedonia | 4.2 | 8.8 | 10.4 | 10.6 | .. | 6.9 | 4.7 |
| Yugoslavia | -13.6 | 3.8 | 9.0 | 8.3 | 2.0 | 5.0 | 8.0[a] |
| Estonia | 9.9 | 12.4 | 13.0 | 12.9 | 13.4 | 13.7 | 13.7[a] |
| Latvia | 5.6 | 4.6 | 7.4 | 9.0 | 13.9 | 13.7 | 11.8[a] |
| Lithuania | 12.6 | 12.4 | 13.2 | 13.9 | 9.9 | 6.5 | 7.1[a] |
| Armenia | 7.4 | 6.3 | 6.9 | 8.5 | 14.2 | 14.0 | 15.2 |
| Azerbaijan | 9.3 | 10.0 | 10.1 | 9.8 | 9.4 | 9.3 | 9.8 |
| Belarus | 18.8 | 10.7 | 6.1 | 8.5 | 18.0 | 16.8 | 18.7 |
| Georgia | 9.7 | 13.2 | 11.4 | 11.0 | .. | 3.3 | 4.3 |
| Kazakhstan | .. | 0.3 | .. | 4.8 | 11.7 | 13.3 | 12.9 |
| Kyrgyzstan | 2.6 | 6.2 | 5.5 | 7.7 | 6.4 | 6.3 | 6.2 |
| Republic of Moldova[c] | -3.0 | 5.1 | 6.4 | 0.1 | 20.6 | 10.8 | 14.2 |
| Russian Federation | 8.1 | 8.4 | 8.7 | 8.7 | 7.3 | 9.5 | 10.1 |
| Tajikistan | -28.8 | -27.0 | -22.1 | -21.2 | .. | -5.7 | -3.6 |
| Turkmenistan | 38.0 | 31.0 | 33.0 | 30.0 | .. | .. | .. |
| Ukraine | 9.2 | 6.5 | 6.2 | 5.6 | .. | 8.8 | 8.9 |
| Uzbekistan | 5.1 | 5.3 | 5.0 | 7.8 | 7.9 | .. | 8.7 |

*Source:* National statistics; CIS Statistical Committee; direct communications from national statistical offices to UNECE secretariat.

*Note:* Retail trade covers mainly goods in eastern Europe, the Baltic states, the Russian Federation, Kazakhstan and Ukraine (for 2000), goods and catering in other CIS countries. The coverage in the monthly statistics is based on current reporting and may differ from the coverage in the annual statistics.

[a] January-August.
[b] January-July.
[c] Registered enterprises for 2000.

---

[24] Since 1996 Hungary has largely maintained a balanced GDP growth path based on expanding external and, to a lesser extent, domestic demand. The exception was in 1998 when net trade contributed negatively to GDP growth.

[25] Hungarian Central Statistical Office, most recent data, GDP (http://test999.ksh.hu/pls/ksh/docs/index_eng.html).

investment.[26] Industrial production has been decelerating in line with aggregate output, its rate of growth dropping to 0.7 per cent in the three quarters of 2001, while the volume of retail sales was declining (table 1.2.4). The rapid weakening of output contributed to the deterioration of the Polish fiscal balance, which, in turn, triggered a dramatic reassessment of the budget deficit projections (see box 1.2.1). This combination of external consolidation and an unfolding fiscal imbalance may be, at least in part, the consequence of a lack of consistency between monetary policy, aimed at curbing domestic demand, and fiscal policy, which assumed and/or targeted a higher rate of growth than was feasible given the monetary policy stance. At the same time, the pace of external consolidation has so far remained unimpressive: the "containing" effect of monetary policy on the current account position has probably been counterbalanced by a real appreciation of the exchange rate, partly resulting from the recent fiscal expansion. In the short run, with domestic demand curbed and an unpromising external environment, the Polish economy is virtually at a standstill. Poland's medium-term macroeconomic stability will depend on the government pressing ahead with structural reforms and privatization, the latter helping both the financing of the public budget as well as increasing FDI inflows to cover current account deficits accompanying higher rates of economic growth.

In the first half of 2001 there was unexpectedly strong growth in the Baltic states; however, closer inspection reveals some country-specific patterns and suggests that this trend may not last for too long. While GDP grew by 5.4 per cent in Estonia, a notable slowdown of output has been taking place since mid-2000. The same was true for industrial sales whose rate of growth halved compared with the first six months of 2000. The principal supply-side contribution to industrial output came from metal processing, textiles and food processing. The communications equipment industry, one of the fast-growing manufacturing branches, suffered from the decision by the Finnish electronic parts maker Elcoteq to phase out the production of components for mobile phone producer Ericsson (see box 1.2.2). Uncertainty about the future is already well reflected on the demand side: the main contribution to GDP growth during the first half of 2001 was the reported build-up of stocks, while total final consumption and gross fixed capital formation jointly contributed only slightly more to GDP growth than was subtracted by the change in net exports.

While Estonia's production and export structure did not allow for terms of trade gains to be reaped from the recent rise in energy prices, a slight real appreciation of the currency and the weakening of demand from western Europe led to some trade diversion to the CIS. In contrast, Latvia and Lithuania, thanks to their involvement in the transit and/or refining of Russian oil, benefited from the energy price increases. At the same time, however, they experienced a larger real currency appreciation due to the pegging of their exchange rates to currencies other than the euro,[27] and this may be adversely affecting their cost competitiveness on European manufacturing markets.

The fastest growing Baltic economy in the first half of 2001 was Latvia with GDP increasing by 8.8 per cent. The most significant source of growth was the important services sector (with a share in GDP of more than 70 per cent): real value added in all the major service branches grew by more than 10 per cent.[28] The growth of manufacturing output also accelerated, led by wood processing, textiles and food processing, primarily due to robust domestic demand as well as growth in exports to CIS countries. Expectations in the economy remained positive, as construction activity remained buoyant (with real value added increasing 5.5 per cent, year-on-year, in the first half of 2001), while private sector investment grew by about 20 per cent.[29]

In addition to strong consumption growth, a 5.1 per cent, year-on-year, increase in Lithuanian gross fixed capital formation during the first half of 2001 was another positive outcome after two years of declining investment. Growth in the first half of the year was largely driven by industry, particularly by a 56 per cent increase in oil refining, which now represents roughly 20 per cent of industrial activity.[30] According to the quarterly national accounts, real value added in services grew by only 1.7 per cent (year-on-year), while in agriculture and construction output declined. The exchange rate peg to the dollar has turned out to be a mixed blessing recently. While benefiting the refining and export of Russian oil, it has proved rather burdensome for the international competitiveness of the rest of the manufacturing sector, since the dollar strengthened considerably against the currencies of Lithuania's mostly European trading partners (see section 1.2(v)).

---

[26] Gross fixed capital formation dropped by 8.4 per cent, year-on-year, in the second quarter of 2001.

[27] The Lithuanian litas is fixed to the United States dollar, but will switch to the euro in February 2002. Latvia's lats is pegged to the IMF's SDR.

[28] In particular, the amount of oil and oil products transported by pipelines increased by 21 per cent.

[29] Bank of Finland, *Russian & Baltic Economies. The Week in Review*, No. 39 (Helsinki), 28 September 2001. Due to the recent growth of investment in Latvia the share of gross fixed capital formation in GDP in nominal terms increased from 15.1 per cent in 1995 to about 21 per cent in the first half of 2001. UNECE Macroeconomic Statistical Database.

[30] UNECE Statistical Database. While the numbers given refer to NACE category DF (manufacture of coke, refined petroleum products and nuclear fuel), there is almost no other activity except oil refining in Lithuania.

### Box 1.2.1

### Fiscal turmoil in Poland

The policy debates accompanying the revision of the Polish 2001 budget in July and especially those during the deliberations on the 2002 budget brought to the surface a problem whose existence was long suspected by some analysts: Poland was facing an escalating fiscal gap which was approaching dangerous proportions. The draft 2002 budget prepared by the Ministry of Finance contained a laconic but alarming message: unless radical fiscal reforms were immediately undertaken, and under conservative growth assumptions, next year's budget deficit could reach some zloty 90 billion, or more than 11 per cent of GDP. Ringing the alarm bell cost the Minister of Finance his post but the re-evaluation of these projections by independent analysts only confirmed this outlook. Although there had been doubts about the actual state of Polish public finances for quite some time (the Monetary Policy Council has repeatedly pointed out that its discontent with the lax fiscal policy stance was one of the main reasons for the tightening of monetary policy since the last quarter of 1999), the sheer proportions of the looming fiscal crisis came as a surprise both to policy makers and to the public. Moreover, during the last several years the reported fiscal deficit in Poland was never considered as a major source of concern.[1]

Although the exact causes of this abrupt deterioration in Poland's fiscal position still need to be analysed in more depth, the academic and policy debate that followed this disclosure has focused on several key factors. An important reason for the emergence of such a large fiscal gap appears to be an ill-fated overlap of structural and cyclical factors in 2001, which had a negative impact on the economy in general and on the fiscal position in particular. These negative effects were compounded by misjudgement and mismanagement that led to a belated response to a looming problem.

In the course of the past several years Poland embarked on an ambitious reform programme including a complete overhaul of the pension system, a reorganization of local government structures, a health care reform and a restructuring of the education system. It has now been acknowledged that the full fiscal implications of these far-reaching reforms – especially given their simultaneous implementation – were not properly assessed at the time they were launched. In addition, it appears that the fiscal coordination of the legislative process during recent years was inadequate: numerous laws were adopted that affected (directly or indirectly) public revenue and expenditure without a comprehensive assessment of their immediate and, especially, their longer-term effects on net revenue.[2] Another structural factor that might have also aggravated the fiscal gap was the automatic adjustment (indexation) of various income-related payments (in particular pensions and various social benefits) for inflation.

The escalation of fiscal liabilities generated by these structural factors coincided with a notable weakening of the Polish economy: in the first half of 2001, the rate of GDP growth slowed to 1.6 per cent, year-on-year. Thus, the official assumption of 4.5 per cent annual GDP growth incorporated in the 2001 budget turned out to be grossly optimistic, with grave consequences for the country's fiscal position. As a large number of expenditure items were fixed in nominal terms, the budget deficit was already starting to rise in the early months of the year: by the end of March it had reached zloty 15.05 billion, or 73.3 per cent of the projected annual figure, and by the end of May it had almost reached the annual target.[3]

Much attention – indeed, not only during the latest debates but already since the first signs of the economy weakening in 2000 – has been focused on the role of the Monetary Policy Council, which sets the course of monetary policy in Poland. There was sharp criticism of what was seen as a too rigid policy stance by the Council, which was responsible for monetary policy being unnecessarily tight. The critics also argued that the monetary authorities were unduly preoccupied with meeting their ambitious disinflation targets while neglecting other serious problems in the economy. The main counter argument of the Council is that in the face of fiscal looseness, cooling down the economy through monetary restraint was the only available option to prevent a further deterioration in the current account and to avoid an exchange rate crisis. Another argument in support of its pursuit of a rapid rate of disinflation was the existence of automatic inflation adjustment mechanisms in a number of public expenditure items (such as pensions and social benefits); given such indexation, a higher inflation rate would imply an even further widening of the fiscal gap.

However, while monetary austerity might have prevented a further deterioration in Poland's external imbalance, in all likelihood it also contributed to the sluggish performance of the Polish economy in 2000-2001 and, in turn, to the cyclical deterioration in the fiscal balance. In terms of the general macroeconomic stance during this period the Polish economy probably received the wrong treatment, which eliminated some of the symptoms but did not address the core problems, namely, the above-mentioned structural factors. These problems cannot be solved with the instruments of monetary policy, or for that matter, of macroeconomic policy alone. They require a wider and more comprehensive policy approach and call for a much greater focus on policy coordination, on the one hand, between the monetary and the fiscal authorities, but on the other, between the legislative and the executive arms of the government as well.

Given the magnitude of the fiscal imbalance and the sluggishness of the Polish economy, Poland's new government, which took office in October, faces formidable problems in searching for ways and means to avert a fiscal and economic crisis. The situation is aggravated by the sharp slowdown in the global and west European economies and the likelihood of negative repercussions for the Polish economy. Time is pressing because a further deterioration in the international environment will cause a further enlargement of the fiscal gap. However, there are no easy solutions and the policy adjustments will inevitably include some highly unpopular measures such as cutting public spending and raising taxes.

> **Box 1.2.1 (concluded)**
>
> **Fiscal turmoil in Poland**
>
> Despite its present problems, Poland's macroeconomic fundamentals appear to be generally sound and the long-run fiscal solvency of the economy has not been questioned. It should also be borne in mind that some of the structural components of the present fiscal imbalance, namely those generated by systemic changes such as the transition to new health care and pension systems, are likely to disappear when the new systems become fully operational. On the other hand, tolerating such a large fiscal deficit as that projected, even if only for a short period, could be an embarrassment as it would send unfavourable signals to the financial markets and could become an obstacle in Poland's accession negotiations with the EU. Thus, if Poland's image as a front-runner in the transition process is to be sustained, the new government will need to initiate and implement a number of painful policy adjustments.
>
> ---
>
> [1] In the period 1997-2000, the consolidated general government deficit varied in the range between 2.7 to 3.8 per cent of GDP. UNECE, *Economic Survey of Europe, 2001 No. 1*, p. 84.
>
> [2] According to the Ministry of Finance, during the 1997-2001 term of the Polish Sejm (which ended in September 2001), some 50 bills were passed with significant effects on the budget. W. Orlowski, "Ways can still be found to avert the present crisis", *Polish News Bulletin Company*, 4 September 2001, quoted in *Reuters Business Briefing*, 4 September 2001.
>
> [3] *Reuters Business Briefing*, 5 May 2001 and PlanEcon, *Monthly Report*, Vol. XVII, No. 15, 24 August 2001.

Developments in the south-east European transition economies were dominated by post-conflict reconstruction in Yugoslavia, the rising tensions in The former Yugoslav Republic of Macedonia and the fallout from the recent Turkish financial crisis. After three years of recession, the Romanian economy started to recover early in 2000, led by industry. Increases in GDP of 4.9 per cent and of industrial production by over 10 per cent during the first half of 2001 largely reflected a strong, broad-based recovery in domestic demand.[31] At the same time, there was an increasingly negative contribution of net exports to GDP growth, raising concerns about a potentially dangerous external imbalance (table 1.2.2). Bulgaria's GDP rose 4.8 per cent during the first half of 2001, the main impetus coming from the continuing strong recovery in fixed investment. Following the general pattern since 1998, gross industrial output was again lagging behind GDP growth, with a mere 1.7 per cent increase during the first six months of this year. The new government has come up with policy measures aimed at invigorating the economy, including a plan to set up a fund for small- and medium-sized enterprises, a lowering of some taxes, as well as a cut in public administration staff, coupled with a rise in electricity and heating prices and increases in various social benefits. Economic growth in Croatia improved in the first half of the year led by a strong performance in manufacturing industry.[32] On the demand side, GDP was driven by consumer spending and investment, largely inventory building, while net trade subtracted 6.7 points from the rate of growth of GDP in the first half of the year due to a surge in imports in the face of a still rather poor export performance.[33] Worsening external conditions and the need for further fiscal tightening, however, point to a slowdown of output in the short term.[34]

Apart from the general worsening of external conditions, The former Yugoslav Republic of Macedonia currently faces a number of additional problems. The beginning of 2001 brought escalating ethnic tensions and political uncertainty, which led to the outburst of an internal military conflict. Although a ceasefire was negotiated in July, the existing tensions are a constant threat to political and economic stability. This uncertainty, as well as the disruption of a number of business activities due to the internal conflict, has had a highly detrimental impact on economic performance. Industrial output – which has been on a downward trend since mid-2000 – fell sharply (by 8.8 per cent, year-on-year) in the first half of this year. The incoming international aid inflow – which so far has been substantial – has been essential for protecting the country from the worst consequences of the past and present conflicts in its region and within its own borders.

---

[31] As a word of caution, the available preliminary Romanian statistics for the first half of 2001 contain some discrepancies. Thus, according to the quarterly national accounts, private consumption in the first half of 2001 increased by more than 7 per cent, year-on-year, while according to the monthly retail trade statistics the volume of retail sales during the same period declined by 2.5 per cent from the same period of the previous year (table 1.2.4).

[32] The output of machinery and equipment was up 13.2 per cent, year-on-year, in this period.

[33] As in the case of Estonia, if the increase in stocks were deducted from total final demand, Croatian GDP would in fact have shrunk during the first half of 2001. However, in contrast to Estonia where statistical discrepancies are explicitly reported in the national accounts, in Croatia these are lumped together with the "changes in stocks".

[34] In accordance with the recommendations of the IMF, the share of consolidated central government spending in GDP will have to be gradually reduced, from 50.2 per cent in 1999 to 37.6 per cent in 2003. IMF Staff Country Report No. 01/50, *Republic of Croatia: 2000 Article IV Consultation and Request for a Stand-By Arrangement. Staff Report; Staff Statement; Public Information Notice and Press Release on the Executive Board Discussion; and Statement by the Authorities of Croatia* (Washington, D.C.), 23 March 2001, p. 14.

### Box 1.2.2

### The Elcoteq case

Thanks to the rapid expansion of a local subsidiary of the Finnish Elcoteq company during recent years, by the beginning of 2001 Estonia had become one of the largest third-country suppliers to the EU of transmission apparatus for radio-telephony (mobile phones, group 8525 of HS) and parts thereof (group 8529 of HS): it ranked 4th and 5th for the two groups, respectively, accounting for 11 and 5 per cent of the corresponding EU markets. These two product groups together made up nearly 40 per cent of Estonia's exports to the EU, which, in turn, accounted for about three quarters of its total exports in the first quarter of 2001. The major export destinations were Finland and Sweden: in 2000 the two countries absorbed 59 and 30 per cent of these exports, respectively. In turn, supplies from Estonia accounted for 40 and 33 per cent of Finland's and Sweden's imports of these goods from the non-EU area.

Recently, however, after losing orders from its major customer Ericsson (which in January decided to outsource the manufacture of its mobile phones to other manufacturers), Elcoteq had to scale down its output and, during the first quarter of 2001, discontinued Ericsson mobile phone production in Estonia and Hungary but continued production for Ericsson in China and Mexico. In May 2001, Elcoteq began production in Tallinn of electronics for base station antennas for a United States company; however, this requires only a small proportion of the available capacity in one of its two Tallinn plants. During the spring of 2001, Elcoteq reduced the workforce in its Tallinn subsidiary from 3,200 to 2,400 people, and it plans to lay off another 400 people by the end of the year.[1]

As a result, according to Eurostat data, in January-May 2001, Swedish imports from Estonia in these two product groups dropped by 65 per cent (year-on-year) in euro value. However, Finnish imports reportedly nearly tripled in value, although the volume growth decelerated noticeably.[2] This helped to contain the deterioration in the year-on-year growth rate of total Estonian export earnings in the second quarter, but in July and August 2001 they fell by 7 and 17 per cent in dollar value, respectively. In volume, Estonia's total exports have been declining since April 2001. With the global economy slowing down and with uncertain prospects for this industry, Estonia's export sector may suffer even more blows in the months to come.

The decision of Elcoteq to close down mobile phone production in its Hungarian plant by mid-year, in addition to ending its manufacture of computer monitors in December 2000, took its toll on Hungarian exports as well, although the impact was not as drastic as that on Estonia owing to the larger size of Hungarian exports and their more diversified structure. Faced with the global weakening of demand and especially of demand for communications equipment, the Elcoteq management has also been considering whether to pull out from Poland by the end of the year, terminating operations at the Wroclaw plant (240 employees) and transferring its manufacturing operations to Hungary. If these plans were to materialize, the new plant currently under construction in Wroclaw would be sold when completed.

---

[1] "Details of Elcoteq adjustment program specified – annual cost savings estimated to total 16 MEUR", *Press Release*, 27 August 2001 and "Elcoteq Network Corporation's Interim Report for January-June 2001", *Press Release*, 3 August 2001 (http://www.elcoteq.fi).

[2] An equally startling discrepancy emerges in the development of export unit values. According to the Estonian Statistics Office, while export unit values for communication network products in group 8525 of the HS rose on average by only 6 per cent between the first and fourth quarters of 2000, they increased nearly threefold between the last quarter of 2000 and the first quarter of 2001. They then dropped by a third in July-August. The exact causes of these fluctuations remain a puzzle but it is most probable that in the wake of the decision to shut down production of mobile phones for Ericsson in March, Elcoteq inflated the transfer prices of these products for strategic reasons before moving the remaining stocks back to its headquarters in Finland.

---

Similarly, neighbouring Yugoslavia is still best described as a postwar economy preoccupied with reconstruction. Recovery is expected to continue from last year, albeit at a reduced rate, especially as industrial output has not proved able to keep up last year's pace. Rebuilding the infrastructure and restructuring industry still require substantial foreign assistance, before private foreign investment will be willing to enter the country.

In Bosnia and Herzegovina, the emphasis is gradually shifting from reconstruction to structural reform. In the first six months of 2001, output development was uneven in the two entities that form the country: while industrial production grew by some 14.6 per cent in the Muslim-Croat Federation, it fell by 9.3 per cent (year-on-year) in Republika Srpska, widening the income disparity between the two.[35] The country still has rather weak export capacities, which result in persistent and sizeable external deficits (table 1.2.2). Private sector activity – now dominating construction, services and the production of foodstuffs and industrial consumption goods – has been the main source of economic growth in still mostly agrarian Albania during the first six months of the year. In contrast, production in state owned industrial firms fell by 20 per cent during the same period (table 1.2.1). Recently, inflows of foreign loans and foreign investment, attracted by the privatization of state owned companies, have joined cash remittances from abroad in financing the

---

[35] According to some estimates, at mid-2001 the average monthly income in Republika Srpska was one third less than that in the Federation. *RFE/RL Newsline*, Vol. 5, No. 186, Part II (Prague), 16 August.

country's chronic external imbalance, adding further impetus to domestic demand.

*(b) Commonwealth of Independent States*

Thanks to a rebound in domestic demand, the CIS as a whole was one of the fastest growing regions in the world in the first half of 2001. At the same time, there are signs that growth is decelerating, partly because of the particular nature of the initial export-driven recovery after the 1998 Russian crisis, which was largely due to subsequent large terms of trade gains. These gains were particularly favourable (at least until mid-2001) for energy and commodity exporters as well as for exporters of some price sensitive bulk commodities (such as chemicals and metals), while all the CIS countries (including net energy importers) benefited from the increase in Russian import demand.[36] Among the energy exporting countries, however, there are signs that the positive terms of trade shock has started to feed into a real appreciation of the exchange rate which threatens to undermine the competitiveness of the non-energy sectors and thus the diversification of their economies. Policy makers seem to be aware of these dangers and have been reacting accordingly, although with varying success.[37]

While GDP growth in Russia remained high (5.1 per cent, year-on-year, during the first half of 2001), it was less than in 2000. With real appreciation of the rouble having gained ground and oil prices having passed their peak, the external stimuli have been gradually weakening. At the same time, strong internal demand has been mostly due to the continuing strength of private consumption, which increased by 9.6 per cent in the first half of 2001. Retail sales were up by 9.5 per cent during the first half but there were signs that the import substitution effect, prompted by the 1998 rouble devaluation, was losing some of its force. The growth of industrial production was broadly based but it also slowed down from the preceding year.[38] Despite its continuing growth, there are some disturbing aspects of fixed capital formation: most of the net increase of investment since 1998 has been concentrated in the energy and transportation sectors, while real investment in the rest of the economy has in fact been declining.[39]

After positive growth was recorded for the first time in a decade in Ukraine in 2000, it strengthened further in 2001: during the first six months GDP increased by 9.1 per cent, year-on-year. However, much of the stimulus so far has been due to a rather narrow export base: in 2000, about a third of manufacturing output consisted of basic metals and fabricated metal products, while exports of steel accounted for roughly 40 per cent of the total value of exports.[40] Reducing the risks from such a narrow specialization in price sensitive products requires the development of a more diversified industrial structure and there were some encouraging signs of that in 2001: during the first six months the growth of manufacturing output was not only very high (up by 23.3 per cent over the same period of 2000) but it was also broadly based.

Belarus – which has not been part of the terms of trade-led expansion – performed significantly below the CIS average in the first half of 2001: GDP rose 3.7 per cent and industrial output by 4.1 per cent. On the demand side, the picture was mixed: investment outlays are reported to have fallen by 3.2 per cent (year-on-year) but private consumption was probably rising.[41] While a recent tightening of monetary policy may have had a negative effect on economic activity, this relatively weak performance in Belarus stems at least partly from the chronic structural problems of its economy. The situation was similar in the Republic of Moldova where GDP increased by 3.1 per cent in the first half of 2001. While preliminary statistics indicate increasing household consumption, falling gross fixed capital formation jeopardizes the potential for stable growth and restructuring.[42]

The relatively high rate of GDP growth in Armenia (6.6 per cent, year-on-year, in the first half of 2001) was largely driven by an upturn in the services sector. Retail trade and personal consumption also picked up strongly.[43] However, as Armenia is a large exporter of processed jewellery and precious stones, it is vulnerable to the effects of the global economic slowdown on the external demand for these items. Georgia's economy grew by 5.5 per cent, year-on-year, in the first half of 2001, having performed worst of all the CIS economies last year. However, its economic situation remains precarious as evidenced by the fact that Georgia was the only CIS economy to report declining industrial output during the

---

[36] Rising energy prices, however, resulted in serious external imbalances in Armenia, Georgia, Kyrgyzstan, the Republic of Moldova and Tajikistan, and may also have been responsible for declining production and distribution of electricity during the first half of this year notably in Armenia and Georgia.

[37] While Azerbaijan has succeeded in avoiding real appreciation by bringing down inflation, Kazakhstan has chosen to subsidize the non-oil industries by channeling part of the increasing investment there.

[38] Oil extraction led the fuels industry with a 6.9 per cent increase (year-on-year) during the first half of 2001. Engineering branches grew above average, while non-ferrous metals output increased by 5.6 per cent during the first six months of this year. There was some stagnation in the production of ferrous metals. Russian Federation Goskomstat, *Sotsial'no-ekonomicheskoe polozhenie Rossii*, January-June 2001 (Moscow), pp. 15-22.

[39] According to some estimates, while real gross fixed investment in fuels and transportation (the latter often related to the former) in the first half of 2001 grew by 132 per cent, compared to the same period of 1998, in the rest of the economy it declined by 10 per cent. Russian-European Center for Economic Policy, *Russian Economic Trends* (Moscow), September 2001, p. 3.

[40] Ukrainian-European Policy and Legal Advice Center, *Ukrainian Economic Trends* (Kiev), June 2001, p. 7.

[41] During the first six months the volume of retail sales was reportedly growing at double-digit rates (table 1.2.4)

[42] TACIS, *Moldova Economic Trends* (Chisinau), April-June 2001, p. 21 (www.economic-trends.org) and *Reuters Business Briefing*, 3 October 2001.

[43] The country has steadily reported levels of aggregate consumption in excess of GDP in recent years. The ratio of consumption to GDP in Armenia reached 107.4 per cent in 2000 and obviously, this does not provide a healthy basis for long-term capital investment and growth. Other low-income transition economies such as Albania, Georgia, Kyrgyzstan, the Republic of Moldova and Tajikistan, also suffer from similar disproportions.

first half of 2001.[44] During most of the past decade, Azerbaijan's economy has been shaped by expanding oil extraction and related activities in construction and transport. The capacity build-up in these areas (so far boosted by FDI) has now come to a temporary halt due to the lack of feasible projects. The continuing strong growth of GDP, which grew 8.4 per cent in the first half of 2001, is now being driven by increasing oil production and, at least until mid-year, relatively high oil prices.[45] Agriculture, which still employs almost half of the total workforce, also contributed to the high rate of GDP growth in 2001, gross agricultural output increasing by 8.2 per cent during the first six months. Consumer demand also remained strong with significant increases in real wages supporting an increase in retail sales of 9.3 per cent.

At 14 per cent and 15 per cent, respectively, the two central Asian oil and gas exporters, Kazakhstan and Turkmenistan, reported the highest GDP growth rates in the ECE region for the first half of 2001.[46] In contrast to Azerbaijan, capacity building in Kazakhstan's energy sector continues to proceed without interruption: while industrial production grew in line with GDP in the first six months of 2001, construction output increased by some 31 per cent, with agriculture lagging far behind the rest of the economy.[47] Real wage increases allowed retail sales to grow by 13.3 per cent – a significant acceleration if compared to the rates of growth in 1999 and 2000 – indicating a surge in private consumption. The rise in imports of machinery and equipment (section 1.2(v)) suggests buoyant investment activity as well, particularly in oil-related sectors. The economy of Turkmenistan is dominated by resource extraction,[48] although strong performance in the textile and food industries also contributed to output growth in 2001. Turkmenistan's export earnings remain constrained by the lack of an independent pipeline network to transport gas to markets outside the CIS.[49]

In the course of 2001 the central Asian commodity exporters, Kyrgyzstan, Tajikistan and Uzbekistan, were already experiencing the negative impact of falling commodity prices. Kyrgyzstan remains dependent on just a few sectors, mainly gold extraction and the related metallurgy as well as agriculture. According to some estimates, almost 2 percentage points of the 6.7 per cent GDP growth during the first half of 2001 are directly attributable to the exploitation and development of the Kumtor gold field;[50] the rest of industrial production has in fact declined. Gross agricultural production in Kyrgyzstan increased by 14.6 per cent, year-on-year, in the first half of 2001 adding to the strength of aggregate output.[51] Uzbekistan, although self-sufficient in oil and gas, exports little of either and has consequently been unable to benefit from the recent terms of trade gains. However, modest but steady growth continued into the first half of 2001. Industrial production was driven by a surge in metal processing and machinery building, largely thanks to increasing capacity utilization and exports by a foreign owned car plant.[52] Tajikistan has a large agricultural sector while its industry is dominated by aluminium and electricity; consequently, its economy is largely dependent on the fluctuating world prices of its main export items, aluminium and cotton. The strong economic upturn, which started in 2000, continued during the first half of 2001 and was broadly-based in the key sectors of the economy.[53] Despite these positive developments, chronic under-investment continues to hamper restructuring in both agriculture and industry.[54] The recent slump in aluminium prices is also likely to have a negative impact on Tajikistan's economy.

(iii) **Costs and prices**

Inflationary pressures weakened in the first eight months of 2001 in most of the transition economies and, by and large, inflation was not a major policy concern in these countries. Imported inflation, a key cause of the setback in the overall disinflationary process in the last two years, subsided significantly as a result of weaker

---

[44] This is due to a large extent to the persistent energy supply disruptions in the country, another indication of its economic fragility.

[45] In the first half of 2001, the share of oil extraction and refining in GDP amounted to 36.9 per cent, while non-oil industry accounted for 6.1 per cent. In 1998, these shares were 13.5 and 9.5 per cent, respectively. TACIS, *Azerbaijan Economic Trends* (Baku), April-June 2001, p. 25 (www.economic-trends.org).

[46] As a word of caution, the reliability of Turkmenistan's official growth statistics, as well as that of the official growth projections (table 1.2.1) have often been questioned by independent analysts due to doubts about the proper deflation of nominal figures.

[47] Statistical Office of Kazakhstan, Sotsial'no-ekonomicheskoe razvitie respubliki Kazakhstan (Almaty), July 2001, p. 8.

[48] Gas, cotton and to a lesser degree oil, together accounted for 83 per cent of export earnings in 2000, while gas alone accounted for over 50 per cent of government revenue and almost 60 per cent of foreign exchange earnings. *Reuters Business Briefing*, 13 September 2001, quoting *Walden Publishing*.

[49] In addition, Ukraine, the largest market for Turkmen gas, owed some $600 million in unpaid gas bills by mid-2001. "Turkmenistan – Review", *Reuters Business Briefing*, 13 September 2001.

[50] *Reuters Business Briefing*, 21 August 2001. The country is the tenth largest gold producer in the world while the Kumtor gold field is responsible for about half of its industrial output.

[51] Interstate Statistical Committee of the CIS, *Main Macroeconomic Indicators, Kyrgyzstan* (http://www.cisstat.com/kir.htm).

[52] TACIS, *Uzbek Economic Trends* (Tashkent), April-June 2001, p. 31 (www.economic-trends.org).

[53] Preliminary data for the first quarter suggest that apart from aluminium there was strong growth in other sectors such as construction, transport and agriculture. IMF Staff Country Report No. 01/115, *Tajikistan: Second Review Under the Third Annual Arrangement Under the Poverty Reduction and Growth Facility and Request for Waiver of a Performance Criterion. Staff Report and News Brief on the Executive Board Discussion* (Washington, D.C.), 27 July 2001, p. 6.

[54] In order to assist at least basic infrastructure development in this, the poorest ECE economy, the EBRD approved in September 2001 a $13 million loan to modernize the telecommunications networks in Dushanbe, Khujand and Qurghon Teppa. *RFE/RL Newsline* Vol. 5, No. 187, Part I (Prague), 3 October 2001, quoting *Interfax News Agency*.

world energy prices and falling prices for other commodities.[55] Import price pressures were further moderated by falling world market prices for manufactures[56] in the face of the global economic slowdown. In addition, in many countries there was real exchange rate appreciation, particularly against the euro, which further reduced the domestic prices of imports.

On the domestic side, food prices, which were an important inflationary factor in 2000 in many transition economies, weakened substantially in 2001 thanks mainly to better than average harvests. As food prices fell considerably faster than usual during the summer, and given their still large weight in the consumption basket of households in these economies, the rate of consumer price disinflation gained further momentum over the summer months. Within the consumer price index, the major upward pressure came from service prices (particularly in Croatia, the Czech Republic, Yugoslavia, Kyrgyzstan and Russia).

Over January-August 2001, consumer price inflation was higher than in the same period of 2000 only in the Czech Republic, the three Baltic states and Russia (table 1.2.5); nevertheless, in Latvia and Lithuania the eight-month cumulative inflation rate remained below 2 per cent. The acceleration in the Czech Republic was mainly due to robust household demand, in turn the result of a combination of expansionary fiscal and income policies and a somewhat loose monetary policy. Relatively large gains in labour productivity, however, partly offset the effect of cost-push factors such as wage growth. In all the other east European countries, inflation rates fell and were even negative in Albania and Bosnia and Herzegovina. In Romania, the rate fell but still remained high at just below 20 per cent, the third highest among all the transition economies after Yugoslavia (32.5 per cent) and Belarus (25.4 per cent). In Poland, household demand weakened significantly as a result of further slowdown in real wage growth and soaring unemployment. This subdued consumer demand combined with the strong zloty and dwindling food prices during summer, lowered the eight-month cumulative inflation rate to 2.6 per cent in August, one of the lowest rates in the region and well below the expectations of the authorities at the beginning of the year. In contrast, in Hungary, a strong recovery in household consumption due to a large rise in real wages, substantial increases in agricultural and tourism incomes, and rapid growth of household credit were the major sources of inflationary pressure. Nevertheless, the strong forint, falling food prices and continuing (although smaller than in previous years) gains in productivity kept the slow rate of disinflation on trend. Recovery or a further strengthening of household demand was also one of the major sources of inflationary pressure in Bulgaria, Slovakia, Slovenia, the Baltic states and most of the CIS economies.

Industrial producer price inflation over January-August 2001 decelerated sharply and was even negative in several countries (table 1.2.5). Slovenia, Latvia and Armenia were exceptions to this pattern but even in these countries, increases in producer prices were not a source of concern. Inflation measured by the producer price index was much lower than that measured by the consumer price index in most transition economies, except Romania, Slovenia and a few small CIS economies. As mentioned above, service prices were the major component of the increase in the consumer price index in 2001. The smaller increase in producer prices than consumer prices may be explained to a large extent by two factors. First, real wages tend to increase not only in sectors with rapid productivity growth (mainly cost-efficient, export-oriented manufacturing branches) but also in less productive sectors (most of the non-tradeable service branches which are much less prone to competition), a phenomenon known as the "Balassa-Samuelson" effect.[57] Second, and probably more important, the experience of most east European countries indicates that consumer prices are usually more sensitive than producer prices to adjustments (usually increases) in administered or controlled prices.

Wage inflation in industry in the first half of 2001 remained relatively strong, rising faster than producer prices except in Bosnia and Herzegovina, The former Yugoslav Republic of Macedonia, Latvia and Lithuania (table 1.2.5). In the two Baltic countries, in fact, nominal wages were nearly 2 per cent lower than in the same period of 2000. In the CIS countries, real product wages, which were shrinking in the previous two years or so, recovered strongly in the first half of 2001 and rose at double-digit rates in most of them. On the other hand, industrial labour productivity grew strongly in the first half of 2001, although less vigorously than in 2000, largely mirroring the slower output growth in many transition economies. Among the east European and the Baltic countries, labour productivity fell only in The former Yugoslav Republic of Macedonia, stagnated in Yugoslavia, and rose at relatively low rates in Slovenia and Estonia. Productivity growth decelerated in the first half of 2001 in Bulgaria, Hungary and Slovakia even though its rate still remained between 5 and 6 per cent (year-on-year). In Russia, the slowdown was considerable, its rate falling from more than 9 per cent in 2000 to less than 4 per cent in the first half of 2001. The best performers, with increases in labour productivity of

---

[55] Over the first nine months of 2001, world market prices for energy commodities (in dollars) increased by only 1.1 per cent, while those for industrial raw materials fell by more than 18 per cent. Over the same period of 2000, these prices had risen by 28 per cent and 1 per cent, respectively.

[56] Developed market economies' export unit values, in dollar terms, for manufactured goods declined by 3-4 per cent, year-on-year, in the first half of 2001. See section 1.2(v) below.

[57] The experience of the transition economies since 1990 provides relatively strong evidence of the presence of the "Balassa-Samuelson" effect. UNECE, "Economic transformation and real exchange rates in the 2000s: the Balassa-Samuelson connection", *Economic Survey of Europe, 2000 No. 1*, chap. 6, pp. 227-239.

## TABLE 1.2.5

**Consumer prices, industrial producer prices and labour costs in industry[a] in the ECE transition economies, 2000-2001**

*(Percentage change)*

|  | \multicolumn{4}{c}{August over previous December} | \multicolumn{8}{c}{Annual average percentage change} |
|---|---|---|---|---|---|---|---|---|---|---|---|---|
|  | \multicolumn{2}{c}{Consumer price index} | \multicolumn{2}{c}{Producer price index} | \multicolumn{2}{c}{Nominal gross wages[b]} | \multicolumn{2}{c}{Real product wages[c]} | \multicolumn{2}{c}{Labour productivity[d]} | \multicolumn{2}{c}{Unit labour costs[e]} |
|  | 2000 | 2001 | 2000 | 2001 | 2000 | 2001[f] | 2000 | 2001[f] | 2000 | 2001[f] | 2000 | 2001[f] |
| Albania | -4.1 | -4.1 | .. | .. | .. | .. | .. | .. | 57.0 | .. | .. | .. |
| Bosnia and Herzegovina | -0.2 | -2.4 | -11.0 | -8.6 | 12.3 | 7.4 | 11.3 | 6.0 | 8.6 | 20.2 | 3.3 | -10.6 |
| Bulgaria [g] | 6.2 | 0.9 | 6.0 | 0.8 | 10.9 | 7.5 | -5.2 | -3.1 | 18.4 | 5.8 | -6.3 | 1.6 |
| Croatia [h] | 5.4 | 3.6 | 5.2 | -2.5 | 6.0 | 9.0 | -3.2 | 2.8 | 3.8 | 8.5 | 2.2 | 0.4 |
| Czech Republic | 3.4 | 4.9 | 3.3 | 0.8 | 7.0 | 7.5 | 1.9 | 3.2 | 8.5 | 7.9 | -1.3 | -0.4 |
| Hungary | 7.2 | 5.8 | 8.4 | 0.5 | 15.0 | 14.9 | 3.2 | 5.2 | 19.7 | 5.8 | -4.0 | 8.6 |
| Poland | 6.0 | 2.6 | 5.1 | 0.4 | 10.9 | 7.2 | 2.9 | 3.7 | 9.5 | 6.7 | 1.3 | 0.5 |
| Romania | 26.4 | 18.9 | 31.2 | 21.5 | 41.7 | 54.1 | -7.6 | 3.7 | 15.5 | 13.6 | 22.7 | 35.7 |
| Slovakia | 6.6 | 6.0 | 6.6 | 3.6 | 9.1 | 10.3 | -0.6 | 1.7 | 12.5 | 5.0 | -3.0 | 5.0 |
| Slovenia | 5.6 | 5.1 | 5.5 | 5.9 | 11.7 | 12.4 | 3.8 | 1.0 | 6.9 | 2.1 | 4.5 | 10.0 |
| The former Yugoslav Republic of Macedonia [h][i] | 7.2 | 2.9 | 5.0 | 2.3 | 5.5 | 5.0 | -3.2 | -0.5 | 8.4 | -2.3 | -2.7 | 7.5 |
| Yugoslavia | 36.2 | 32.5 | 55.5 | 22.9 | 97.8 | 128.1 | -3.7 | 4.6 | 14.8 | 0.2 | 72.3 | 127.6 |
| Estonia | 2.7 | 3.7 | 3.5 | 2.1 | 10.6 | 13.9 | 5.4 | 7.6 | 9.0 | 1.0 | 1.5 | 12.8 |
| Latvia | 0.7 | 1.7 | 0.4 | 1.5 | 15.1 | -1.8 | 14.2 | -2.9 | 1.0 | 8.2 | 14.0 | -9.2 |
| Lithuania | 0.7 | 1.7 | 2.2 | -1.9 | 1.2 | -1.7 | -14.1 | -2.6 | 9.2 | 21.8 | -7.3 | -19.3 |
| Armenia | -3.1 | 0.3 | -1.0 | 4.1 | 14.1 | 12.5 | 14.5 | 12.7 | 15.6 | 3.3 | -1.3 | 8.9 |
| Azerbaijan | -0.2 | -0.8 | 6.4 | -2.7 | 15.0 | 24.4 | 5.1 | 28.4 | 11.0 | 20.0 | 3.6 | 3.6 |
| Belarus | 67.1 | 25.4 | 107.1 | 20.4 | 201.9 | 119.5 | 5.7 | 3.9 | 8.2 | 5.6 | 179.1 | 107.9 |
| Georgia | 0.4 | 0.5 | 4.6 | 2.1 | .. | .. | .. | .. | .. | .. | .. | .. |
| Kazakhstan | 5.2 | 3.4 | 8.5 | -10.1 | 25.9 | 25.4 | -8.8 | 17.6 | 22.2 | 23.2 | 3.0 | 1.8 |
| Kyrgyzstan | 4.7 | 1.6 | 14.9 | 5.9 | 21.7 | 24.5 | -7.6 | 8.6 | 14.2 | 6.7 | 6.6 | 16.6 |
| Republic of Moldova | 13.9 | 1.1 | 19.4 | 4.6 | 32.3 | 28.5 | -1.0 | 9.1 | -2.0 | 11.4 | 35.0 | 15.4 |
| Russian Federation | 12.6 | 13.4 | 23.0 | 9.7 | 42.5 | 45.3 | -2.7 | 16.6 | 9.2 | 3.7 | 30.6 | 40.1 |
| Tajikistan | 23.8 | 7.3 | 14.6 | 5.0 | 30.7 | 42.8 | -6.0 | 3.8 | 20.9 | 14.9 | 8.0 | 24.3 |
| Turkmenistan | .. | .. | .. | .. | .. | .. | .. | .. | .. | .. | .. | .. |
| Ukraine | 18.6 | 3.3 | 14.1 | 1.3 | 30.2 | 39.0 | 7.8 | 23.2 | 19.6 | 21.0 | 8.9 | 14.8 |
| Uzbekistan [i] | 9.1 | 9.0 | 31.6 | 6.6 | 39.0 | 54.6 | -13.8 | 2.6 | .. | .. | .. | .. |

*Source:* UNECE secretariat estimates, based on national statistics and direct communications from national statistical offices.

[a] Industry = mining + manufacturing + utilities.

[b] Average gross wages in industry except in Bosnia and Herzegovina – net wages in industry; in Bulgaria, Estonia, Lithuania and all CIS economies – gross wages in total economy; in The former Yugoslav Republic of Macedonia and Yugoslavia – net wages in total economy.

[c] Nominal gross wages deflated by producer price index.

[d] Gross industrial output deflated by industrial employment.

[e] Nominal wages deflated by productivity.

[f] January-June 2001 over January-June 2000.

[g] June over previous December for purchasing price index.

[h] Retail price index for consumer prices.

[i] June over previous December for consumer and producer prices.

20 per cent or more, were Bosnia and Herzegovina, Lithuania, Azerbaijan, Kazakhstan and Ukraine. As a result of these developments, industrial unit labour costs, with few exceptions, remained broadly in check and in real terms they increased only in The former Yugoslav Republic of Macedonia, Yugoslavia, Estonia and some CIS countries, particularly in Russia. This suggests that producers' operating profits in most transition economies were subject to only mild downward pressures during the first half of 2001.

Preliminary quarterly national accounts data indicate that household consumption growth continued to be weak or had lost steam in the first half of 2001 in some of the east European countries (Bulgaria, Poland and Slovenia) and Estonia. On the other hand, significantly higher real household incomes, due to rising real wages and some easing of the situation on the labour markets, boosted private consumption throughout the CIS except Tajikistan and, probably also, Georgia, as suggested by the retail trade data.[58]

In the short term, unless another external shock raises input costs, a further slowdown in core inflation

---

[58] See table 1.2.4 above. In the absence of national accounts data for the CIS economies except Russia, retail trade data were used as a proxy for developments in private consumption in the first half of 2001.

rates (that is, excluding increases in administrative prices) can be expected. The major source of domestic pressure on industrial producer prices in the rest of 2001 may come from unit labour costs due to the expected slowdown in export-led growth in output and productivity. However, this may be partly offset by the expected deterioration in the labour markets, which is likely to ease wage pressures. At the same time, the increase in global economic uncertainties and the weakening growth in household incomes can be expected to check consumption demand in the near future. Consequently, the inflation forecasts and targets incorporated in the budgets for 2001, some of which seemed to be rather ambitious when they were set, now seem to be within reach, and in a few cases, the outturn can be expected to be even lower.

(iv) **Labour markets**

In the first half of 2001 there were some signs of improvement in the labour markets of some transition economies but, with the deepening of the global economic downturn, these appear to be coming to an end. The decline in employment decelerated in eastern Europe, with several countries reporting actual increases. For the first time, employment increased in the CIS region as a whole. Nevertheless, in many cases the demand for labour continued to be weak and, on average, rates of unemployment remained very high. One possible reason for the relatively weak response of employment to output growth is the specific pattern of industrial restructuring which combines rationalization of labour in existing firms with the adoption of modern, highly productive technologies in newly emerging ones. On balance this results in a sluggish demand for labour despite the robust output growth. However, the effect of changes in output on labour demand is asymmetric: thus if growth in the transition economies starts to weaken due to the global slowdown, there is a risk that employment will continue to fall and at an accelerating rate in late 2001 and 2002.

Unemployment declined in many countries during the first eight months of the year, and has done so on the largest scale since mid-1998, when in most transition economies it rose sharply in the wake of the Russian financial crisis. Nevertheless, the situation on the labour markets is rather heterogeneous among countries and subregions and several economies still have high and persistent rates of unemployment, which present a major challenge for economic and social policy. Moreover, in some cases (Bulgaria, the Czech Republic and Romania) a fall in unemployment in the first eight months of 2001 occurred in tandem with falling employment, suggesting departures from the labour force. It should also be mentioned that the negative social consequences of the tense situation on the labour markets in some countries are further exacerbated by the continued decline in the numbers eligible for unemployment benefits[59] as well as the regional distribution of unemployment. Average levels of unemployment mask considerable regional disparities in a number of countries.[60]

In eastern Europe, the decline in employment which started in 1999 and continued throughout 2000, decelerated in the first half of 2001 in the region as a whole (to 1 per cent from about 2 per cent in the same period of 2000) (table 1.2.6). In addition to Hungary and Slovenia, where relatively large increases in employment (around 1 per cent) were maintained for a fourth consecutive year, employment increased also in Albania and Slovakia. Employment was basically stagnant in Romania and Yugoslavia, but elsewhere it continued to fall, although in Bulgaria and the Czech Republic at a much slower rate than in the recent past. In the former this reflected higher labour demand in small businesses, while in the latter it was mainly due to net job creation in manufacturing.

In the 12 months to August 2001, unemployment in the region declined in the majority of countries for the first time since the second half of 1998 (table 1.2.6). There were large falls in Albania, Romania and Bulgaria[61] and further improvement in the Czech Republic, Hungary and Slovenia. After peaking at nearly 20 per cent in January 2001, the unemployment rate in Slovakia fell substantially during the following months. Nevertheless, at close to 18 per cent in August, it was still slightly higher than a year earlier and the highest among the EU accession candidates. In the rest of the region joblessness continued to rise. In Croatia, a persistently high and rising unemployment rate reached a new record of just below 23 per cent in March.[62] Although the rate declined somewhat in the summer, in August it was nearly one percentage point higher than a year earlier.

---

[59] In countries such as Albania, Bulgaria, Poland and Slovakia where unemployment rates were 15 per cent and higher in August 2001, some two thirds or more of the unemployed were no longer eligible for unemployment benefits.

[60] Interregional disparities in unemployment within individual transition economies largely reflect the countries' inherited industrial structures, which under central planning were, geographically, highly concentrated. The spatial patterns that emerged were sometimes quite irrational, paying little attention either to transport cost (for industry) or to natural conditions (for agriculture). As a result, the whole of central and eastern Europe faces serious problems with regional unemployment: for example, in Poland, the former state farms in the north have given rise to very high regional unemployment rates. As an illustration of the existing regional disparities, the numbers below indicate the ranges between the lowest unemployment rates (usually located in capital districts), the average national rates and the highest rates (located in depressed regions) in selected countries (the countries are listed in the order of their average national rates, all numbers are in per cent and refer to August 2001): Latvia: 3.6-7.8-26.8; the Czech Republic: 2.4-8.5-21.4; Lithuania: 5.5-12.1-26.2; Poland: 4.3-16.0-35.7; Bulgaria: 4.5-16.7-53; Slovakia: 6.1-17.8-24.9.

[61] As in Romania employment was somewhat stagnant and in Bulgaria it continued to decline; these falls suggest, *inter alia,* departures from the labour force.

[62] The statistics of registered unemployment in Croatia appear to be biased upwards. Thus, in the second half of 2000, the unemployment rate according to the labour force survey stood at 17 per cent as opposed to 21.3 per cent according to the unemployment register. Similar caution is advised when interpreting the unemployment data in all the successor states of the former SFR of Yugoslavia. For a more detailed discussion of the reasons for this discrepancy see UNECE, *Economic Survey of Europe, 2001 No. 1*, pp. 134, 136.

TABLE 1.2.6

**Total employment and registered unemployment in the transition economies, 1998-2001**
*(Percentage change over the same period of preceding year, per cent of labour force, end of period)*

|  | Employment[a] |  |  | QI-QII |  | Unemployment August |  |  |  |
|---|---|---|---|---|---|---|---|---|---|
|  | 1998 | 1999 | 2000 | 2000 | 2001 | 1998 | 1999 | 2000 | 2001 |
| **Eastern Europe** | 0.4 | -1.9 | -0.9 | -2.0 | -1.0 | 11.7 | 13.8 | 14.6 | 14.9* |
| Albania | -2.0 | -1.8 | 0.3 | -1.7 | 1.2 | 16.8 | 18.1 | 17.4 | 15.1 |
| Bosnia and Herzegovina [b] | 5.9 | 3.1 | 1.1 | 1.1 | -0.8 | 37.9 | 39.3 | 39.4 | 39.7 |
| Bulgaria | -0.2 | -2.1 | -4.7 | -10.0 | -4.6 | 10.8 | 13.6 | 18.0 | 16.7 |
| Croatia | 6.6 | -0.4 | -1.8 | -0.8 | -1.0 | 17.0 | 19.1 | 20.8 | 21.5 |
| Czech Republic | -1.6 | -3.6 | -2.3 | -2.3 | -0.6 | 6.4 | 9.0 | 9.0 | 8.5 |
| Hungary | 1.4 | 3.1 | 1.0 | 0.7 | 1.0 | 9.0 | 9.4 | 8.8 | 8.1 |
| Poland | 2.3 | -2.7 | -0.5 | -3.6 | -3.1 | 9.5 | 11.9 | 13.9 | 16.2 |
| Romania [c] | -1.9 | -0.6 | -0.1 | -0.5 | -0.3 | 8.7 | 10.9 | 10.1 | 8.1 |
| Slovakia [c] | -0.3 | -3.0 | -1.4 | -2.5 | 1.5 | 13.8 | 18.2 | 17.4 | 17.8 |
| Slovenia | 0.2 | 1.8 | 1.3 | 2.0 | 1.1 | 14.2 | 13.3 | 11.7 | 11.1 |
| The former Yugoslav Republic of Macedonia | -2.9 | 1.8 | -1.3 | 0.4 | -3.4 | 41* | 43* | 44* | .. |
| Yugoslavia [d] | -0.1 | -8.2 | -2.7 | .. | -0.2 | 27.0 | 27.2 | 26.8 | 27.8 |
| **Baltic states** | -0.5 | -1.2 | -2.0 | -1.7 | -2.6 | 5.8 | 8.3 | 9.5 | 9.8 |
| Estonia [e] | -1.3 | -4.1 | -0.9 | -2.1 | 0.7 | 4.2 | 6.3 | 6.0 | 7.3* |
| Latvia | 0.6 | -0.5 | – | 0.6 | -0.1 | 7.4 | 9.8 | 8.1 | 7.7 |
| Lithuania | -0.8 | -0.5 | -3.7 | -3.0 | -5.4 | 5.4 | 8.1 | 11.8 | 12.1 |
| **CIS** | -1.1 | -0.4 | .. | -0.2 | 0.6 | 8.4 | 8.6 | 7.0 | 6.2 |
| Armenia | -2.5 | -2.9 | -1.6 | -1.0 | -0.6 | 8.7 | 11.3 | 11.4 | 10.0 |
| Azerbaijan | 0.2 | – | – | – | 0.1 | 1.4 | 1.2 | 1.1 | 1.3 |
| Belarus | 1.1 | 0.6 | – | 0.1 | -2.6 | 2.3 | 2.1 | 2.1 | 2.5 |
| Georgia | 2.2 | .. | .. | .. | .. | 3.6 | 5.3 | .. | .. |
| Kazakhstan | -5.3 | -0.4 | 1.6 | -3.3 | 8.1 | 4.0 | 3.6 | 4.0 | 2.8 |
| Kyrgyzstan | 0.9 | 3.5 | 0.2 | 0.2 | -0.1 | 3.7 | 3.1 | 3.1 | 3.2 |
| Republic of Moldova | -0.2 | -9.0 | 1.3 | 3.0 | -1.0 | 2.0 | 2.4 | 2.1 | 1.9 |
| Russian Federation [f] | -1.5 | 0.5 | 0.6 | 0.7 | 1.1 | 11.6 | 11.9 | 9.8 | 8.2 |
| Tajikistan | 0.3 | -3.3 | 0.5 | -0.7 | -3.4 | 3.5 | 3.3 | 3.0 | 2.6 |
| Turkmenistan | 1.3 | 3.8 | .. | .. | .. | .. | .. | .. | .. |
| Ukraine | -1.1 | -2.3 | -2.5 | -2.4 | -1.6 | 3.6 | 4.1 | 4.2 | 3.7 |
| Uzbekistan | 1.4 | 1.0 | 1.1 | 1.0 | .. | 0.4 | 0.5 | 0.5 | .. |
| *Memorandum items:* |  |  |  |  |  |  |  |  |  |
| CETE-5 | 1.2 | -2.0 | -0.6 | -2.1 | -1.1 | 9.4 | 11.7 | 12.6 | 13.8 |
| SETE-7 | -0.7 | -1.8 | -1.2 | -1.7 | -0.8 | 15.0 | 16.8 | 17.5 | 16.5* |
| Russian Federation [g] | .. | .. | .. | .. | .. | 2.4 | 1.9 | 1.4 | 1.4 |
| Former-GDR | .. | .. | .. | .. | .. | 17.2 | 17.6 | 17.0 | 17.1 |

*Source:* National statistics; direct communications from national statistical offices to UNECE secretariat.

[a] Annual average unless otherwise stated. Regional quarterly aggregates of employment exclude Yugoslavia, Georgia, Turkmenistan and Uzbekistan.

[b] Figures cover only the Muslim-Croat Federation. Data for Republika Srpska are not available.

[c] Labour force survey employment data.

[d] Since 1999 excludes Kosovo and Metohia.

[e] Unemployment: until October 2000 – job seekers, thereafter – registered unemployed as percentage of the labour force.

[f] Unemployment figures are based on monthly Russian Goskomstat estimates according to the ILO definition, i.e. including all persons not having employment but actively seeking work.

[g] Registered unemployment.

Given the expected slowdown of GDP growth and the need to implement austerity measures in order to reduce macroeconomic imbalances, any radical improvement of the situation is unlikely this year. In Poland, an already difficult situation deteriorated further in the first eight months of 2001, reflecting the sharp slowdown in output growth coupled with the ongoing restructuring of unprofitable industries and the continued pressure of newcomers on the labour market. In August, the unemployment rate was 2.3 percentage points higher than a year earlier (the largest increase among all the transition economies) and reached a six-year high at 16.2 per cent of the labour force. As economic growth is expected to remain weak, and with a large influx of baby-boom graduates entering the labour market in the course of the year, national analysts estimate that the unemployment rate could reach a record 17-18 per cent of the labour force by the end of 2001.[63] The highest unemployment rates in the region were again in Bosnia and Herzegovina and probably in The former Yugoslav Republic of Macedonia, in both cases the rates standing at around 40 per cent.

Despite a robust economic performance in the Baltic states in 2000-2001, so far there have been no radical improvements in their labour markets, although the situation differed considerably between the three countries. In Estonia, the fall in employment bottomed out at the end of 2000 and in the first half of 2001 it grew by nearly 1 per cent, the first increase since 1997. The improvement was mainly due to manufacturing industry where employment started to grow rapidly in 2000 and continued to do so in the first two quarters of 2001.[64] In Latvia, where output growth was even stronger than in Estonia, the demand for labour remained weak and employment was actually flat during the first half of the year. This is partly explained by the fact that the growth of the Latvian economy was largely due to the increased transit of oil and oil products. The rapidly growing Lithuanian economy has not invigorated the country's labour market either. On the contrary, the situation has deteriorated considerably: the decline in employment accelerated in the first half of 2001 with a fall of over 5 per cent. Most of the job losses were in manufacturing, energy and agriculture, a reflection of the continuing restructuring of the enterprise sector and the country's specific pattern of growth (see section 1.2(ii)).

In August 2001, the average unemployment rate in the Baltic region stood at nearly 10 per cent,[65] slightly higher than a year earlier (table 1.2.6), although, again, the situation differed considerably among the countries. In Latvia, the downward trend, observable since 1999, continued in the first eight months of 2001. But the situation continued to deteriorate in Lithuania: unemployment there has been rising steadily since August 1998 and reached a record of more than 13 per cent in March 2001. In 1998, unemployment rates were very similar in all three economies but by August 2001 the rate in Lithuania was considerably higher than in the other two Baltic states and more than double what it was in 1998, just before the Russian crisis.[66] The official figures suggest that unemployment also increased in Estonia in the 12 months to August 2001, but this may largely reflect a change introduced in October 2000 to ease the eligibility conditions.[67]

As a result of the widespread and strong economic recovery in the CIS countries, there was some improvement in employment in the region as a whole.[68] In Russia, it rose by more than 1 per cent, reflecting the strength of the continuing recovery. The growth in employment continued for the third consecutive year after the August 1998 financial crisis, although its rate decelerated in the second quarter of 2001. As employment in large and medium-term enterprises – which still account for more than 60 per cent of total employment – broadly stagnated, most of the expansion was concentrated in small enterprises. The main sectors providing new jobs were services, but there was also a substantial increase, of nearly 2 per cent, in industry. In the first two quarters of 2001, there was also a large increase in employment in rapidly growing Kazakhstan with labour demand increasing in construction, and particularly in services. Employment remained flat in Azerbaijan and Kyrgyzstan and in the other countries for which data are available continued to decline, although in some cases at lower rates than in 2000 (except in Belarus and Tajikistan, where the falls were larger than in 2000).

Goskomstat's estimates of unemployment based on data from the labour force survey indicate a marked fall in unemployment in Russia, the rate declining steadily during the first eight months of 2001 to just above 8 per cent in August, 1.6 percentage points less than a year earlier. The situation also improved in Ukraine where according to the latest labour force survey, the unemployment rate fell to 10.5 per cent in June 2001, 0.9 percentage points lower than a year earlier.[69] According to rough estimates for the first half of 2001, the total number of unemployed in the CIS region (i.e. all those who are out of work and searching for a job, not just those who are registered) amounted to some 12 million persons, down from 13.5 million in the same period of 2000.[70] The main

---

[63] In August some 2,300 enterprises declared that they were planning some 81,000 layoffs in the near future. *Polish News Bulletin Company*, 24 September 2001, as quoted by *Reuters Business Briefing*.

[64] Despite the continued strong growth of manufacturing employment, there were signs of a sharp deceleration in the first half of 2001. The rate of growth declined from 12.9 per cent (year-on-year) in the first quarter to 5.5 per cent in the second. The situation may be further aggravated by the adjustment programme recently announced by Elcoteq Network Corporation (box 1.2.2).

[65] In interpreting the unemployment figures in these countries, one should take into consideration that the official figures for registered unemployment in the Baltic states tend to underestimate the actual levels of unemployment. In the second quarter of 2001, unemployment rates derived from labour force surveys were at 12.4 per cent in Estonia, 13.5 per cent in Latvia (May) and 16.6 per cent in Lithuania (May), whereas the registered rates were 7.5, 7.9 and 12.6 per cent, respectively.

[66] This partly reflects the fact that the Lithuanian economy has taken longer to recover from the effects of the Russian crisis.

[67] Although data from the labour force survey are not yet available for the third quarter of 2001, the data for the first half of the year do not indicate any increase in unemployment. On the contrary, according to the labour force surveys, the unemployment rate declined from 14.2 per cent in the first quarter of 2001 to 12.4 per cent in the second; these rates were also lower than the corresponding rates a year earlier, by 0.6 and 0.8 percentage points, respectively.

[68] Recently there has been a deterioration in the current statistical reporting of the situation on the labour markets of the CIS countries, which was inadequate in the first place. Thus, quarterly employment statistics are no longer available for several CIS countries. As for unemployment, registered unemployment data, often improperly measured, are the only series available for most of the CIS countries. Moreover, several countries (Georgia and Uzbekistan, in addition to Turkmenistan) recently stopped reporting unemployment statistics altogether.

[69] Direct communication to the ECE secretariat from the national statistical office.

[70] CIS Statistical Committee, *Statistika SNG, Statistical Bulletin*, No. 14, July 2000, p. 50 and No. 14, July 2001, p. 68.

contributors to this development were Russia, Ukraine and most likely Kazakhstan, where the labour market has improved since 2000. The average unemployment rate based on these figures was 9.4 per cent (compared with 10 per cent in 2000).

It is increasingly difficult to assess the unemployment situation in the rest of the CIS countries due to the scanty and unreliable statistics. The official figures indicate few changes in the 12 months to August 2001. Although registered unemployment rates remain very low, they do not reflect the true level of unemployment as a large proportion of the jobless, although willing to work, do not register for various reasons.[71]

### (v) International trade

#### (a) Trade of eastern Europe and the Baltic states

Contrary to widespread expectations, the foreign trade of the east European and Baltic countries maintained its momentum throughout the first six months of 2001, expanding at nearly the same rate (11-13 per cent) in current dollar value as in 2000.[72] However, growth in volume, although in double digits, weakened significantly in 2001: the region's exports and imports grew some 13 and 10 per cent in volume, respectively, against 21 and 15 per cent in 2000.[73] The terms of trade slightly worsened for a number of countries in the region, as dollar import prices increased slowly or were stagnant on average, whereas average export dollar prices declined, mainly under the influence of manufactured goods.[74] The region's aggregate merchandise trade deficit increased by one billion dollars in the first half of 2001, widening to $21.4 billion. This was in spite of a noticeable improvement in the Polish and Slovene trade balances.

The 13 per cent rise in the aggregate dollar value of east European and Baltic exports in the first six months of 2001 resulted partly from the diversified product structure and flexibility of the export sector in a number of these countries; the Czech Republic, Poland and Slovenia markedly increased their export earnings over those of 2000 (table 1.2.2). While the growth of exports in 2000 was primarily boosted by buoyant western import demand, in the first half of 2001 the latter faltered; but at the same time some of the transition economies benefited from the ongoing recovery of demand in the Russian, CIS and south-east European markets.[75] In particular, Croatia, Slovenia and the Baltic states were able to increase significantly their exports to the CIS.[76] However, the actual increase in total exports would not have been possible had there not been a steady growth of exports to the developed market economies in spite of the weakening of their overall import demand (table 1.2.7).[77]

The markedly increased presence of east European and Baltic producers on specific western markets is a result of their ongoing integration into the production networks of multinational companies and of improved cost competitiveness (see table 1.2.5 on changes in unit labour costs). According to Eurostat data, the share of the east European and Baltic countries in total EU imports increased from 4 per cent to 4.5 per cent between the first quarters of 2000 and 2001, and their share of EU imports originating from third countries rose to nearly 11 per cent (from 10 per cent in January-March 2000).[78] In the first three months of 2001, there were particularly large gains in EU market share for vehicles, spare parts and components (chapter 87 of HS) produced or assembled in CEFTA-7 countries, which rose from 21 to 28 per cent of extra-EU imports. This development reflects the deepening production fragmentation in major western automotive companies, often based on cheaper unit labour costs, and geographical proximity and other logistical factors. There were similar developments

---

[71] The proportion of those who do not register varies in different countries between 50 to 80 per cent of total unemployment. For a more detailed discussion see UNECE, *Economic Survey of Europe, 2001 No. 1*, p. 137. Ad hoc estimates suggest that actual unemployment rates in some CIS countries were much higher than those in table 1.2.6. These estimates suggest that the rate of unemployment in Georgia was around 12 per cent on average in the first half of 2001; in Kazakhstan at 12.7 per cent in April; and in the Republic of Moldova 8.8 per cent in the first quarter of 2001. National Bank of Georgia, *Bulletin of Monetary and Banking Statistics*, No. 6 (Tbilisi), 2001, p. 5; *Sotsial'no-economicheskoe razvitie Respubliki Kazakhstan, yanvar'-aprel' 2001 goda* (Alma Aty), May 2001, p. 93, TACIS, *Moldova Economic Trends*, April/June 2001, p. 42 (www.economic-trends.org).

[72] The continued weakness of the euro – in which most of the region's trade is denominated – relative to the dollar depressed export and import values expressed in dollars, but not as much as in 2000: in euros, east European and Baltic trade rose by 19-21 per cent in the first six months of 2001, following a 29-31 per cent rise in 2000.

[73] These are ECE secretariat estimates, based on preliminary data from several east European and Baltic countries (see table 1.2.8).

[74] World market prices in dollars for energy resources stagnated on average in the first six months of 2001 compared with January-June 2000, but starting in June they also dipped down; world commodity prices excluding energy fell markedly at the same time. Export unit values (in dollars) for manufactured goods declined, year-on-year, by 3-4 per cent in developed market countries (United Nations Statistical Division, *Monthly Bulletin of Statistics On-line*, query results for trade-manufactured goods exports, unit value indices in dollars, 1990=100). According to national statistics, export dollar unit values for manufactured goods also declined in Hungary but remained virtually unchanged in the Czech Republic and Poland. On the unprecedented increase in the export prices of Estonia see footnote *d* to table 1.2.8 and box 1.2.2.

[75] Russia's imports from non-CIS countries rose by more than 45 per cent in volume in the first half of 2001. This trend seems to have continued into the second half.

[76] In January-June 2001, Slovenia recorded a rise of 63 per cent, and Croatia and the Baltic states of 40 per cent, year-on-year, in their export dollar values to the CIS region. In contrast, exports from Bulgaria and Romania declined, while those from Hungary remained stagnant.

[77] According to Eurostat data, in January-March 2001 total EU imports rose by 3 per cent in current dollar value and those from third countries by nearly 6 per cent, while imports from east European and Baltic countries surged by more than 15 per cent, year-on-year. The pattern seems to have continued through April-May, when the dollar value of extra-EU imports to 12 EU countries (data for Denmark, Greece and Spain were not yet available at the moment of writing) shrank by 1 per cent, year-on-year, but imports from the east European and Baltic countries grew by 9 per cent. European Commission/Eurostat, *Intra- and Extra-EU Trade Flows, Monthly Data*, No. 8, 2001.

[78] European Commission/Eurostat, op. cit.

TABLE 1.2.7

**Foreign trade of the ECE transition economies by direction, 1999-2001**

*(Value in billion dollars, growth rates in per cent)[a]*

| | Exports | | | | Imports | | | |
|---|---|---|---|---|---|---|---|---|
| | Value | Growth rates | | | Value | Growth rates | | |
| Country or country group[b] | 2000 | 1999 | 2000 | 2001[c] | 2000 | 1999 | 2000 | 2001[c] |
| **Eastern Europe,** to and from: | | | | | | | | |
| World | 132.9 | -0.6 | 12.8 | 12.8 | 172.6 | -2.1 | 11.1 | 11.2 |
| ECE transition economies | 26.7 | -17.0 | 13.5 | 12.3 | 38.2 | -0.8 | 27.7 | 12.2 |
| CIS | 4.9 | -41.2 | 14.3 | 8.7 | 18.0 | -3.4 | 50.5 | 12.5 |
| Baltic states | 1.2 | 1.3 | 15.3 | 26.1 | 0.5 | 31.9 | 50.8 | 3.1 |
| Eastern Europe | 20.5 | -9.2 | 13.2 | 13.6 | 19.8 | 0.6 | 11.8 | 11.8 |
| Developed market economies[d] | 96.7 | 4.8 | 12.2 | 13.5 | 115.0 | -3.7 | 6.1 | 9.3 |
| European Union | 88.4 | 4.6 | 11.7 | 13.8 | 99.7 | -3.6 | 4.8 | 9.3 |
| Developing economies | 8.3 | 1.8 | 18.9 | 10.2 | 17.3 | 8.5 | 13.3 | 26.5 |
| **Baltic states,** to and from: | | | | | | | | |
| World | 8.9 | -11.6 | 24.5 | 18.9 | 13.0 | -13.6 | 15.5 | 13.1 |
| ECE transition economies | 2.5 | -34.7 | 18.7 | 23.7 | 4.4 | -10.5 | 23.4 | 15.3 |
| CIS | 0.9 | -50.9 | 0.9 | 39.7 | 2.7 | -13.9 | 36.8 | 14.1 |
| Baltic states | 1.2 | -12.3 | 28.2 | 15.2 | 0.8 | -6.3 | 7.8 | 14.3 |
| Developed market economies | 6.1 | 5.6 | 25.9 | 15.4 | 7.7 | -15.1 | 9.5 | 4.0 |
| European Union | 5.5 | 5.4 | 26.6 | 16.1 | 6.7 | -15.4 | 10.5 | 3.1 |
| Developing economies | 0.3 | 8.3 | 50.6 | 20.2 | 0.9 | -13.6 | 38.3 | 105.9 |
| **Russian Federation,** to and from: | | | | | | | | |
| World | 103.0 | 2.2 | 41.3 | 5.4 | 33.9 | -30.5 | 12.0 | 24.2 |
| Intra-CIS | 13.8 | -21.8 | 28.8 | 3.7 | 11.6 | -26.3 | 39.6 | 18.3 |
| Non-CIS economies | 89.2 | 7.9 | 43.5 | 5.7 | 22.3 | -32.0 | 1.5 | 27.2 |
| ECE transition economies | 17.9 | 6.3 | 61.1 | .. | 2.4 | -45.3 | 15.5 | .. |
| Baltic states | 4.9 | 26.0 | 73.7 | .. | 0.3 | -56.2 | 12.2 | .. |
| Eastern Europe | 12.9 | 0.9 | 56.8 | .. | 2.1 | -43.0 | 16.1 | .. |
| Developed market economies[d] | 49.6 | 4.2 | 37.6 | .. | 15.4 | -31.9 | 2.9 | .. |
| European Union | 36.9 | 7.1 | 48.4 | .. | 11.1 | -28.9 | -0.4 | .. |
| Developing economies | 21.7 | 19.4 | 44.5 | .. | 4.4 | -24.5 | -9.2 | .. |
| **Other CIS economies,** to and from: | | | | | | | | |
| World | 41.0 | -4.7 | 34.4 | 12.8 | 36.9 | -17.1 | 34.4 | 8.5 |
| Intra-CIS | 15.1 | -21.2 | 38.0 | 16.6 | 20.4 | -18.3 | 38.0 | 8.8 |
| Non-CIS economies | 25.9 | 7.9 | 32.4 | 10.5 | 16.5 | -15.9 | 32.4 | 8.1 |
| **ECE transition economies,** to and from: | | | | | | | | |
| World | 285.8 | -0.7 | 25.2 | 10.4 | 256.4 | -9.8 | 12.5 | 12.5 |

*Source:* National statistics and direct communications from national statistical offices to UNECE secretariat; for the Russian Federation, State Customs Committee data; for other CIS economies, CIS Statistical Committee.

*Note:* There were changes in the methodology of foreign trade reporting in several east European and Baltic economies in 1999-2000. In 1999, Poland changed its customs declaration system increasing substantially the coverage. In 2000, Estonia in its basic trade statistics switched to a "special trade" reporting system; this change, however, is not yet reflected in the Baltic states aggregate above. For details on prior-1998 changes see UNECE, *Economic Bulletin for Europe*, Vol. 48, 1996 and Vol. 49, 1997.

[a] Growth rates are calculated on values expressed in dollars.

[b] For country groups see table 1.2.1.

[c] January-June over same period of 2000.

[d] "Developed market economies" exclude Turkey and include Australia, New Zealand and South Africa.

elsewhere in the machinery and equipment sector, but also in other manufacturing sectors (furniture for instance), leading to marked increases in the presence of the east European and Baltic countries in specific product markets in the EU. Thus, exports of mechanical and electrical machinery and equipment and their parts (chapters 84 and 85 of HS) from east European and Baltic countries currently account for 10 and 12.5 per cent of extra-EU imports, respectively, while furniture, bedding, etc. (chapter 94 of HS) account for 42 per cent of extra-EU imports in that sector. However, some of the east European and Baltic countries, being small open economies, are now increasingly dependent on a limited range of export products and/or markets, which implies a high degree of risk should the industry or market fail. The recent major reshuffling of the Elcoteq production site in Estonia is a case in point (see box 1.2.2).

Export growth in eastern Europe and the Baltic states had slowed down by mid-2001 (chart 1.2.1) as weakening business and consumer confidence in western countries led to a fall in demand for most investment goods and some consumer durables, in particular for electrical machinery and communications network products and new vehicles.

TABLE 1.2.8

**Changes in the volume of foreign trade in selected transition economies, 1997-2001**
(Per cent)

| | Exports | | | | 2001[a] Jan.-Mar. | 2001[a] Jan.-Jun. | Imports | | | | 2001[a] Jan.-Mar. | 2001[a] Jan.-Jun. |
|---|---|---|---|---|---|---|---|---|---|---|---|---|
| | 1997 | 1998 | 1999 | 2000 | | | 1997 | 1998 | 1999 | 2000 | | |
| Croatia | .. | 11.6 | -2.7 | -1.3 | 0.7 | .. | .. | -4.4 | -4.2 | 3.7 | 16.6 | .. |
| **CETE-4** | 18.4 | 12.2 | 8.4 | 21.5 | 16.9 | 14.1 | 18.4 | 15.3 | 7.7 | 15.0 | 10.3 | 8.3 |
| Czech Republic | 14.2 | 14.4 | 9.7 | 20.2 | 17.1 | 14.9 | 7.6 | 8.5 | 5.8 | 19.2 | 16.5 | 14.7 |
| Hungary | 29.9 | 22.5 | 15.9 | 21.7 | 16.3 | 14.2 | 26.4 | 24.9 | 14.3 | 20.8 | 15.6 | 13.4 |
| Transition economies | 25.2 | 4.7 | -9.3 | 22.5 | 17.8 | 20.4 | 5.3 | 12.1 | 6.0 | 17.1 | 16.2 | 6.6 |
| European Union | 33.6 | 24.1 | 20.6 | 21.3 | 17.8 | 14.6 | 29.7 | 23.8 | 14.6 | 14.4 | 12.3 | 11.1 |
| Poland | 13.7 | 2.3 | 2.0 | 25.3 | 19.2 | 15.0 | 22.0 | 14.3 | 4.4 | 10.8 | 5.2 | 2.5 |
| Transition economies | 35.8 | -5.0 | -9.3 | 25.1 | 18.9 | 18.3 | 13.5 | 12.6 | 7.8 | 16.7 | 7.9 | 3.3 |
| European Union | 11.9 | 8.5 | 5.4 | 26.8 | 20.4 | 15.7 | 25.2 | 16.2 | 4.1 | 10.5 | 3.6 | 2.1 |
| Slovenia[b] | 11.9 | 8.1 | 3.7 | 11.4 | 10.0 | 7.7 | 10.5 | 11.3 | 9.2 | 4.1 | -1.4 | -0.3 |
| **Baltic states** | 20.3 | 8.7 | -7.1 | 25.9 | 15.6 | 12.3 | .. | 13.4 | -9.3 | 15.9 | 19.2 | 15.7 |
| Estonia[c,d] | 29.2 | 16.7 | 0.9 | 41.1 | 20.4 | 0.3 | .. | 13.1 | -9.0 | 35.0 | 27.4 | 18.8 |
| Latvia | 20.3 | 10.2 | -2.1 | 13.6 | 9.0 | 10.9 | .. | 21.3 | -3.2 | 5.1 | 12.7 | 10.1 |
| Lithuania | 12.8 | 1.3 | -16.3 | 19.2 | 14.8 | 23.0 | 24.6 | 9.0 | -13.0 | 7.4 | 16.6 | 16.6 |
| **Total above** | .. | 11.9 | 6.8 | 20.9 | 16.2 | .. | .. | 14.0 | 5.6 | 14.5 | 11.5 | .. |
| Russian Federation | 1.8 | -0.3 | 9.4 | 10.2 | 0.2 | 1.4 | 21.1 | -11.0 | -15.6 | 29.2 | 27.2 | 36.7 |
| Non-CIS | 1.8 | -0.6 | 11.3 | 9.9 | 1.7 | 1.9 | 31.7 | -8.4 | -19.4 | 28.6 | 35.3 | 47.0 |
| CIS | 1.9 | 0.8 | 1.5 | 12.5 | -8.0 | -0.7 | -1.8 | -18.4 | -4.8 | 30.6 | 10.4 | 16.3 |

*Source:* UNECE secretariat calculations, based on national foreign trade statistics.

*Note:* Dollar values of exports and imports in 2000 used for weights in calculating aggregate growth rates.

[a] Over same period of 2000.

[b] Changes in volumes for Slovenia are derived from indices of export and import unit values and changes in trade values as reported in *Statistical Yearbook of the Republic of Slovenia 2000* and *Monthly Statistical Review of the Republic of Slovenia*, Volume L, No. 6 (Ljubljana), 2001.

[c] Changes in volumes for Estonia are derived from indices of export and import prices and changes in trade values as reported in the monthly bulletin *Estonian Statistics*. Since Estonia switched to the "special trade" reporting system as from 1 January 2000, volume indices in this table were calculated on export and import values under the new reporting system and differ from the previously published trade data for Estonia in the statistical appendix and in other text tables of this Survey.

[d] According to the Estonian statistical office, export prices had risen 13 and 43 per cent year-on-year in the first and second quarters of 2001. For details see box 1.2.2.

Decelerating (year-on-year) growth rates prevailed throughout the region (except for Latvia and Lithuania) in April-June, but in July and August export performance was rather mixed. In July 2001, Croatian, Hungarian and Estonian exports shrank by 2-7 per cent, while Czech and Slovene exports picked up after a slump in June; in August, there was an even larger decline in exports from Estonia (down 17 per cent, year-on-year), while for most other countries, according to preliminary data, there were important increases in the dollar value of their exports (in the range of 9-15 per cent).[79]

In the first half of 2001 there was also a notable rise in the region's imports, although the aggregate dollar value of east European and Baltic imports increased at a slower pace than exports, mainly reflecting the low rate of growth of Polish imports. In fact, Polish imports, after surging 14 per cent in January, dropped slightly below the previous year's monthly dollar values in February and again in May onwards. This reflected the squeeze on domestic demand by the tight monetary policy, although the strengthening of the zloty in real effective terms probably helped to keep imports afloat. As a result, Poland's merchandise trade deficit shrank to $7.1 billion in January-June 2001 from nearly $9 billion in the first half of 2000. There was also a substantial improvement in the trade balance of Slovenia, where imports remained nearly flat throughout January-August as domestic demand was weak and the tolar depreciated slightly in real effective terms.[80]

---

[79] The strong export growth in Latvia and Lithuania through the second quarter of 2001 partly reflected continued expansion of their exports to the CIS and other transition economies. Lithuanian exports to the EU also grew rapidly (30 per cent, year-on-year) in the first half of 2001, but this growth was very narrowly based: some three quarters of the increase came from exports of oil and its products from Mazeikiai Nafta, a development which hardly has longer-term prospects. Already in July the export growth rate had decelerated noticeably and in August it was just 7 per cent, year-on-year. At the same time, the growth of exports of mineral products, including oil and electricity, was sharply reduced from over 50 per cent in January-June to below 4 per cent in July-August (year-on-year).

[80] In The former Yugoslav Republic of Macedonia, imports declined by a quarter in the first half of 2001 according to the Macedonian Central Bank, mainly because of the high import values in the base period when there was an influx of goods destined for refugees and for Yugoslavia which was then still under the Milosevic regime.

## CHART 1.2.1
**Monthly dollar exports and imports in selected transition economies, 2000-2001**
*(Year-on-year indices)*

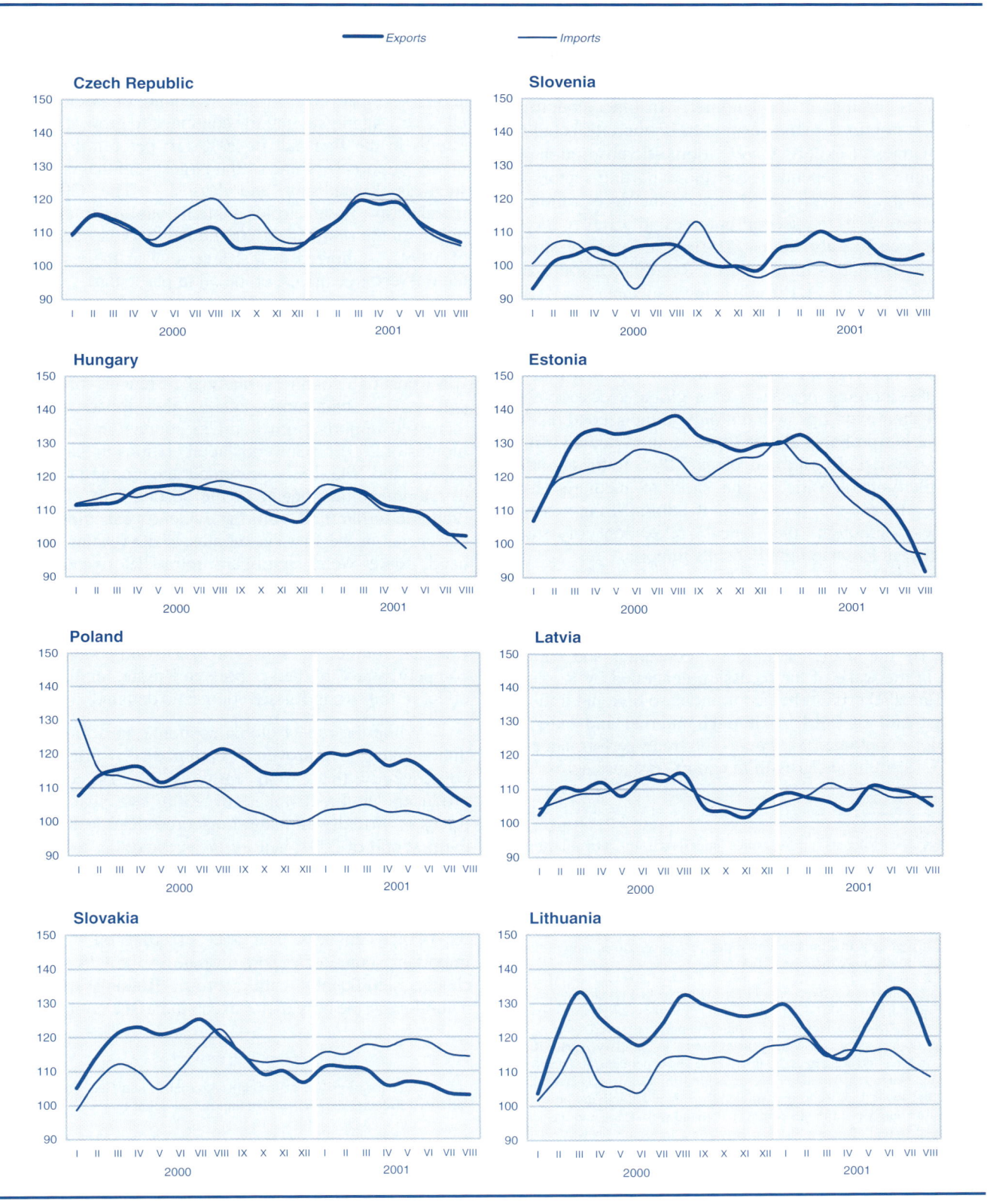

*Source:* UNECE secretariat calculations, based on national statistics.
*Note:* Monthly indices against same month of previous year are based on three-month moving average of monthly values in current dollars.

These were not, however, across-the-board developments: backed by recovering domestic demand and reduced trade protection,[81] import growth in Croatia, Romania and Slovakia, and to a lesser extent in Bulgaria, not only accelerated in the second quarter of 2001, but also exceeded that of exports, resulting in a considerable increase in their merchandise trade deficits.[82] In addition, import growth in these countries continued at double-digit rates throughout the summer. In other countries, however, import demand seemed to be more affected by the dynamics of exports than by domestic absorption – a clear sign of the dependency of the exporting sectors on imported inputs. This could explain the deceleration in import growth in the second quarter and a rather mixed performance in July-August 2001 – in spite of strong domestic demand – in the Czech Republic, Estonia and Hungary (chart 1.2.1), Estonia being a special case.[83]

Although the slackening of west European import demand had no immediate impact on the rise in east European and Baltic export revenues in the first half of 2001, the looming prospects of a global recession will hamper their export growth not only in the second half of this year but probably also in 2002. Import demand from commodity-rich Russia and other CIS countries is also most likely to subside in the short to medium term, mainly under the pressure of diminishing export revenues, although the recent relaxation of trade protection in Russia might boost its imports.[84]

### (b) International trade of the CIS countries

The continuing economic rebound in the CIS area – driven by the rising demand for consumer and capital goods in the wake of the recession generated by Russia's 1998 financial crisis – helped to increase regional trade in the first half of 2001. Although internal and external conditions continued to support the trade performance of many CIS countries, growth in export revenues across the CIS slowed markedly following the stabilization of crude oil prices and the weakening of other natural resource prices. In the first half of 2001, the dollar value of aggregate merchandise exports from the CIS increased by 8 per cent (table 1.2.2) down from 49 per cent in the corresponding period of 2000. In most CIS countries (except Georgia, Kyrgyzstan and Tajikistan) total exports rose, with increases of almost 60 per cent in Azerbaijan and 20-30 per cent in the Republic of Moldova and Ukraine. Aggregate imports continued to grow at the same rate as in the first half of 2000 (16 per cent), reflecting higher imports by some of the largest regional economies such as Kazakhstan and Russia. The CIS area's merchandise trade surplus – after doubling to about $33 billion in the first half of 2000 relative to the same period of 1999 – stabilized at this level due to the levelling of natural resource prices, crude oil in particular.

Currency movements also had a notable impact on CIS trade flows in the first half of 2001. The prevailing trend of real appreciation of national currencies in the CIS (most CIS countries' national currencies rose by 2-15 per cent in real terms against the dollar) supported increased imports from non-CIS sources in the first six months of 2001.[85] Currency appreciations in Kazakhstan, Russia and Ukraine represented the continuation of a development that began in the second half of 2000, a dramatic turnaround from the evident weakening of their currencies in the first six months of 2000. On the other hand, there were significant intra-CIS exchange rate movements. Given the substantial real appreciation of the rouble against the dollar, the currencies of all the CIS countries depreciated sharply in real terms against the Russian rouble.[86] This gave a solid boost to the competitiveness of their exports to Russia, reinforcing the effect of the strong Russian domestic demand.

While the robust global economic performance had dramatically improved commodity prices and consequently the export earnings of CIS commodity producers in the first half of 2000, one year later the prices of virtually all natural resources had declined.[87] In the first half of 2001, year-on-year, the dollar prices of all base metals decreased: aluminium and copper were down by 2-3 per cent and the price of nickel by almost a third. The gold price was lower by 7 per cent while cotton sold for 14 per cent less. The price of crude oil – the most important source of export revenue in the CIS – did not change significantly. In contrast, Russia's exports of natural gas – the only major exception to the commodities'

---

[81] For some details on measures undertaken in 2001 to further liberalize trade see UNECE, *Economic Survey for Europe, 2001 No. 1*, p. 148.

[82] According to preliminary data, merchandise trade deficits in relation to GDP in the first half of 2001 amounted up to 24 per cent in Croatia, 12 per cent in Romania and 8 per cent in Slovakia, some by 4-5 percentage points more than their 2000 levels.

[83] In Estonia, gross fixed capital formation picked up strongly in the second quarter of 2001 (11 per cent, year-on-year) and private consumption also rose. However, Estonian imports, after registering a 21 per cent dollar value growth, year-on-year, in January-April 2001, declined in May and were actually down by 6 and 4 per cent, respectively, in July and August. The major factor behind this was the sharp fall in imports for inward processing, which more than halved in these two months. In Hungary, although domestic demand was strong and the forint appreciated in real effective terms, a slowdown in import growth became obvious after May and in August dollar imports were down by 5 per cent, year-on-year. In the Czech Republic, in contrast, after falling by 3 per cent in June, imports picked up in the months that followed.

[84] Russia has announced import tariff reductions and unification for roughly 400 commodities as from 1 October 2001 effective for nine months. The tariff reductions will mainly affect imports of technological equipment. *Interfax News Agency, Weekly Business Report*, 3 October 2001, as reported by *Reuters Business Briefing*, 3 October 2001.

[85] The exceptions were Armenia, Azerbaijan and Georgia, where the national currencies depreciated slightly in real dollar terms.

[86] In most cases the real depreciation was in the range of 15-25 per cent except for the currencies of the Republic of Moldova and Ukraine, which depreciated slightly less (by 9 and 4 per cent, respectively).

[87] In the third quarter of 2001, commodity prices continued to decline reflecting deteriorating world demand and the uncertainty created by the 11 September terrorist attacks on the United States. In the wake of the terrorist attacks, gold benefited from its role as a safe haven. Gold averaged $284 per ounce for the month of September, up by almost $12 from August.

declining price trend, due to the specific pricing mechanisms in the gas sector – fetched over 50 per cent more relative to the first half of 2000.[88]

In general, CIS producers responded to lower metal prices by decreasing the volumes shipped. Oil exporters, however, increased shipments with the additional exports being accommodated by new pipeline capacity. For example, in the first half of 2001 the volume and value of Kazakhstan's oil exports increased by about 20 per cent, year-on-year – owing to accelerated production ahead of the opening of a major new pipeline – but the volumes of the country's major metal exports were either flat (copper and zinc) or declined (aluminium and steel products). As a result, the revenues from exports of metals declined by up to 26 per cent compared to the first half of 2000. Similarly, in Russia, virtually all metals exports declined in volume, but the country continued to ship more crude oil and oil products. In Azerbaijan, despite a modest increase in oil production, which now accounts for about 90 per cent of total exports, the combination of a 50 per cent rise in export volumes and higher prices led to a significant increase in total export revenues.[89] In contrast, Tajikistan's shipments of aluminium increased – the metal accounts for more than half of the country's exports – but much lower exports of electricity reduced the value of its total exports. Overall, as a result of the interplay between export prices and quantities shipped, on average, the earnings that accrued to CIS commodity exporters such as Azerbaijan, Kazakhstan, Russia and Tajikistan in the first six months of 2000 probably remained the same as in 2001.

In some other CIS countries, total exports increased in response to solid CIS-wide economic growth and the resulting increase in import demand. The Russian economy grew by 5.1 per cent in the first half of the year and this growth boosted directly and indirectly the export performance of those countries that rely heavily on CIS trade such as Armenia, the Republic of Moldova, Ukraine and, to a lesser extent, Belarus. Armenia and the Republic of Moldova increased their shipments of agricultural and food products while Ukraine boosted its major exports such as steel and chemicals. In Georgia and Kyrgyzstan total exports fell slightly because of lower demand in Turkey (Georgia's important trade partner) and lower CIS sales of electricity from Kyrgyzstan.

Although total imports into the CIS increased notably in the first half of 2001 (by 16 per cent, year-on-year) performance varied among countries. Compared with the first half of 2000, a number of CIS economies reported large increases in the dollar value of their total imports: 7 to 17 per cent in the Republic of Moldova, Tajikistan, Ukraine and Uzbekistan and between 24 and 46 per cent in Georgia, Kazakhstan, Russia and Turkmenistan. Among the countries with the largest increases in imports – Kazakhstan (46 per cent) and Turkmenistan (38 per cent) – imports of machinery and equipment accounted for most of the rise. These two countries were also the fastest growing CIS economies in the first half of 2001. In Kazakhstan, for example, imports of machinery and equipment, mostly related to investment in infrastructure for the gas and oil industry, increased by 65 per cent to over $900 million, while the value of imported steel pipes doubled. Similarly, the value of Russian imports of machinery and equipment increased by 27 per cent, year-on-year, in the first half of 2001, again almost exclusively due to fixed investment in oil-related sectors (section 1.2(ii)). In contrast, lower imports of machinery and equipment were behind the fall in total imports into Armenia, Azerbaijan and Kyrgyzstan (despite the high GDP growth in these countries) and the same was probably true of Belarus.[90] The falls in the total imports of these countries ranged between 4 and 17 per cent.

In the first half of 2001, the dollar value of CIS exports to non-CIS countries increased by 7 per cent (table 1.2.9). Traditional commodity exporters such as Kazakhstan, Kyrgyzstan, Russia, Tajikistan and Turkmenistan all increased exports in value by between 2 and 6 per cent, the differences reflecting variations in the importance of crude oil and oil products as well as the differential movements of commodity prices. Azerbaijan was the major exception among the natural resource exporters with its non-CIS exports rising by almost three fourths, the result of a 64 per cent increase in the volume of crude oil exports. Turkmenistan also increased its sales of crude oil and oil products – reportedly by 40 and 20 per cent, respectively – its total non-CIS exports increasing by 6 per cent. Similarly, Kyrgyzstan increased its sales of gold by about 15 per cent in value, which contributed to the overall increase in its non-CIS exports of 5 per cent. The Republic of Moldova increased its food and textile exports (to the EU in particular) and Ukraine's steel and chemical sales to non-CIS countries both rose by 15 per cent. Russia's non-CIS exports increased slightly in volume together with a 3 per cent increase in average export prices. In the remaining CIS countries, non-CIS exports were flat, except in Georgia where they fell 10 per cent and Armenia (-2 per cent), mainly because of much lower exports of jewellery and precious stones to the EU.

The dollar value of CIS imports from non-CIS countries rose in the first six months of 2001 by 19 per cent (table 1.2.9). These often consisted of imports of

---

[88] Russian Federation Goskomstat, *Sotzial'no-ekonomicheskoe polozhenie Rossii* (Moscow), various issues.

[89] The large increase in exports of crude oil from Azerbaijan probably came from inventories. The price for the country's oil exports was also reported to be 16 per cent higher than in the first six months of 2000.

[90] In Armenia total imports fell despite a 13 per cent increase in investment outlays. In Azerbaijan, which has passed a peak in its spending on oil and gas infrastructure, total imports fell. In Kyrgyzstan, total imports also fell as investment expenditures fell by one fifth.

### TABLE 1.2.9

**CIS countries' trade with CIS and non-CIS countries, 1999-2001**
*(Value in million dollars, growth rates in per cent)*

|  | Export 2000 | Export 2001[a] | Import 2000 | Import 2001[a] | Trade balances 1999 | Trade balances 2000 | Trade balances 2001[a] |
|---|---|---|---|---|---|---|---|
| **Armenia** | | | | | | | |
| Non-CIS | 29.5 | -1.9 | 13.9 | -12.3 | -449 | -484 | -196 |
| CIS | 30.1 | 48.6 | -7.3 | 3.0 | -131 | -100 | -43 |
| **Azerbaijan** | | | | | | | |
| Non-CIS | 110.3 | 73.5 | 12.1 | -16.4 | 8 | 713 | 760 |
| CIS | 11.4 | -27.6 | 15.4 | 25.6 | -114 | -140 | -126 |
| **Belarus** | | | | | | | |
| Non-CIS | 27.9 | 0.3 | 3.8 | -18.9 | -98 | 450 | 417 |
| CIS | 23.0 | 6.9 | 39.9 | -9.6 | -667 | -1 548 | -420 |
| **Georgia** | | | | | | | |
| Non-CIS | 48.0 | -10.1 | 27.0 | 19.5 | -246 | -285 | -141 |
| CIS | 26.9 | 9.8 | 10.0 | 33.7 | -118 | -111 | -76 |
| **Kazakhstan** | | | | | | | |
| Non-CIS | 64.6 | 1.8 | 9.9 | 43.3 | 2 011 | 4 455 | 1 673 |
| CIS | 60.2 | 16.0 | 72.6 | 47.5 | -106 | -367 | -409 |
| **Kyrgyzstan** | | | | | | | |
| Non-CIS | 9.8 | 5.3 | -24.9 | -23.5 | -70 | 42 | 49 |
| CIS | 13.1 | -17.6 | 15.2 | -10.4 | -76 | -91 | -36 |
| **Republic of Moldova** | | | | | | | |
| Non-CIS | -7.0 | 15.2 | 49.9 | 15.1 | -134 | -321 | -159 |
| CIS | 8.7 | 40.2 | 7.7 | 19.6 | 12 | 15 | 34 |
| **Tajikistan** | | | | | | | |
| Non-CIS | 9.8 | 2.4 | -22.3 | 43.9 | 225 | 295 | 130 |
| CIS | 18.7 | -31.6 | 8.7 | 5.7 | -200 | -186 | -168 |
| **Turkmenistan** | | | | | | | |
| Non-CIS | 71.4 | 5.9 | 10.0 | 15.7 | -300 | 100 | -200 |
| CIS | 165.3 | 17.8 | 36.0 | 100.8 | -10 | 620 | 220 |
| **Ukraine** | | | | | | | |
| Non-CIS | 21.0 | 14.8 | 15.9 | 11.3 | 3 227 | 4 159 | 2 286 |
| CIS | 38.3 | 38.5 | 19.2 | 4.8 | -3 491 | -3 542 | -1 692 |
| **Uzbekistan** | | | | | | | |
| Non-CIS | -7.1 | -1.0 | -20.0 | 10.1 | – | 290 | -10 |
| CIS | 20.0 | 3.6 | 40.0 | 4.1 | 200 | 90 | 60 |
| ***Total above*** | | | | | | | |
| Non-CIS | 32.4 | 10.5 | 7.1 | 8.1 | 4 174 | 9 414 | 4 608 |
| CIS | 38.0 | 16.6 | 30.8 | 8.8 | -4 701 | -5 361 | -2 657 |
| **Russian Federation** | | | | | | | |
| Non-CIS | 43.5 | 5.7 | 1.5 | 27.2 | 40 244 | 66 959 | 30 366 |
| CIS | 28.8 | 3.7 | 39.6 | 18.3 | 2 364 | 2 137 | 937 |
| **CIS total** | | | | | | | |
| Non-CIS | 40.8 | 6.8 | 3.8 | 18.9 | 44 418 | 76 373 | 34 974 |
| CIS | 33.4 | 10.2 | 33.8 | 12.1 | -2 337 | -3 224 | -1 720 |

*Source:* CIS Statistical Committee; UNECE estimates for Turkmenistan and Uzbekistan.

[a] January-June.

machinery and equipment to upgrade the capital stock in the oil, natural gas and metal sectors. This was the case in varying degrees in Kazakhstan, Russia and Turkmenistan. Russian non-CIS imports also rose in volume by 47 per cent helped by lower unit import values. In addition to increased imports of machinery and equipment, the strong growth in Russian private consumption led to more imports of consumer goods. In contrast, reduced non-CIS imports in Azerbaijan, Kyrgyzstan and probably Belarus were linked to low spending on capital goods in Azerbaijan and negative investment rates in the latter two countries.

In the first half of 2000, the value of intra-CIS trade increased by some 11 per cent. The relatively strong and extended economic recovery in Russia has increased demand for both capital and consumer goods from neighbouring economies. Increased CIS exports to Russia were also helped by the real appreciation of the rouble against the currencies of the other CIS countries. In the first half of 2001, Russia's imports from the CIS increased by 16 per cent in volume, benefiting the many countries that rely on Russia for a large part of their foreign sales. Armenia and the Republic of Moldova increased their sales of agricultural and food products and the solid performance of Ukrainian CIS exports was also due to sales to Russia. Ukraine's key export items, ferrous metals and chemicals, increased by 11 and 6 per cent, respectively, with a large proportion going to Russia. Kazakhstan also benefited from higher demand for raw materials in the CIS, its CIS exports, mainly crude oil and coal, increasing by 16 per cent. Turkmenistan's CIS exports increased by 18 per cent following the resumption of gas sales to Ukraine at the end of 2000 (up by 24 per cent in volume).

### (vi) Current accounts and external financing

Despite the deterioration in the international economic climate, the external financial situation of most transition economies remained stable or improved in the first half of 2001. The current account deficits of the east European and Baltic states were generally easily financed by increasing capital inflows, and foreign currency reserves increased (which was the case in many CIS countries as well). The access of creditworthy transition economies to the international financial markets improved and in some cases credit ratings were upgraded. This included the long-term (sub-investment grade) debt rating of Russia, thanks to its better internal and external finances and continuing structural change. However, the current account deficits of a number of countries increased to worrisome levels, particularly as signs of slower export growth became more widespread. Several transition economies continue to face serious financial constraints, generally because foreign investment has been discouraged by slow economic reform, political uncertainty and/or doubts about the sustainability of their debt.

### *(a) Current account developments*

In the first half of 2001, the combined current account of the transition economies was in surplus by $9 billion, down from $12 billion in the comparable period of 2000 (table 1.2.10).[91] This large surplus is due mainly to Russia, which has benefited from relatively high international fuel prices since 1999, while most of the other countries have current account deficits. In general, there are large differences among individual economies in their external performance.

---

[91] Unless otherwise noted, all comparisons and growth rates for the first half of 2001 are relative to the same period in 2000. All data in this section derive from national balance of payments statistics.

## TABLE 1.2.10

**Current account balances of the ECE transition economies, 2000-2001**

*(Million dollars, per cent)*

|  | Million dollars | | | Per cent of GDP | |
|---|---|---|---|---|---|
|  |  | January-June | | | Jan.-Jun. |
|  | 2000 | 2000 | 2001 | 2000 | 2001 |
| **Eastern Europe** | -19 024 | -10 349 | -11 291 | -5.0 | -5.7 |
| Albania | -163 | -81 | -147 | -4.4 | -7.2 |
| Bosnia and Herzegovina | -909 | -400* | -400* | -21.6 | -17.1 |
| Bulgaria | -702 | -432 | -270 | -5.9 | -4.8 |
| Croatia | -399 | -677 | -1 420 | -2.1 | -14.7 |
| Czech Republic | -2 273 | -690 | -1 131 | -4.5 | -4.3 |
| Hungary | -1 496 | -832 | -726 | -3.3 | -3.0 |
| Poland | -9 946 | -5 611 | -4 375 | -6.3 | -5.2 |
| Romania | -1 359 | -581 | -1 337 | -3.7 | -8.1 |
| Slovakia | -713 | -156 | -785 | -3.7 | -8.1 |
| Slovenia | -612 | -285 | -57 | -3.4 | -0.6 |
| The former Yugoslav Republic of Macedonia | -113 | -124 | -194 | -3.2 | -11.0 |
| Yugoslavia | -339 | -480* | -450* | -4.2 | -9.0 |
| **Baltic states** | -1 483 | -626 | -623 | -6.3 | -5.2 |
| Estonia | -315 | -118 | -148 | -6.3 | -5.6 |
| Latvia | -493 | -193 | -191 | -6.9 | -5.3 |
| Lithuania | -675 | -315 | -283 | -6.0 | -5.0 |
| **CIS** | 47 638 | 23 159 | 21 118 | 13.9 | 11.4 |
| Armenia | -278 | -155 | -100* | -14.5 | -14.2 |
| Azerbaijan | -168 | -34 | 100* | -3.2 | 4.3 |
| Belarus | -220 | -258 | 240 | -2.1 | 4.6 |
| Georgia | -262 | -128 | -180* | -8.7 | -12.4 |
| Kazakhstan | 743 | 609 | -551 | 4.1 | -5.5 |
| Kyrgyzstan | -77 | -32 | –* | -5.9 | – |
| Republic of Moldova | -121 | -57 | -38 | -9.4 | -6.4 |
| Russian Federation | 46 317 | 22 843 | 21 149 | 18.4 | 15.2 |
| Tajikistan | -62 | -37 | -80* | -6.3 | -21.2 |
| Turkmenistan | 100* | -50* | -240* | 2.3 | -9.6 |
| Ukraine | 1 481 | 292 | 845 | 4.7 | 5.0 |
| Uzbekistan | 184 | 164 | -28 | 1.4 | -0.6 |
| **Total above** | 27 131 | 12 183 | 9 205 | 3.6 | 2.3 |
| *Memorandum items:* | | | | | |
| **CETE-5** | -15 040 | -7 574 | -7 073 | -5.2 | -4.6 |
| **SETE-7** | -3 984 | -2 775* | -4 218* | -4.6 | -9.9 |

*Source:* UNECE secretariat, based on national balance of payments statistics; IMF, *Staff Country Reports,* for Bosnia and Herzegovina, Yugoslavia, Tajikistan and Uzbekistan (www.imf.org); for Azerbaijan: TACIS, *Azerbaijan Economic Trends* (Baku) (www.economic-trends.org).

*Note:* UNECE secretariat estimates of the current account of Bosnia and Herzegovina and Yugoslavia (January-June 2001) reflect their reported changes in merchandise trade in that period. Estimates for Armenia, Azerbaijan, Georgia, Kyrgyzstan and Tajikistan also incorporate first quarter 2001 current account results. The estimate for Tajikistan are one half of the balance projected for the full year by the IMF. The current account of Turkmenistan has been estimated using the estimated merchandise trade balance (custom basis) and assuming no change in the other current account items since 1998.

In eastern Europe, the combined current account deficit increased to over $11 billion in the first half of 2001, some 5.7 per cent of aggregate GDP. The dollar value of east European exports of goods and services continued to expand at a rate of 11-12 per cent in the first half of 2001, thanks in part to the faster growth of services. As noted above, however, the expansion of merchandise exports slowed as the year progressed, in contrast to what the year-on-year changes suggest. At the same time, the rate of growth of imports of goods and services quickened from 8 to 11 per cent in the first half of the year. The surplus on services continued to recover after the sharp decline in the late 1990s, but this has only partially offset the deterioration of the merchandise trade balance. Current account balances were also adversely affected by moderately larger income outflows and smaller inward transfers.

Despite the general weakening of the external environment, only about one half of the east European and Baltic countries experienced a widening of their current account deficits (although some of these deficits are quite large). Among the latter are Romania and Slovakia where the impact of slowing exports of goods and services has been amplified by an acceleration of domestic demand and imports. Domestic demand was also the dominant factor behind increased deficits in Croatia and the Czech Republic, despite the fact that Czech exports actually accelerated during the first half of the year. Dynamic growth of exports of goods and services were instrumental in reducing the current account deficit of Poland, tight monetary policy having kept import growth in check. A larger surplus on services explains the improvement in the current account balance of Hungary.[92] As a percentage of GDP, the aggregate imbalance of the south-east European transition economies was about twice as large as that of central Europe (table 1.2.10).

On average some 73 per cent of the combined current account deficit of eastern Europe and nearly all of that of the Baltic states was covered by net inflows of FDI (table 1.2.11). There are, however, considerable intercountry differences in the relative importance of this source of financing.

The traditionally large current account surplus of *Russia* declined slightly, to $21 billion in the first half of 2001 (from $22.8 billion in the same period of 2000). Although the growth of exports of goods and services quickened somewhat to 6 per cent, imports increased by 20 per cent, spurred by domestic demand and the real appreciation of the rouble. The same factors also boosted the demand for foreign travel, which accounts for the rise in the services deficit. However, net interest payments declined markedly, chiefly because of higher interest receipts on growing foreign exchange reserves.

The expansion of Russian demand continued to stimulate exports of goods and services from the other CIS countries. Fragmentary data indicate that around one half of them reported an improvement in their current account balances.[93] Belarus and Ukraine were in surplus, but their imports of goods and services remain below the

---

[92] Although the current account of Hungary moved into surplus in July and August, a deficit is expected for the full year.

[93] UNECE secretariat estimates are explained in table 1.2.10. Balance of payments data are no longer available for Turkmenistan and those for Uzbekistan are only released with a considerable lag.

### TABLE 1.2.11

**Foreign direct investment in the ECE transition economies, 2000-2001**

*(Million dollars, per cent)*

|  | FDI inflows [a] |  |  | Net FDI/ current account [b] |  |
|---|---|---|---|---|---|
|  |  | January-June |  |  | Jan.-Jun. |
|  | 2000 | 2000 | 2001 | 2000 | 2001 |
| **Eastern Europe** | 20 279 | 7 532 | 8 803 | 102 | 73 |
| Albania | 143 | 50 | 101 | 88 | 69 |
| Bosnia and Herzegovina | 150 | 70* | 80* | 17 | 20 |
| Bulgaria | 1 002 | 286 | 275 | 143 | 101 |
| Croatia | 926 | 678 | 388 | 225 | 27 |
| Czech Republic | 4 595 | 2 242 | 2 317 | 197 | 201 |
| Hungary | 1 700 | 906 | 1 245 | 78 | 123 |
| Poland (cash basis) | 8 294 | 2 737 | 2 748 | 82 | 61 |
| Romania | 1 025 | 274 | 313 | 76 | 25 |
| Slovakia | 2 075 | 130 | 714 | 289 | 84 |
| Slovenia | 176 | 40 | 189 | 18 | 287 |
| The former Yugoslav Republic of Macedonia | 170 | 99 | 373 | 149 | 192 |
| Yugoslavia | 25 | 20* | 60* | 7 | 13 |
| **Baltic states** | 1 173 | 530 | 729 | 74 | 98 |
| Estonia | 387 | 146 | 260 | 103 | 117 |
| Latvia | 408 | 191 | 128 | 81 | 63 |
| Lithuania | 379 | 193 | 341 | 56 | 112 |
| **CIS** | 5 329 | 2 152 | 3 464 | .. | .. |
| Armenia | 104 | 81 | 90* | 37 | 90 |
| Azerbaijan | 129 | 67 | 20* | 77 | .. |
| Belarus | 88 | 52 | 32 | 40 | .. |
| Georgia | 131 | 63 | 40* | 50 | 22 |
| Kazakhstan | 1 246 | 369 | 1 372 | .. | 249 |
| Kyrgyzstan | -2 | 14 | 25* | .. | .. |
| Republic of Moldova | 128 | 72 | 104 | 105 | 278 |
| Russian Federation | 2 714 | 1 004 | 1 188 | .. | .. |
| Tajikistan | 22 | 17 | 10* | 35 | 13 |
| Turkmenistan | 100* | 50* | 50* | .. | 14 |
| Ukraine | 595 | 333 | 515 | .. | .. |
| Uzbekistan | 75 | 32 | 17 | .. | 62 |
| **Total above** | 26 782 | 10 214 | 12 996 | .. | .. |
| *Memorandum items:* |  |  |  |  |  |
| **CETE-5** | 16 839 | 6 055 | 7 213 | 106 | 94 |
| **SETE-7** | 3 440 | 1 477* | 1 590* | 86 | 38 |

*Source:* UNECE secretariat, based on national balance of payments statistics; IMF, *Staff Country Reports*, for Bosnia and Herzegovina, Yugoslavia, Tajikistan and Uzbekistan (www.imf.org); for Azerbaijan: TACIS, *Azerbaijan Economic Trends* (Baku) (www.economic-trends.org).

*Note:* Estimates for Bosnia and Herzegovina, Yugoslavia and Tajikistan are one half of the corresponding IMF projections for the full year. Estimates for Armenia, Azerbaijan, Georgia and Kyrgyzstan reflect first quarter 2001 results and quarterly changes in 2000.

[a] Million dollars. Inflows into the reporting countries.

[b] Per cent. In this table (..) indicates that the current account balance was positive (and/or net FDI negative).

---

faster growth of imports of goods and services (and the shift in the current account to deficit) seems to reflect mainly larger purchases of oil and gas field equipment and services, both of which are associated with a surge in FDI. Concern has been expressed about the sustainability of the debt burdens of five low-income CIS countries – Armenia, Georgia, Kyrgystan, the Republic of Moldova, and Tajikistan.[95] In the first half of 2001, only the Republic of Moldova posted a relatively rapid growth of exports of goods and services (one of the conditions for debt sustainability) of 23 per cent. In Armenia export growth was considerably weaker. In the other countries exports fell, increasing the already large current account deficits of Georgia and Tajikistan.

### (b) Capital flows and FDI

Despite increasing signs of a global economic downturn and nervousness about emerging financial markets, net inflows on the capital and financial account into eastern Europe increased by nearly 30 per cent in the first half of 2001 (table 1.2.12). FDI remained the leading source of finance (as it did in the Baltic states), but portfolio investment increased sharply (table 1.2.13). The latter reflects external bond issues (see below) and foreign investments in local securities markets.[96] Net inflows of long-term finance were small, but there was another marked (net) outflow of short-term funds.[97] Inflows of unrecorded capital (the "errors and omissions" item in the balance of payments statistics) rose in the first half of 2001 to $3.7 billion (about equally divided between central and south-east Europe). Croatia and Romania reported the largest increases in capital inflows, consisting of modest amounts of FDI, external bond issues and, in the case of Romania, large inflows of short-term funds (including unrecorded flows which are often presumed to be short term). In both countries foreign exchange reserves also rose significantly.

Fragmentary data indicate that most CIS countries received less capital in early 2001, or experienced net outflows. In Belarus, Russia and Ukraine outflows were fuelled by current account surpluses (although in Ukraine FDI was a significant source of funds as well). Although the types of transaction underlying these outflows differ

1997-1998 levels. Both countries have been subject to severe financing constraints in recent years, although for Ukraine these appear to be easing.[94] In Kazakhstan, the

---

[94] Ukraine has been excluded from the international financial markets but it has continued to service outstanding bonds. In 2001 it obtained a Paris Club rescheduling, regained access to IMF/World Bank funds and, very recently, negotiated a rescheduling of its debt arrears to Russia's Gazprom.

[95] UNECE, *Economic Survey of Europe, 2001 No. 1*, pp. 162-163; IMF and World Bank, "Armenia, Georgia, Kyrgyz Republic, Moldova and Tajikistan: external debt and fiscal sustainability", 7 February 2001, background paper prepared jointly by the European II Department of the IMF and the Europe and central Asia region of the World Bank (www.imf.org/external).

[96] During the course of 2001, foreign investors reduced their exposure to eastern Europe (as they did to emerging markets in general), which partly explains the decline in local stock market prices. However, the Russian stock exchange was one of the best performing in the world.

[97] Virtually all east European countries reported a build-up of short-term assets abroad, the total amounting to over $4 billion in the first half of 2001.

## TABLE 1.2.12

**Net capital flows into the ECE transition economies, 2000-2001**
*(Million dollars, per cent)*

| | Capital and financial account flows[a] Million dollars | | | Capital flows/GDP | | Changes in official reserves[b] (million dollars) | |
|---|---|---|---|---|---|---|---|
| | | Jan.–Jun. | | | Jan.–Jun. | | Jan.–Jun. |
| | 2000 | 2000 | 2001 | 2000 | 2001 | 2000 | 2001 |
| **Eastern Europe** | 25 051 | 12 272 | 15 775 | 6.6 | 8.0 | 6 028 | 4 484 |
| Albania | 295 | 93 | 218 | 7.9 | 10.7 | 132 | 72 |
| Bosnia and Herzegovina | 923 | 450* | 410* | 21.9 | 17.5 | 14 | 10* |
| Bulgaria | 1 111 | 488 | 106 | 9.3 | 1.9 | 409 | -164 |
| Croatia | 981 | 958 | 1 946 | 5.2 | 20.1 | 582 | 527 |
| Czech Republic | 3 092 | 1 420 | 1 644 | 6.1 | 6.2 | 819 | 513 |
| Hungary | 2 557 | 982 | 1 794 | 5.6 | 7.5 | 1 060 | 1 068 |
| Poland | 10 565 | 4 846 | 5 041 | 6.7 | 6.0 | 619 | 666 |
| Romania | 2 287 | 943 | 2 496 | 6.2 | 15.2 | 928 | 1 159 |
| Slovakia | 1 537 | 948 | 578 | 8.0 | 6.0 | 824 | -206 |
| Slovenia | 790 | 362 | 662 | 4.4 | 7.2 | 178 | 605 |
| The former Yugoslav Republic of Macedonia | 349 | 223 | 279 | 10.6 | 17.1 | 236 | 85 |
| Yugoslavia | 566 | 560* | 600* | 7.0 | 12.0 | 227 | 150* |
| **Baltic states** | 1 754 | 815 | 667 | 7.5 | 5.6 | 271 | 44 |
| Estonia | 437 | 117 | 41 | 8.7 | 1.6 | 122 | -107 |
| Latvia | 511 | 196 | 224 | 7.2 | 6.2 | 18 | 33 |
| Lithuania | 806 | 503 | 402 | 7.1 | 7.1 | 131 | 118 |
| **CIS** | -30 313 | -13 994 | -13 299 | -8.8 | -7.2 | 17 325 | 7 759 |
| Armenia | 298 | 140 | 69* | 15.5 | 9.9 | 19 | -30* |
| Azerbaijan | 442 | -20 | -29* | 8.4 | -1.3 | 274 | 70* |
| Belarus | 295 | 336 | -176 | 2.9 | -3.3 | 76 | 64 |
| Georgia | 242 | 105 | 190* | 8.0 | 13.1 | -20 | 10* |
| Kazakhstan | -602 | -673 | 777 | -3.3 | 7.8 | 141 | 226 |
| Kyrgyzstan | 98 | 42 | -10* | 7.5 | -1.9 | 21 | -10* |
| Republic of Moldova | 167 | 72 | 35 | 13.0 | 5.9 | 46 | -3 |
| Russian Federation | -30 307 | -13 522 | -13 980 | -12.1 | -10.1 | 16 010 | 7 170 |
| Tajikistan | 91 | 35 | 69* | 9.2 | 18.3 | 29 | -11* |
| Turkmenistan | .. | .. | .. | .. | .. | .. | .. |
| Ukraine | -1 083 | -465 | -463 | -3.4 | -2.7 | 398 | 382 |
| Uzbekistan | -153 | -274 | -2 | -1.1 | – | 31 | -30 |
| **Total above** | -3 508 | -908 | 3 143 | -0.5 | 0.8 | 23 624 | 12 288 |
| *Memorandum items:* | | | | | | | |
| **CETE-5** | 18 540 | 8 557 | 9 719 | 6.4 | 6.3 | 3 500 | 2 646 |
| **SETE-7** | 6 512 | 3 715 | 6 056 | 7.5 | 14.2 | 2 528 | 1 839 |
| Russian Federation[c] | -21 040 | -10 687 | -9 996 | -8.4 | -7.2 | 16 010 | 7 170 |

*Source:* UNECE secretariat, based on national balance of payments statistics; IMF, *Staff Country Reports,* for Bosnia and Herzegovina, Yugoslavia, Tajikistan and Uzbekistan (www.imf.org); for Azerbaijan: TACIS, *Azerbaijan Economic Trends* (Baku) (www.economic-trends.org).

*Note:* Estimates for Bosnia and Herzegovina, Yugoslavia and Tajikistan are one half of the relevant IMF projections for the full year. Estimates for Armenia, Azerbaijan, Georgia and Kyrgyzstan reflect first quarter 2001 results and quarterly changes in 2000.

[a] Includes errors and omissions; excludes changes in official reserves.
[b] A negative sign indicates a decrease in reserves.
[c] Excluding errors and omissions.

weakens. Although this scenario remains likely, data for the first half of 2001 show FDI inflows continuing to rise (table 1.2.11). This growth presumably reflects the long-term nature of FDI and projects in the pipeline. Also, large privatizations involving foreign strategic investors have continued in a number of countries. Such deals, often involving national monopolies, seem to be less subject to the business cycle than greenfield investments. Inflows of FDI increased in most countries – the large energy-related investment in Kazakhstan is noteworthy – but in some they fell short of initial expectations due to delays or postponements of privatizations. FDI inflows have remained relatively small in Russia, and steadily increasing outflows have made the country a net exporter of FDI since 2000.

International bond issues by the transition economies were modest in the first three quarters of 2001, remaining at around $5 billion. The creditworthy countries in the area had only a limited need for such funds thanks to FDI inflows (over one half of the transition economies, mainly the CIS, remain excluded from the bond markets). Although investor sentiment towards emerging markets was generally negative, lenders were willing to discriminate among countries according to their perceived credit quality. In fact the Czech Republic, Hungary, Poland and Slovenia are no longer viewed by bond investors as emerging market risks.[98] Sovereign borrowers accounted for the bulk of the euro-denominated bonds that were issued, mainly in the first half of the year. Hungary borrowed •1 billion to refinance maturing debt, while Poland issued a •750 million bond largely for benchmarking purposes. However, in Croatia (a •750 million emission) and Lithuania (•200 million) the funds were intended to finance budget deficits. Romania, which had acute financial difficulties in 1999, consolidated its access to the international markets with a •600 million bond. Corporate borrowing remains far below the pre-rouble crisis levels and has been largely limited to blue chip companies.[99] The transition economies benefited from investors seeking to diversify their portfolios and, as a result, all of the issuing countries successfully broadened their investor bases. Following the terrorist attacks in the United States on 11 September, several sovereign bonds and initial public equity offerings (IPOs) planned by the transition economies (and other issuers as well) have been postponed. However, the city of Moscow was able to issue a bond in late October suggesting that the international market remains receptive to the debt issues of some of these countries.

among countries, it appears that in all three there were various degrees of capital flight. This was particularly true of Russia, but its official reserves rose sharply nonetheless, to nearly $38 billion in September (about $34 billion excluding gold).

A decline in FDI inflows into the transition economies has been expected because investors tend to postpone new investment when the global economy

---

[98] These countries' long-term bonds are also attractive because they offer the possibility of a so-called "convergence play": the holders of the bond hope to profit from the expected convergence of the bond yield to the lower yields prevailing in the euro area.

[99] For example, bonds were issued by Poland's telecommunications company TSPA (•500 million) and gas monopoly PFNiG (•800 million in October) and by Kazakhstan's oil transport firm KazTransOil ($150 million).

TABLE 1.2.13

**Net capital flows into eastern Europe, the Baltic states and selected members of the CIS, by type of capital, 1999-2001**

(Billion dollars)

|  | Eastern Europe[a] |  |  |  | Baltic states |  |  |  | Three European CIS[b] |  |  |  |
|---|---|---|---|---|---|---|---|---|---|---|---|---|
|  |  |  | Jan.-Jun. |  |  |  | Jan.-Jun. |  |  |  | Jan.-Jun. |  |
|  | 1999 | 2000 | 2000 | 2001 | 1999 | 2000 | 2000 | 2001 | 1999 | 2000 | 2000 | 2001 |
| Capital and financial account | 24.4 | 21.9 | 9.1 | 12.1 | 2.2 | 1.6 | 0.7 | 0.6 | 0.1 | -0.6 | -0.4 | -0.2 |
| Capital and financial account[c] | 28.7 | 25.1 | 12.3 | 15.8 | 2.2 | 1.8 | 0.8 | 0.7 | -1.1 | -0.6 | -0.1 | -0.6 |
| of which: |  |  |  |  |  |  |  |  |  |  |  |  |
| FDI | 18.8 | 19.4 | 7.4 | 8.3 | 1.0 | 1.1 | 0.5 | 0.6 | 1.0 | 0.8 | 0.5 | 0.6 |
| Portfolio investment | 2.6 | 2.3 | 2.2 | 5.5 | 0.8 | 0.1 | 0.4 | 0.1 | -0.2 | -0.1 | – | -0.2 |
| Medium-, long-term funds | 4.9 | 4.8 | 2.9 | 0.6 | 0.5 | – | – | -0.1 | 1.0[d] | -1.2[d] | -1.4[d] | -0.4[d] |
| Short-term funds | -2.9 | -5.4 | -3.7 | -3.1 | -0.1 | 0.5 | -0.2 | -0.1 | -1.8 | -0.2 | 0.5 | -0.2 |
| Errors and omissions | 4.3 | 3.1 | 3.1 | 3.7 | – | 0.1 | 0.1 | 0.1 | -1.2 | – | 0.3 | -0.4 |
| *Memorandum item:* |  |  |  |  |  |  |  |  |  |  |  |  |
| Short-term investment[e] | 4.1 | – | 1.6 | 6.1 | 0.6 | 0.6 | 0.3 | 0.1 | -3.3 | -0.3 | 0.8 | -0.8 |

*Source:* UNECE secretariat estimates, based on national balance of payments statistics.

[a] Excludes Bosnia and Herzegovina and Yugoslavia.
[b] Belarus, Republic of Moldova and Ukraine.
[c] Including errors and omissions.
[d] Includes short-term funds for Belarus.
[e] Portfolio investment, short-term funds and errors and omissions.

## 1.3 The short-term outlook

### (i) The western market economies

The short-term economic outlook for the ECE region, and indeed for the world economy as a whole, has become exceptionally uncertain since the terrorist attacks in New York and Washington on 11 September. At the time of writing this *Survey* it is impossible to know how long the military action in Afghanistan will last, what kind of responses it will trigger and whether the conflict can be contained. The current public fears of anthrax in the United States have, moreover, raised the spectre of "bio-terrorism" that risks accentuating the sense of insecurity caused by the events in September and further depressing business and consumer confidence. But all this remains to be seen.

The current trends in the world economy are pointing to a worst-case scenario in which the simultaneous weakening of economic activity in the major economies and all major regions has strengthened deflationary forces which, if allowed to continue, could push the global economy into deep recession.

Predictions of a quick recovery from the cyclical downturn in the United States have so far proved to be wrong. Although many forecasts now assume that a recovery will start in the second half of 2002, there are no definite signs that this will indeed be the case. The main reasons for scepticism about this are the considerable internal and external imbalances that have accumulated in the United States economy in the period 1995-2000 (see section 1.1 above). The orderly unwinding of these imbalances will take time and their presence will tend to reduce the effectiveness of the considerable stimulus from monetary and fiscal policy.[100]

The corporate sector needs to eliminate a large amount of spare capacity before business investment can be expected to respond to the lower interest rates and assume its traditional role of supporting economic recovery. In a similar vein, private households need to restructure their balance sheets and to reduce high debt levels in view of deteriorating prospects in labour markets and significant stock market losses. As a result, the outlook for fixed investment and private consumption is not clear and this leaves the prospects for a recovery in 2002 in doubt.

As the United States was the main engine of global economic growth in 1995-2000, it is clear that a prolonged weakness of the United States economy will have serious consequences for other regions, notably the developing countries. This is particularly so for its major trading partners in Asia and Latin America and also Canada. In Japan, real GDP is expected to decline not only in 2001 but also next year as well. These developments will, in turn, weaken further the growth prospects for the European economies.

Against this background, there has been a marked lowering of growth forecasts for the United States, Japan and the euro area for both this year and next compared with those made before the attacks in early September (table 1.3.1). The figures for 2002 should, of course, be regarded with great caution – they reflect inevitably rough estimates of the impact that the attacks and the

---

[100] UNECE, *Economic Survey of Europe, 2001 No. 1*, chap. 1.

### TABLE 1.3.1

**Changes in the consensus forecasts of economic growth in 2001 and 2002**

*(Percentage change over previous year)*

|  | Survey data 10 Sept. 2001 2001 | 2002 | Survey data 8 Oct. 2001 2001 | 2002 | Change[a] 2001 | 2002 |
|---|---|---|---|---|---|---|
| United States | 1.6 | 2.7 | 1.0 | 1.2 | -0.6 | -1.5 |
| Canada | 1.8 | 2.7 | 1.4 | 1.7 | -0.4 | -1.0 |
| Euro area | 1.9 | 2.4 | 1.7 | 1.8 | -0.2 | -0.6 |
| France | 2.4 | 2.5 | 2.0 | 1.8 | -0.4 | -0.7 |
| Germany | 1.1 | 2.1 | 0.9 | 1.5 | -0.2 | -0.6 |
| Italy | 2.1 | 2.4 | 1.9 | 1.5 | -0.2 | -0.9 |
| United Kingdom | 2.1 | 2.6 | 2.1 | 2.1 | – | -0.5 |
| Japan | 0.1 | 0.5 | -0.5 | -0.4 | -0.4 | -0.9 |

*Source:* Consensus Economics Inc., *Consensus Forecasts* (London), various issues.

[a] Percentage points.

steeper downturn in the global economy might have on expectations and confidence, and hence on the spending behaviour of households and enterprises. At the same time, however, they illustrate the considerable downside risks facing the global economy given the possibility of a simultaneous contraction of output in the euro area, Japan and the United States and the ensuing negative multiplier effects.

By far the largest reductions in the forecasts of economic growth are those for the United States, especially for 2002, but expectations have also been reduced substantially for other countries and the euro area as a whole. Although many forecasts now assume that a recovery will start in the second half of 2002, the overall strength of economic activity is seen to remain rather weak for the year as a whole. In the United States, the consensus is for a continued decline of business investment and a marked slowdown of consumer spending, which had been the main source of growth until the second quarter of 2001. Business investment has collapsed against a background of deteriorating prospects for sales and profits. Private consumption expenditures can be expected to weaken under the impact of a deteriorating labour market and increasing strains in households' balance sheets. Both business and consumer confidence will be depressed by the heightened uncertainty and this may lead to a postponement of spending plans. Real GDP in the United States is now expected to increase by only 1 per cent in 2001 followed by only 1¼ per cent in 2002.

In the euro area, the rate of economic growth is likely to be only some 1¾ per cent this year, down from 3.4 per cent in 2000 and a full percentage point less than the European Commission's forecasts in the spring 2001.[101] On current forecasts, the average growth rate of the euro area in 2002 is unlikely to be much better than this year.

The same holds for the aggregate of the 15 European Union countries (table 1.3.2). The average outcome is strongly influenced by the pronounced cyclical weakness in Germany, the largest western European economy. The German government now expects an annual rate of economic growth of only ¾ of a percentage point in 2001, down from 3 per cent in 2000. Growth is expected to strengthen only slightly in 2002, largely a reflection of the continued modest growth of the global economy.

In the euro area the virtual stagnation of economic activity in the second quarter of 2001 increases the probability of a contraction in the third quarter. The sharp fall of the *ifo* business climate index, a key cyclical indicator, in September 2001 signalled a rapid deterioration of business prospects in Germany. This is likely to be the case in other member countries as well. Given the weak economic conditions in the euro area and the global economy at large and forecasts of rapidly declining rates of inflation, the ECB now has room for a further loosening of monetary policy. As regards fiscal policy, at minimum the automatic stabilizers should be allowed to operate fully in order to avoid procyclical effects. The ECB has argued in favour of continued fiscal discipline in the three larger economies of the euro area in line with the letter of the Stability and Growth Pact. But it is difficult to reconcile the two policies: demanding fiscal discipline in a cyclical downturn and at the same time pursuing an overly cautious monetary policy – either with each other or with the current needs of the European economy.[102]

It was accepted in the early years of EMU, that countries with large budget deficits – i.e. close to 3 per cent or above – would need time to achieve the desired medium-term fiscal position of "close-to-balance or in surplus" and that in the event of a severe cyclical downturn in the early years of EMU they could therefore be forced into a procyclical fiscal stance by the rules of the Stability and Growth Pact. The probability of such a risk was, of course, unknown but it was assumed that a sustained cyclical upturn would enable the medium-term fiscal targets to be reached. But actual developments have not turned out that way. The euro area is now on the edge of recession in a very worrisome global context. One of the major principles of the Pact, namely that fiscal discipline is a precondition for eventual fiscal flexibility,[103] is therefore hardly adequate for the current economic situation, which was not envisaged when the Pact was designed. It can, of course, be argued that countries have not done enough to consolidate their fiscal positions, although this is not only a matter of choice but also of economic circumstance. But there is no point in crying over spilt milk. What is required now in view of European and global economic conditions is a more flexible interpretation of the Pact (e.g. in the form of a temporary derogation). The reasons for this should and must be explained to the general public and the

---

[101] Commission of the European Communities, *European Economy*, Supplement A, No. 3-4 (Luxembourg), March/April 2001.

[102] UNECE, *Economic Survey of Europe, 2001 No. 1*, chap. 1.

[103] M. Artis and M. Buti, "'Close-to-balance or in surplus': a policy makers guide to the implementation of the Stability and Growth Pact", *Journal of Common Market Studies*, Vol. 38, No. 4, November 2000, pp. 563-591.

### TABLE 1.3.2

**Real GDP in the ECE market economies, 1999-2002**

*(Percentage change over previous year)*

|  | 1999 | 2000 | 2001[a] | 2002[a] |
|---|---|---|---|---|
| France | 2.9 | 3.1 | 2.0 | 1.8 |
| Germany | 1.8 | 3.0 | 0.8 | 1.5 |
| Italy | 1.6 | 2.9 | 1.9 | 1.5 |
| Austria | 2.8 | 3.3 | 1.8 | 2.3 |
| Belgium | 2.7 | 4.0 | 1.8 | 2.2 |
| Finland | 4.0 | 5.7 | 2.1 | 2.7 |
| Greece | 3.4 | 4.1 | 3.7 | 3.7 |
| Ireland | 10.8 | 11.5 | 5.5 | 4.8 |
| Luxembourg | 7.5 | 8.5 | 5.6 | 4.2 |
| Netherlands | 3.7 | 3.5 | 1.3 | 1.6 |
| Portugal | 3.3 | 3.2 | 1.9 | 2.1 |
| Spain | 4.0 | 4.1 | 2.7 | 2.3 |
| *Euro area* | 2.6 | 3.4 | 1.7 | 1.8 |
| United Kingdom | 2.3 | 3.0 | 2.1 | 2.1 |
| Denmark | 2.1 | 3.2 | 1.1 | 2.1 |
| Sweden | 4.1 | 3.6 | 1.6 | 2.0 |
| *European Union* | 2.6 | 3.3 | 1.8 | 1.9 |
| Cyprus | 4.5 | 4.9 | 4.5 | 4.6 |
| Iceland | 4.3 | 3.6 | 1.5 | 1.7 |
| Israel | 2.6 | 6.2 | 1.0 | 3.0 |
| Malta | 4.0 | 4.3 | 4.3 | 4.3 |
| Norway | 1.1 | 2.3 | 1.2 | 1.9 |
| Switzerland | 1.5 | 3.4 | 1.6 | 1.5 |
| Turkey | -5.0 | 7.2 | -6.0 | 3.0 |
| **Western Europe** | 2.2 | 3.5 | 1.4 | 1.9 |
| Canada | 5.1 | 4.4 | 1.4 | 1.7 |
| United States | 4.1 | 4.1 | 1.0 | 1.2 |
| **North America** | 4.2 | 4.2 | 1.0 | 1.2 |
| Japan | 0.8 | 1.5 | -0.5 | -0.5 |
| **Total above** | 2.8 | 3.5 | 0.9 | 1.3 |
| *Memorandum items:* | | | | |
| 4 major west European economies | 2.1 | 3.0 | 1.6 | 1.7 |
| Western Europe and North America | 3.2 | 3.8 | 1.2 | 1.6 |

*Source:* OECD, *National Accounts of OECD Countries* (Paris) various issues; Eurostat, New Cronos Database; national statistics; Consensus Economics Inc., *Consensus Forecasts* (London), various issues.

[a] Forecasts.

actors in financial markets. The possibilities for a coordinated fiscal policy response to offset the weakening of private sector demand should also be explored.

The costs of a further weakening of global growth and a failed recovery in the United States are potentially so high that governments should take measures to avoid such an outcome. In view of the increasingly large downside risks to the global economic outlook, there is now an urgent need for a coordinated policy response, including multilateral measures, to ensure a sustainable recovery, avoid disruption to the liquidity needs of developing countries and guarantee the inflow of other funds to support their economies.[104]

The current situation is especially worrisome because of the presence of factors that have been posing serious risks for the world economy for quite some time but which have become more acute in a context of sharply diminished growth expectations. Apart from the fragility of the financial sector in Japan and the risk of emerging market crises, this refers especially to the considerable domestic and external imbalances in the United States economy. The reduction of these imbalances is a necessary condition for a new sustainable upswing. There is now a much greater risk that the inevitable adjustment costs to be borne by the rest of the world (mirrored in the reduction of its current account surplus) will be abrupt rather than gradual; sudden and large changes in exchange rates and in the direction of international capital flows would greatly increase the risk of international financial turmoil and of even larger disruptions to global economic activity.

In the United States, both monetary and fiscal policy has shifted to a significantly more expansionary stance. It can be expected that the Federal Reserve will lower interest rates further if there are no definite signs of a recovery emerging in the near future. Hopes for a rapid and strong rebound of economic activity in the United States in the course of next year (the so-called V-shaped recovery) – with concomitant benefits for the rest of the world – are pinned on the potential stimulus to domestic demand associated with the expansionary stance of fiscal and monetary policies. But such a domestic demand-led recovery in the United States economy could turn out to be a mixed blessing for the world economy because it would only postpone the inevitable reduction of the large domestic and external imbalances and actually increase the risk of abrupt and disruptive adjustment.

Sustained growth in the rest of the world, especially in western Europe, would create the best environment for a smoother adjustment in the United States.[105] More generally, western Europe needs to face the fact that the strengthening of global economic growth forces cannot be left to the United States alone. The orderly reduction of economic imbalances in the United States requires sustained growth in the rest of the world, especially in western Europe and Japan. In Japan, huge fiscal imbalances and official interest rates close to zero have left economic policy with little room for manoeuvre, although there is still some scope for further monetary easing to reverse deflationary pressures and stimulate domestic demand. In western Europe, the need for greater reliance on domestically generated growth now requires a more deliberate shift toward expansionary economic policies in the euro area. Such action, which would support the expansionary stance of United States economic policy, has now become rather urgent, and it would be in line with the advocacy of multilateral action to sustain the global economy.

---

[104] UNCTAD, *Global Economic Trends and Prospects* (Geneva), 1 October 2001 (UNCTAD/GDS/Misc.21); IMF, "Statement of the managing director on the situation of the world economy and the fund responses", *News Brief No. 01/98* (Washington, D.C.), 5 October 2001 (www.imf.org).

[105] UNECE, *Economic Survey of Europe, 2001 No. 1*, p. 5.

## (ii) The transition economies

The short-term outlook for the ECE transition economies is also surrounded by considerable uncertainty. During the first half of 2001 most of these economies escaped the immediate impact of the global economic slowdown, but this is unlikely to continue because the synchronous global downturn – unprecedented in recent history – has increased considerably the downside risks. Several major factors have to be taken into account when judging the short-term economic outlook for the transition economies:

- Most of the east European and Baltic countries are small, open economies with intense trade links with the European Union, and their recent performance indicates that they are increasingly synchronized with the EU trade cycle. Thus, the notable weakening of output and import demand in the euro area will inevitably have an adverse effect on east European and Baltic exports and may have a negative impact on capital flows to this region, especially as privatization runs its course;

- The global economic downturn is already causing a considerable weakening of oil and commodity prices and this is likely to continue at least in the short run. Although not all commodity exports may be immediately affected (with oil standing somewhat apart),[106] the combined effect on export revenue of declining activity, increasing uncertainty and falling prices is likely to be negative for both commodity and oil exporters;

- In addition, some of the central Asian economies in the CIS are exposed to political risks due to the increasing tensions in the region following the terrorist assault on the United States. A prolonged military conflict in Afghanistan may have both direct and indirect negative economic consequences for the whole central Asian region.

Given the considerable external threats, it is not realistic to expect that the transition economies as a whole will be able to maintain their present rates of economic growth; moreover, increasing output volatility in individual countries can be expected in the rest of 2001 and the beginning of 2002. A general slowdown of output growth thus seems most likely in the short run; at the same time, given the differential impact of external factors, there may be large differences in the performance of individual transition economies. The outlook for the transition economies as a whole as well as for individual countries will also depend crucially on the duration and depth of the global downturn.

The current economic situation presents a serious challenge for all the transition economies but the immediate risks for eastern Europe and the Baltic area appear to be greater than those for the CIS. While the central European and the Baltic economies have made substantial progress towards establishing functioning market economies and strengthening their institutions, they are extremely dependent on trade with western Europe. In addition, some of them (particularly the Baltic states) are very small economies where even a single event can have a considerable impact on overall performance (as was the case with the closure of the Elcoteq operation in Estonia – box 1.2.2). Most of the south-east European transition economies – which are generally lagging behind in the reform process – still suffer from chronic weaknesses while at the same time also being more exposed to west European markets. In the present circumstances this combination increases their susceptibility to external disturbances. Thus, if the west European downturn intensifies, the consequences for eastern Europe and the Baltic area could be considerable. The manufacturing sector of these transition economies, which relies heavily on exports, is most exposed to such risks and would be the first to be affected; however, the negative repercussions would be wide-ranging and overall economic performance can be expected to suffer as well.

Contrary to the situation in previous cases of global turmoil, the CIS as a whole appears to be less exposed to the immediate impact of the current global economic downturn. In the present circumstances, the fact that intra-CIS trade still accounts for a significant share of the total trade of most of these countries provides some insulation from weakening demand in the rest of the world. Hence, the continuing strong recovery in Russia and in some of the other large economies in the Commonwealth (Ukraine and Kazakhstan) provides a source of growth for the rest of the CIS.

The escalating global economic risks and the ensuing threats for the transition economies have not so far triggered major revisions to the official growth forecasts. At the moment of writing this *Survey*, there were few indications that policy makers were preparing for a notable weakening of growth in the coming months.[107] It was only after the terrorist attacks on the United States on 11 September that some warnings were made about the possible economic fallout in the region. Nevertheless, as of early autumn, and despite some modest downward revisions of forecasts,[108] governments remained generally optimistic about their short-term

---

[106] The increase in uncertainty about the future, which is detrimental for investment, manufacturing output and exports, may have different implications for oil and commodity exporters: for example, there may be a rise in "precautionary demand" for oil but not necessarily for cotton. Thus, while declining activity combined with increasing uncertainty will most likely negatively affect real demand for most commodities, their joint effect on the real demand for oil is a priori unclear.

[107] The exception of Poland stands out as a special case, as the deterioration in output and the lowering of forecasts were mostly driven by domestic factors.

[108] Slight downward revisions were made in some countries in this period: apart from Poland, these included Bulgaria, Croatia, Hungary, Slovenia, Estonia and Latvia.

prospects despite the persistent lowering of output forecasts in western Europe. Thus, according to the official October forecasts for the year 2002, aggregate GDP was still expected to grow by some 3.5 per cent in eastern Europe and by close to 5 per cent in both the Baltic states and the CIS (table 1.2.1).

These fairly optimistic forecasts suggest that governments in most transition economies see few serious risks from the present global downturn. Consequently, with few exceptions, there is little sign that they are considering specific policy responses to an eventual slowdown in their economies. One important reason for this may be the fact that policy makers in most transition economies have fairly limited degrees of freedom to design and implement such measures. Widespread constraints on policy in the transition economies are the existing macroeconomic imbalances: current account deficits (table 1.2.2) as well as fiscal deficits are already quite high in a number of countries. Moreover, these imbalances are likely to increase because of various factors such as different rates of adjustment of exports versus imports in response to the global downturn, insufficient flexibility of product or factor markets, and a possible inconsistency of policy in a changing environment.

Another policy constraint stems from the rigid monetary regimes to which some countries adhere (such as the currency boards in Bosnia and Herzegovina, Bulgaria, Estonia and Lithuania or the fixed exchange rate regime in Latvia). Such regimes preclude the use of monetary policy to cushion the economy from an external shock. In principle, a rigid monetary regime such as a currency board implies that the external shock is fully absorbed by the real economy which, in turn, requires a sufficient degree of flexibility on the labour and product markets (this in fact is a precondition for the smooth functioning of a currency board arrangement). To put it differently, an aggravation of the current economic situation will provide a test not only of the degree of flexibility of markets in these countries but also of the stability of their monetary regimes in the event of an external shock.

Among the advanced reformers, Hungary and Slovenia, which in recent years have enjoyed balanced growth and adhere to more flexible monetary regimes, appear to have more room for engaging in activist policies in response to a deteriorating external environment. With both external and internal balances broadly in check, a moderately expansionary policy would seem to be the proper response to an external shock.[109] However, even in these countries the current macroeconomic equilibrium is rather delicate and much policy caution would be needed to avoid upsetting it. Slovakia and to a lesser extent the Czech Republic are facing some difficulties in keeping their macroeconomic balances under control: the latter is currently coping with a fiscal shock caused by the bailout of several large banks, while the former has been trying to scale down an ambitious public investment programme that has led to a dangerous escalation of both external and domestic imbalances. The policy response in such cases will have to be much more cautious and selective, avoiding measures that could further destabilize the macroeconomy. Poland is in an even weaker position due to its acute fiscal crisis (box 1.2.1); possible counter-cyclical measures may only be envisaged in the context of an internal restructuring of an already shrinking government budget. The three Baltic states as well as Bulgaria adhere to rigid monetary regimes which, as noted above, leave very little, if any, degrees of freedom to policy makers to undertake counter-cyclical measures. At the same time, as indicated by past experience (for example in the aftermath of the Russian crisis in 1998), they are highly susceptible to external shocks. These economies are now exposed to a double risk: first, the extent to which they may be affected by an external shock and, second, whether the real economy can absorb it without jeopardizing macroeconomic stability. Most of the remaining south-east European transition economies also have limited room for manoeuvre due to their overall macroeconomic instability.

The immediate downside risks for the CIS are probably not so high; as already noted, as long as the Russian economy continues to grow the direct negative repercussions of the global downturn on the rest of the CIS will be partly offset. In fact, while the Russian authorities envisage some slowdown in the rate of growth in the short run (the draft budget for 2002 submitted to the *Duma* assumes a 4.3 per cent rate of GDP growth), the prevailing expectation is that output will remain relatively strong in the medium term.[110]

The most important external influence on the continuation of recovery, not only in Russia but in the whole of the CIS, is the direction of world commodity prices. These economies can probably weather a temporary drop in world oil prices and, actually, some governments in the region have been preparing for such a turn. Thus, some of the CIS oil exporting countries (Azerbaijan, Kazakhstan and Russia) have already instituted, or are in the process of doing so, special off-budgetary reserve funds, which accumulate some of the windfall revenue in periods of boom. When conditions

---

[109] Hungary is one of the few transition economies where a policy stimulus (in particular, by stimulating home construction and investment in infrastructure) is actually being discussed as a precautionary measure to offset any adverse impact of the global slowdown on the Hungarian economy. *Reuters Business Briefing*, 28 September 2001.

[110] According to the medium-term economic programme approved by the Russian government in August, the average annual rate of growth of GDP in the period 2002-2004 is expected to be in the range of 3.5 to 4.5 per cent. *Interfax News Agency* as reported in *Reuters Business Briefing*, 9 August 2001.

reverse, some of the accumulated resources may be used as a cushion against the shock. However, as this is a relatively recent development, the reserve funds have not yet accumulated sufficient resources to cope with a prolonged period of distress. Thus, a deep and lasting fall in commodity prices will inevitably have a direct detrimental effect on the economic performance of all the commodity exporters, including Russia, and, indirectly, further negative repercussions due to any weakening in the Russian economy.

Given the depth and scope of the present synchronous global downturn, the transition economies obviously cannot isolate themselves from the external shock or cope with its negative consequences on their own. Economic policy in the transition economies should obviously try to counter weakening cyclical growth forces, but the extent to which this is possible will vary among countries. In general, such measures will only be able to complement and strengthen the policy responses to recessionary forces in western Europe and overseas. It goes without saying that a long-lasting negative shock could jeopardize reform efforts in the transition economies and prolong still further the time required to reduce the gap between their per capita income levels and those of the more developed part of the continent.

# PART TWO

## CREATING A SUPPORTIVE ENVIRONMENT FOR BUSINESS ENTERPRISE AND ECONOMIC GROWTH: INSTITUTIONAL REFORM AND GOVERNANCE

*PAPERS FROM THE ECE SPRING SEMINAR, MAY 2001*

# INTRODUCTION AND SUMMARY OF DISCUSSION

## Economic Analysis Division, UNECE

The UNECE's fourth Spring Seminar focused on two related issues that have become an increasing preoccupation of economic policy makers and researchers over the last decade, namely, the role of institutions in determining the performance of a market economy and the problem of corruption.[111] Given the regional perspective of the UNECE and the fact that all of the countries of central and eastern Europe and the former Soviet Union are among its members, the papers and discussion focus largely on the problems of the transition economies although these are not by any means the only countries to suffer from institutional weakness and corruption.

Although institutions and corruption have become more prominent in policy discussions and research since the mid-1990s this actually represents a re-emergence – or rediscovery – of questions that were being raised and debated by development economists in the 1950s and 1960s but which, for various reasons, were put aside in the 1980s. When the communist regimes of eastern Europe and the Soviet Union collapsed in 1989-1991 neo-liberalism was in the ascendancy in North America and western Europe (and therefore in the major international financial institutions) and thus the emphasis in policy advice was heavily weighted towards the swift liberalization of domestic prices and international trade – and privatization. Rapid price and trade liberalization implied a large degree of structural adjustment in economies where resource allocation had previously been determined by planners' preferences rather than comparative advantage and market prices. But it was assumed that the transformation of property rights through privatization would unleash domestic entrepreneurial energies which, together with foreign direct investors eager to be in at the start of a new source of market growth, would accelerate the transition to a market economy and trigger a period of sustained growth that would eventually allow the former centrally planned economies to start "catching up" with the living standards of western Europe.

Despite the considerable progress made in many aspects of the transition process since 1989, only a small handful of central European economies have shown any tendency to conform to this scenario and to start converging on the levels of GDP per head in western Europe. In general most of the transition economies – and there are 27 members of UNECE with economies in transition – have continued to fall further behind the income levels of western Europe.[112] The slowness of, and setbacks to, the transition process in all but a few economies have given rise to a search for explanations and these have ranged from a "lack of political will" through poor governance to corruption. For the most part, however, these "non-market" factors are all aspects of the institutional infrastructure, broadly conceived, or, as some development economists prefer to describe it, the social infrastructure, which is understood to include not only property rights and market institutions but also political and social structures, and economic and political cultures.

Although many of the supporters of the neo-liberal approach were aware of the importance of institutions for the efficient operation of a market economy, for the most part they shared the optimism of the neo-classical school that institutional and other impediments to the reallocation of resources would not prove a serious obstacle to investment and sustained growth as long as "prices were right" and private enterprise given its head. Much of the required institutional development could therefore be postponed to a longer-run "second stage" of reforms. However, an earlier, structuralist school of development economists, which includes Gunnar Myrdal the first Executive Secretary of the ECE,[113] argued that because of institutional shortcomings and coordination failures such reallocation could be blocked or would occur only very slowly and not sufficiently to

---

[111] The Spring Seminar was held on 7 May 2001. The choice of topic (or topics) for the Seminar reflects the preferences of ECE member governments for the suggestions made by the secretariat and by participants at the previous year's Seminar. If there is no consensus, or if member governments are equally happy with more than one proposal, the final decision is made by the Executive Secretary of ECE.

[112] UNECE, "Catching up and falling behind: economic convergence in Europe", *Economic Survey of Europe*, 2000 No. 1, pp. 155-187.

[113] Myrdal was the ECE's first Executive Secretary, from 1947 to 1957. On his approach to economic development, see G. Myrdal, *Economic Theory and Under-Developed Regions* (London, Duckworth, 1957) and *Asian Drama. An Inquiry into the Poverty of Nations* (New York, Pantheon, 1968). Myrdal was one of the first economists to draw attention to the problems of corruption and to dismiss the idea that it was a necessary "lubricant" for a cumbersome system – on this, see below. Others sharing this scepticism about the spontaneity of the reallocation process in response to the "right prices" include Paul Rosenstein-Rodan, Ragnar Narkse, Albert O. Hirschman, W. Arthur Lewis, W.W. Rostow, and Alexander Gershenkron. However, Myrdal and Hirschman probably attached the most importance to institutions and social capabilities as preconditions for sustained growth.

allow the take-off into sustained growth. They therefore saw a crucial role for the state (government) in creating the appropriate institutions for a market economy and in overcoming the obstacles to economic recovery. Whether or not the governments of the transition economies in the early 1990s were capable of playing an active development role, this was not a message that was likely to find much of an audience in the political and ideological atmosphere of the 1980s or in the immediate post-1989 euphoria over the collapse of communism.

Nevertheless, the "institutional hiatus"[114] remains in most of the transition economies, although its size varies considerably: progress in closing it is most advanced in those central European countries which are in the first wave of candidates for accession to the EU but it remains large in most of south-east Europe and in the CIS. But the problem with an institutional hiatus is that if the state is unable to move rapidly enough to shape an infrastructure that will encourage legitimate and productive activity, other forces will move in to take its place – property rights will still be defined and enforced, but by gangsters with guns rather than by law and judges. In practice, the institutional framework in a transition economy will often be a mixture of old and new institutions and practices, a mix that tends to provide many opportunities for arbitraging and rent-seeking. The lack of effective systems of governance and of public administrations with the skills appropriate for a market-based economy should help to explain not only differences in economic performance among the various transition economies but also differences in the prevalence of corruption and the extent to which economic activity has been criminalized. The "institutional hiatus" is thus the leitmotiv of this seminar.

Some of the economists subscribing to the structuralist view of development would argue that the neo-liberal or Washington consensus is an appropriate basis for policy advice to fully-fledged, developed market economies, but not for developing or transition economies where the adjustment processes are hindered or blocked by institutional inadequacies.[115] However, in the opening paper of the Spring Seminar, **Professor William Lazonick** of the University of Massachusetts Lowell and INSEAD (The European Institute of Business Administration), argues that the mainstream theory of the market economy – the general equilibrium model – is "fundamentally flawed" because it fails to account for the innovative enterprise and therefore is unable to provide an explanation of the process of economic development. Lazonick recognizes the importance – economic but also social and political – of efficient product and factor markets, but his key point is that such markets are an outcome of a successful development process rather than preconditions for it. Thus, to create a market economy and reap its economic and social benefits, policy must identify and put into place the institutions and structures that will eventually produce well-functioning markets for labour, capital, goods and services.

Lazonick argues that the key social unit or actor responsible for the allocation of resources and the generation of innovations is not the market but the modern business corporation. This comes about because the process of innovation is: (i) collective – it requires the combined services of a large number of people with different skills and responsibilities; (ii) it is cumulative, in the sense that previous decisions and conditions constrain the possibilities of present action; and (iii) the process is inherently uncertain. These characteristics stand in sharp contrast to the mainstream theory of the market economy where allocative decisions are: (i) individual, in that the decisions of independent actors have no influence on one another; (ii) reversible, in the sense that past decisions place no constraint on the future; and (iii) optimal, in that individual decisions are made subject to the prevailing conditions of technology and the market over which they have no influence.[116] If the process of resource allocation is characterized by the first set of conditions then the mode of allocation "must be organizational, developmental and strategic", a mode which can be adopted by organizations but not by markets. The innovative enterprise is in turn characterized by three social conditions, namely, organizational integration, financial commitment and strategic control, although their actual form and content in any given economy will

---

[114] The "hiatus" arises because while the old system of coordination could be rapidly destroyed, its replacement takes much longer to be put into place. R. Kozul-Wright and P. Rayment, "The institutional hiatus in economies in transition and its policy consequences", *Cambridge Journal of Economics*, Vol. 21, No. 5, September 1997, pp. 641-661.

[115] Thus, Irma Adelman has stated that "once growth and savings habits are firmly entrenched in the entrepreneurial and household sectors ... the appropriate role of the government ... should change to that prescribed by the current neo-liberal ... Washington consensus". But "had [this consensus] been enforced on the east Asian miracle countries during the fifties, sixties, and early seventies, there would not have been an east Asian miracle". *Comments on Angus Maddison's Paper,* presented to the Development Thinking and Practice Conference (Washington, D.C.), 3-5 September 1996. Adelman, together with Cynthia Taft Morris, has carried out a number of pioneering studies in comparative development history taking explicit account of the interdependence of economic, social

and political structures. I. Adelman and C. Morris, *Society, Politics and Economic Development – A Quantitative Approach* (Baltimore, Johns Hopkins University Press, 1967) and C. Morris and I. Adelman, *Comparative Patterns of Economic Development, 1859-1914* (Baltimore, Johns Hopkins University Press, 1988). Temple and Johnson have shown recently that forecasts of long-run growth made in the 1960s would have been considerably improved had Adelman and Morris' index of social capability not been ignored. The forecasts failed, for example, to foresee the Asian miracle and the prospects of sub-Sahara Africa were judged to be better than South-East Asia's. J. Temple and P. Johnson, "Social capability and economic growth", *The Quarterly Journal of Economics*, August 1988, pp. 965-990.

[116] M. O' Sullivan, "The innovation enterprise and corporate governance", *Cambridge Journal of Economics*, Vol. 24, No. 4, 2000.

depend on the interaction between the "prevailing institutional (financial, employment and regulatory, and organizational (cognitive, behavioural and strategic) conditions ...".

Lazonick's conclusion is that the mainstream, general equilibrium theory of the market economy is a seriously misleading model on which to base policy recommendations to transition and developing economies because it views as "market imperfections" the very institutional and organizational conditions that are crucial for the emergence of the innovative enterprise and economic development. The alternative recommendations will depend on the social and historical context of each individual country, but he would stress the enhancement of people's skills rather than labour mobility, the need for financial commitment rather than stock markets, and in general giving priority to enterprise development. Lazonick also stresses the importance of the developmental state, since the state is the other principal form of organization capable of engaging in an allocative process which is collective, cumulative and uncertain.

Lazonick's analysis clearly places him in the tradition of the structuralist school of development economists, discussed above, but his call for a more realistic analysis of the nature and role of the business enterprise and entrepreneurship in a modern economy also places him in a heterodox tradition of economic growth analysis which can be traced through the work, *inter alia*, of Schumpeter, Galbraith, Abramovitz and Robinson.[117]

***Professor Victor Polterovich*** (Central Economics and Mathematics Institute, Moscow) opened the discussion of Lazonick's stimulating paper which he thought failed to acknowledge that the neo-classical framework had taken into account, over the last decade or so, many of the important elements that Lazonick accused it of ignoring, such as research and development, learning by doing, and so on. Nevertheless, he agreed that these elements are not well integrated into modern growth theory and that a dogmatic interpretation of neo-classical principles had led to serious policy mistakes in the transition economies: for example, the belief that privatization alone would be sufficient to improve their economic performance. Polterovich agreed that the role of the state was especially important in preventing an economy falling into an "institutional trap" where missing or ineffective institutions led to a persistent failure of coordination among economic agents. He also supported a developmental role for the state. But the effectiveness of the state in carrying out these functions will depend to a large extent on the prevailing civic culture which, however, tends to be "much more inertial than the political and economic organization of a society".

***Eugen Jurzyca*** (Institute for Economic and Social Reforms, Bratislava, and Member of the Board of the National Bank of Slovakia) also agreed with much of Lazonick's paper and thought that it was misleading to assume that an economy's performance would be positively correlated with the degree to which its markets were perfect. He thought that the theory of the innovative enterprise needs to be developed further and he agreed that despite examples of failure the state can play a developmental role. But he queried whether the theories of the market economy and of the innovative enterprise were really in opposition rather than being complementary. Markets should serve as mechanisms of cooperation between hierarchically organized entities and he thought that, historically, they had developed simultaneously rather than the one being a precondition or end product of the other.

In the general discussion many of the speakers found Lazonick's paper stimulating and refreshing, and one participant from a central bank in eastern Europe expressed "pleasant surprise" at its being presented at such an international forum. There was a lot of support for his emphasis on the importance of enterprise development but many speakers asked for more policy prescriptions and some, while supporting the idea of a developmental role for the state, queried whether this was currently feasible[118] in most of the transition economy countries. On the question of policy prescription, Lazonick's basic conclusion is that much higher priority must be given to the development of innovative enterprises, but his insistence on history and national context must imply that individual policies are specific to each national situation. This is why, as Professor Hare notes in the next paper, there is such a large diversity of practice for any given institution considered "normal" for a market-type economy.

In the second paper presented to the Seminar ***Professor Paul Hare*** of the Heriot-Watt University, Edinburgh, asks what sorts of institutions are necessary for a market economy to function effectively and how institutional change can be brought about. He surveys the actual progress in institution building that has occurred in the transition economies since the early 1990s and then investigates the extent to which the various indicators of institutional change are linked to intercountry differences in economic performance.

---

[117] J. Schumpeter, *The Theory of Economic Development*, (Cambridge, MA, Harvard University Press, 1934); J. Galbraith, *The New Industrial State* (Boston, Houghton Mifflin, 1967); M. Abramovitz, "Catching up, forging ahead and falling behind", *The Journal of Economic History*, Vol. XLVI, No. 2, June 1986, pp. 385-406; J. Robinson, "What are the questions?", *Journal of Economic Literature*, Vol. XV, No. 4, December 1977, pp. 1318-1339. On the importance of developing the social capabilities for entrepreneurship in the transition economies of eastern Europe and the former Soviet Union, see R. Kozul-Wright and P. Rayment, loc. cit.

[118] Similar doubts were expressed by Jan Svenjar at last year's Spring Seminar. UNECE, *Economic Survey of Europe, 2000 No. 2/3*, p. 63.

Hare's approach is first to ask what are the key indicators of a vibrant economy and then to identify those institutions that support and strengthen them. The key indicators in his view are: high rates of domestic saving and investment; high rates of new business formation; a sufficient but not excessive degree of competition; and arrangements for the orderly exit of failed enterprises.

High rates of domestic saving and investment, which are macroeconomic variables, are necessary conditions for sustained economic growth and it is a delusion to suppose that foreign direct investment can compensate for a failure to mobilize domestic savings. The institutional requirements to raise savings are complex and broad-ranging: they include a network of institutions able to attract and manage small deposits from individuals and business; commercial banks able to select projects and monitor loans; institutions providing a range of financial services such as insurance, pension funds, etc; a diversified set of financial assets for investors to hold (government and corporate bonds, equities, unit trusts, etc.); effective financial regulation of the financial sector; and appropriate protection of investors and savers. Hare recognizes that all of these add up to a very tall order for many transition economies and he suggests that either group of them could cooperate to set up joint or regional markets for such financial products or that they should simply use those already established in other countries.

Hare's other three indicators refer to the microeconomy, essentially the business enterprise sector. Although there are differences of focus and emphasis, his conclusion that the next round of privatization must "ensure that firms are mostly held by owners capable of supplying new funding, new technology and competent management" overlaps with Lazonick's strategic priority of promoting the innovative enterprise. But Hare pays more attention to the macroeconomic conditions for enterprise development – price stability, prudent fiscal policy, etc. – and to the institutional reforms that influence both the performance of enterprises and the relations between them.

Reviewing the empirical literature on the relations between various aspects of institutional change and intercountry variations in economic performance, Hare concludes that sound macroeconomic policies are certainly necessary for sustained growth but that institutional change in the widest sense is absolutely vital. This conclusion derives both from studies investigating the links between macroeconomic performance and broad indicators of structural reform and from more micro-investigations of the institutional barriers to business activity.

Hare observes that although the lessons for transition economies are not straightforward "it is clear that weak property rights and poor market signals are major parts of the story". Without reasonably reliable market signals, enterprise restructuring will be poorly focused and likely to disappoint in the delivery of productivity growth. Nevertheless, he is frank in admitting that the current state of research into the relationships between the general institutional framework, enterprise governance and performance does not provide a basis for precise policy conclusions.

He also stresses the important role of the state in institution building – setting the rules of the game – and he sees a need for coherent industrial trade and competition policies; but he is also concerned by the practical point that many of the states with economies in transition "are weak, incompetent or corrupt". In such conditions there is no point in urging them to pursue complex and sophisticated policies which will simply create more opportunities for rent-seeking, bribery and corruption. Thus, for example, a high welfare/high tax model might be more favourable for enterprise restructuring than a low welfare/low tax version, but the former may simply not be feasible in countries with weak administration and unable to collect sufficient tax revenue. In other words, many countries are caught in an institutional trap and the problem is how to find a way out.

One way forward is for transition economies to replicate the institutional structure of the EU: this is a ready-made, operational model and copying it would reduce the costs of designing all the necessary institutions from scratch. This is probably more feasible for the leading group of accession candidates, most of which started the transition process in much better initial conditions with respect to both institutional and economic development and which are now receiving large amounts of assistance in their attempts to comply with the *acquis communautaire*. Moreover, the political desire for EU membership supplies a strong incentive for the state to promote institutional change. Nevertheless, Hare points out that the institutional structures of the EU "evolved to suit a group of relatively developed economies with high incomes, accustomed to steady but quite slow rates of economic growth" and he questions whether, given the stage of development of most transition economies, all aspects of the EU's institutions and practices are appropriate for them. Here, he comes close to the central point stressed by William Lazonick.

The question of what actually drives institutional change is raised by **Danica Popovic** (Belgrade University) in her comments on Paul Hare's paper: there are a lot of suggestions but the underlying impulse is still a mystery. Hare's statement that good institutions are public goods and, as such, undersupplied, is contradicted by contractual theorists who see the creation of the state and other institutions as a collective response to market demand. Popovic agrees that weak or missing institutions can be devasting for economic performance, but she notes that the link between good institutions and good performance is not that close and is anyway subject to substantial time lags. These lags may deter governments from implementing difficult reforms whose results will only appear after the next election. Popovic thought Hare's suggestion that there could be "too much competition" was interesting but queried how it could be translated into a practical yardstick. On the issue of safety nets and restructuring she emphasized the need for proper sequencing: governments must give high priority to creating a good environment for the creation of new firms otherwise they will be tempted to try to protect those which should really go out of business.

Both Popovic and *László Csaba* (Central European University, Budapest), the other discussant of Hare's paper, took issue with his statement that the "notion that FDI can be brought in on a large scale where domestic savings are insufficient is a widely held myth, but a myth nevertheless". Popovic argued that attempts to raise savings at all costs would weaken consumption and growth: a better approach is to create a credible reform programme which will attract FDI, which in turn will stimulate higher levels of both consumption and savings. Csaba also stresses the importance of FDI as a pacesetter for the enterprise sector and as an indicator of the underlying rate of institutional improvement. However, the disagreement on this issue is perhaps more apparent than real: the two discussants emphasize the dynamics of FDI as part of a virtuous circle of credible reform efforts and improving performance, while Hare's emphasis on the need to raise domestic savings is one of the key factors in getting economic development underway.

Laszlo Csaba thinks it an overstatement to say that the importance of institutions was recognized from the start of the transition: the arguments tended to be polarized between those calling for radical policies with little knowledge of the context in which they would be implemented and those appealing to institutional problems as a cover to hold back the entire process of transformation. Policies were introduced without careful attention to initial conditions and institutional legacies, and this has proved seriously counterproductive in the less developed transition economies. "State failure" has been largely ignored "in most analytical and policy papers" and yet in his view this is the cause of market failure. The idea that the minimal state would encourage private initiative to come up with better institutional alternatives has proved to be mistaken: the results have instead been perverse and the resulting fuzziness of institutions has led to state capture and, as a consequence, "a network-dominated economy in which inherited and newly acquired power positions are used for rent-seeking and the creation of monopolies irrespective of the formal legislation". How can this situation be reversed and a virtuous circle of sound institutions and improving economic performance be got underway? Csaba agrees with Hare's emphasis on a "bottom-up" approach, focusing on the performance of the enterprise sector and with particular emphasis on the encouragement of new firms and the exit of failures. He also agrees that, given its present weakness in many transition economies, the state cannot be expected to undertake very complicated or too detailed tasks of institution building and restructuring, but he is more optimistic about the possibilities of obtaining institutional knowledge and disciplines from international markets, foreign direct investors and the EU. He admits, however, that the fundamental question of how countries learn and absorb such knowledge is largely unexplored, but this almost certainly leads to the issues of culture and the more intangible elements in the institutional structure emphasized by Victor Polterovich.

Professor Hare's stimulating paper led to a lot of questions from the floor but the principal ones concerned his remarks on moderate competition, the need to boost domestic saving, and his emphasis on new institutions. On competition he admitted it was difficult to define a "moderate degree" but a pragmatic balance has to be struck somehow between excessive competition which kills off innovation, especially in its infancy, and too little which also stifles the drive to innovate. On the need to boost savings, there was really little disagreement with Danica Popovic on this – he was not downplaying the role of domestic consumption, simply emphasizing the need for a sharp increase in domestic investment. On institutions a number of speakers argued that the problem was not so much that there was a lack of them but that those that were in place did not function properly. Hare thought this was a doubtful distinction since they often required such radical reform that effectively new bodies were created.

The problem of corruption was frequently raised throughout the first two papers and their discussion, usually in the context of weak or missing institutions and in the company of rent-seeking and other forms of socially unproductive activity, but the afternoon session of the Seminar focused directly on this problem with papers presented by Shang-Jin Wei and Antoni and Bartlomiej Kaminski.

Corruption, like the poor, has always been with us, but over the last 10 years or so there has been a marked upsurge of both policy efforts to combat it on a global scale and of empirical research into its effects on economic performance. Like all new issues, the concern with corruption has an historical tail reaching back in this case to the 1950s and 1960s. A few economists had already drawn attention to the demoralizing effects of corruption in the 1950s,[119] but it was Gunnar Myrdal who highlighted the issue in his

---

[119] T. Mende, *South-East Asia Between Two Worlds* (London, Turnstile Press, 1955).

major study of economic development in Asia.[120] Myrdal complained that, at the time he was writing, corruption was virtually a taboo subject for researchers and analysts working on development problems; when it was occasionally mentioned, usually by American authors (Europeans remaining silent on the subject), it was generally to argue that bribery was a harmless and necessary lubricant for cumbersome administration. Myrdal argued the contrary, that corruption was one of the causes of cumbersome administration and a major constraint on development.

Not only research into corruption but also national policy initiatives and the development of international cooperation to combat it increased steadily in the 1970s and 1980s. Although a small number of countries had introduced legislation in the 1950s to deal with corruption the major influence in changing international attitudes came from the United States where, in the wake of the Watergate scandal and the creation of a Presidential task force to examine the overseas behaviour of United States corporations, the Foreign Corrupt Practices Act was passed in 1977. This made it illegal for United States citizens and United States-based companies to bribe foreign officials. Further progress in the 1970s and 1980s was gradual but uneven: in the United Nations, for example, ECOSOC set up a number of working groups and commissions to consider transnational bribery but their proposals were never translated into General Assembly resolutions. Nevertheless, the work that was done in these years provided a basis for the rapid expansion of research and of anti-corruption activity in the 1990s. The collapse of communism in eastern Europe and the Soviet Union and the end of the cold war was the trigger for this: the United States, no longer constrained by cold war political and strategic considerations, began to be much more active in seeking a level playing field for United States corporations. At the same time the considerable publicity given to a series of corruption scandals, not least the *Mani Pulite* investigations in Italy in 1992, heightened public awareness of the extent of corruption and increased support for action to combat it. There was also growing awareness of the scale of international money laundering which had increased considerably in the wake of the financial deregulation of the 1980s; although by no means exclusively related to corruption, it nevertheless heightened popular perceptions of global malfeasance and of the need to do something about it.

Transparency International was created in 1993 and, together with other international organizations and NGOs, contributed to a rapid expansion in the supply of basic data and information about the nature and extent of corruption, and this has supported the expansion of academic research on the subject. The OECD Convention on Combating Bribery of Foreign Officials was adopted in 1997 and by mid-2001 34 countries had ratified it.[121] Thus, the development of international norms against corruption, including institutional and legal instruments as well as a shift in public attitudes, are now in a relatively advanced state of development.[122]

In order to design appropriate policies to combat corruption, and to build and maintain popular support for them, it is important to establish its actual costs in retarding or frustrating efforts to create market economies and build democratic institutions. Myrdal's call in 1970 for a research programme "to establish the general nature and extent of corruption in a country, its incursion upon various levels and branches of economic life, and any trends that are discernible" has now met with a strong response by academic economists and by a number of international organizations (intergovernmental and NGOs), producing various indexes of the extent of various dimensions of bribery and corruption.

**Shang-Jin Wei** (Harvard University and Brookings Institution) is one of the leading academic researchers in this field and in his paper he confronts directly the issue of whether corruption eases or obstructs the process of transition and economic development and he explores the channels through which its effects are transmitted. He reaches the same conclusion as Myrdal: corruption has a strongly negative effect on economic development. Synthesizing the available body of empirical work, to which he has made a large contribution, Wei shows that corruption has a negative effect on the investment-GDP ratio and discourages foreign direct investors. Instead of offering substantial tax incentives to foreign investors, Wei argues that it would be more effective to reduce the level of corruption. Although corruption acts like a fiscal disincentive to enterprise activity, it is actually more damaging than an explicit tax: bribes are secretive and arbitrary, and the implicit contract is not enforceable in the courts. As for the argument that bribery may "grease" the wheels of bureaucracy, detailed data for a large sample of enterprises and countries show that managers spend more of their time

---

[120] G. Myrdal, *Asian Drama. An Inquiry ...*, op. cit., especially Vol. II chap. 2. See also, idem, *The Challenge of World Poverty* (London, Penguin Books, 1970), especially chap. 7. Myrdal's definition of corruption followed that of the Indian Government's *Report of the Committee on the Prevention of Corruption* (the "Santhanam report") (New Delhi), 1964, namely, "the improper or selfish exercise of power and influence attached to a public office or to the special position one occupies in public life", although Myrdal insisted that it should also include the activity of the bribers. This is essentially the same definition used by the World Bank and in the two papers discussed here.

[121] See (www.oecd.org//daf/nocorruption/annex2.htm). See also the comments below of Mark Pieth, on the increasing incidence of national legislation against bribery.

[122] This paragraph draws largely on J. McCoy and H. Heckel, "The emergence of a global anti-corruption norm", *International Politics*, Vol. 38, March 2001, pp. 65-89. See also M. Pieth, "International cooperation to combat corruption", in K. Elliott (ed.), *Corruption and the Global Economy* (Washington, D.C., Institute for International Economics, 1997).

negotiating with officials, not less, when they have to offer bribes. Although the direction of causality is sometimes ambiguous, Wei concludes from the empirical work that there is a clear and significant effect of corruption on the overall rate of GDP growth.

Corruption, however, does not only affect a country's *rate* of growth but also its composition. The *quality* of output is reduced by a misallocation of resources which can occur in a number of ways. If the allocation of credit and other forms of finance is based on personal or political relationships rather than the competence of managers and the viability of projects, funds will tend to go to riskier projects and thus increase the volatility of investment and of output growth. Corruption will also tend to skew public investment away from socially desirable sectors (health, education) to those where bribes are more easily obtained (public construction projects), and to divert spending from running and maintenance costs to new equipment. The net effect is to lower the productivity of public investment and to weaken the country's infrastructure.

Wei also stresses that the negative effects of corruption on productivity and output growth, and the deterrent effect on foreign investors, imply that a country will tend to fall further behind in an increasingly globalized world while at the same time leaving it more vulnerable to the financial risks of global financial markets.

What can be done to combat and control corruption? Wei emphasizes that its root causes are to be found in the incentive structure and the surrounding institutional environment in a country, and he distinguishes between the conditions and activities which create the opportunities for corruption and the incentives which encourage people to make use of them.

The key source of opportunities for corruption, by definition, lies in the scale and nature of government activity in the economy. Although the size of government spending and corruption are positively related, the key features involve the nature of government activity: a lack of transparency and excessive complexity of regulations; industrial, commercial and other policies which allow large margins of discretion to officials; and generous levels of foreign aid determined by geopolitical rather than economic motives. The point is not to deny a role for government and for regulation but to stress that the risks of increasing the opportunities for corruption should be assessed when policies are being designed. To reduce the temptation of civil servants to take bribes, Wei stresses the importance of meritocratic appointment and promotion and of paying them wages that are comparable to "the best private sector alternatives". Other elements of a broad-based anti-corruption strategy include increasing the probability of detection and prosecution of corrupt officials, encouraging the emergence of grassroots watchdog organizations, a free press, and democratic institutions willing and able to protect "whistle blowers" and dismiss the corrupt. In addition, international pressure can be applied through the lending criteria of the international financial institutions and by international agreements to make the bribery of foreign officials illegal.

The key battle against corruption, however, has to be a domestic one but, as is clear from the above, the necessary reforms are wide-ranging, expensive and will often carry high political risks for governments. Moreover, such reforms will inevitably have to confront problems of culture and the inertia of history and existing institutions – and, as Victor Polterovich remarked, these are among the most difficult obstacles to overcome. Recognizing this, Wei suggests that one way forward could be to introduce a "special governance zone", an enclave within a country where a comprehensive set of reforms can be introduced ahead of the rest of the country. The political risks would be less than a national programme, but success would have a powerful demonstration effect on the rest of the country.

***Ivo Bićanić***, of the University of Zagreb, warmly welcomed the application of more rigorous economic and econometric analysis to corruption and stressed the importance of Wei's conclusion that corruption was never justified. Bićanić, however, drew attention to the prevalence of "small-scale corruption" in many transition and developing economies: payments for queue jumping, or to officials to overlook this or that relatively minor breach of this or that regulation. The sums involved were generally small, but the behaviour was widespread and accepted as part of the daily routine of getting by in "untidy economies". To a large extent a legacy of the 1980s, when small entrepreneurs had to learn how to manage with a collapsing socialized sector, these corrupt practices have persisted (in part because of an understaffed and underfunded public sector) and are now firmly rooted in the social capital. Eliminating this pervasive small-scale corruption is essential for the success of transition and for sustained economic growth, but it is extremely difficult to do so (and more difficult than combating "grand corruption"), not least because so much of the population is engaged in it. Moreover, Bicanic saw little possibility of obtaining outside help for dealing with this sort of corruption – so again, the key effort has to be domestic, however difficult it may be.

***Krassen Stanchev*** (Institute for Market Economics, Sofia), concluded from Wei's paper that honesty was the best policy because it lowered transaction and enforcement costs and enlarged the degree of economic freedom. Extending the scope of economic freedom – especially by diminishing the "licence requiring state" – was key because, although Wei does not address the issue as such, there is a close inverse correlation between measures of economic freedom and the scale of corruption. Measures to enlarge economic freedom – for example, the removal of government controls over prices, profits and contracts – increase transparency and lower the costs of investment. But Stanchev stressed the importance of getting a critical number of policies right: this was more important than worrying too much about whether corruption held back reforms or whether poor reforms encouraged corruption. There was no point in liberalizing prices if the state continued to control entrepreneurship (through the grant of licences to establish new businesses) and investment. Stanchev illustrates his point with examples from Bulgaria between 1990 and 1997.

In the general discussion of Wei's paper a number of questions were raised about the measurement of corruption, the interpretation of corruption indexes, and the direction of causality between corruption and growth. It was recognized that there were problems here: the various measures of corruption generally measured the perception of corruption than the phenomenon itself, although they tended to provide similar rankings of countries.[123] Despite some evidence that causality works in both directions, Wei said that the studies which had addressed this issue concluded that the stronger link was from governance to economic outcome. Wei agreed with a number of comments that corruption tended to fall disproportionately on small- and medium-sized enterprises that were less able to bear the burden. This was clearly an important factor to bear in mind when considering how to promote enterprise development and entrepreneurship.

There was also general agreement with Wei's list of proposals for reducing corruption. It was clear that any strategy would have to be pursued over a number of fronts and that those offering bribes had to be dealt with as severely as those accepting them. But it was also pointed out that many individual businesses see themselves as victims of corruption – and in the absence of collective action to stamp out bribery it was difficult for an individual enterprise to act without incurring significant loss. International organizations could also contribute, in addition to the ways already mentioned above, by exerting peer pressure in the manner of OECD on its member governments. There was also favourable comment on Wei's suggestion of making a start against corruption with special governance zones – these were certainly worth trying.

*Antoni Kaminski* (Polish Academy of Sciences) and *Bartlomiej Kaminski* (University of Maryland) broaden the discussion of institutions and governance by insisting that the institutional organization of the economy and the polity are intimately related in a liberal democracy and have to be treated together.

Although the choices to be made between alternative constitutional arrangements have a crucial influence on the general policy environment and the quality of both decision-making and its accountability, the authors argue that in the aftermath of the collapse of communism there was virtually no serious discussion about the shape of the post-communist political architecture. Most of the crucial choices were made haphazardly or by default. It was thus a major failure of the reformers not to have paid "as much attention to the design of the political system as they did to the problems of privatization and economic transformation". This was a particularly unfortunate omission as post-communist societies, especially those in the CIS, were "poorly endowed with the institutions and human capital capable of handling either the economic or political dimension of transition".

Kaminski and Kaminski argue that corruption is typically the product of weakness or breakdown in the organization of social activity and specifically in the structure of public institutions. "All purposeful social action depends for its success on the will of participants to conform to the rules adopted by the organization or the norms and modes of behaviour prevailing in society". Corruption thus belongs to the broad category of consequences that arise when formal institutional rules are no longer observed and when people pursue their activities without regard to community obligations or purposes. Thus the creation of a rule-governed polity involves three interrelated areas: the choice of institutional design, the political culture of a country and the maturity of civil society.

Among the strategic choices to be made by the post-communist countries in the early 1990s, the first was whether to go for a radical shock or to instigate a more gradual process of reform. The two authors are firm in their conclusion that this was a false dilemma because gradualism was never "a viable policy approach to first-generation reforms". Those countries, mainly in the CIS and parts of south-east Europe, that adopted gradualism simply "created fertile ground for systemic rent-seeking and ... for vested interests opposing liberalizing reforms not only in the economy but also in the political realm". This seems to play down the importance of initial conditions and the inherited institutional legacy mentioned already, and which varied considerably among the transition economies. The authors admit that "the liberal-democratic transition in the CEECs was, in cultural and psychological terms, easier than in the CIS".

Of the constitutional choices the key ones concern the electoral system and the distribution of power between the legislative, the executive and the judiciary. On the electoral system, the authors prefer the majoritarian or first-past-the-post system because it is more likely to produce stable and accountable governments – and there is evidence that it leads to less

---

[123] One participant thought the definition should be broadened. In his own country, in central Asia, he observed that top officials overseeing specific areas of economic activity did not take bribes from the private sector because they had their own private businesses in the same sphere and restricted competition to their benefit. The definition used by Myrdal, however, would seem to be able to cope with this. In the paper by Kaminski and Kaminski (below) this would presumably be seen as an extreme case of "state capture" by private interests.

corruption than proportional representation (PR). But this conclusion rests on the assumption of a strong civil society which prevents majorities from acting undemocratically towards minority and ethnic groups. Hence there may have to be some trade-off between the stability properties of "first-past-the-post" and the broader representation supplied by PR since the latter "may be indispensable for the emergence of democracy supported by a buoyant civil society".

As for the choice between presidential and parliamentary systems of government, most of the CIS countries and the former constituents of the former SFR of Yugoslavia (except Slovenia) have adopted the former, and the CEECs the latter, but the implications for corruption are unclear. Parliamentary governments appear to provide greater stability and "seem to coincide with lower levels of corruption" but as a generalization this is debatable. Parliamentary government has often led to unstable government and political parties degenerating into competing cliques.[124] On the other hand, presidential systems do not appear to be very capable of pushing through reforms. Corruption and instability can flourish under either system if institutions, including public administration, are generally weak and where social discipline and support for the rule of law is wanting. The Kaminskis' paper discusses these various aspects of institutional weakness, and the consequent capture of the state by narrow interest groups, and this analysis provides the basis for some of their suggestions for combating corruption. Here they share the emphasis of Hare and Wei on the need for laws and regulations to be as simple as possible, transparent and neutral, and on minimizing the degree of discretion available to the bureaucracy in their application. The volume of regulation should be kept as low as possible and every effort made to curtail the penetration of economic activity by politics. Public administration must be reformed and made more professional and meritocratic, and also more accountable and resistant to the influence of powerful pressure groups. The electorates must also be encouraged to develop a louder "voice" and this can be helped by strong and articulate media. In other words, there has to be considerable change in prevailing political and social cultures.

**Mark Pieth** (University of Basle and Chairman of the OECD Working Group on Bribery in International Business Transactions) raised the question of social solidarity in making institutions effective: if the distribution of income was seen as unjust, such solidarity would be weak and corruption was likely to flourish. Picking up an earlier comment, he dismissed the idea that companies should be regarded as the "victims" of corruption: they were the "supply side" of the corruption story and were very quick to offer bribes when it suited them. This underlined the need for international standards and conventions which outlawed the giving of bribes: this could be done through a mixture of legislation and peer pressure. Cutting off the overseas supply of bribes into corrupt polities would make it more difficult for local elites to pay off their supporters and allies.

***Jorge Braga de Macedo*** (President of the Development Centre, OECD, Paris) stressed the multi-faceted nature of the corruption problem and the need to tackle it on a wide front and employing a range of disciplines. He was very sceptical about the direct causality between corruption and growth and would instead put much more emphasis on factors such as openness, income levels and so on, which influenced both variables. He also disagreed with the implicit suggestion in the Kaminski paper that the only way to get reforms underway was to have a crisis: gradualism did have advantages over radical reform, particularly if it reduced the risk of policy reversal. The speed at which different reforms can be implemented varies a lot but one should pay more attention to sustainability rather than speed. In general, however, he shared the view that globalization would improve governance via the pressures that were created as a result of greater openness.

Braga de Macedo emphasized the heterogeneity of the transition economies, both in their initial conditions in the early 1990s and in their present state, and this was taken up by several speakers in the general discussion. The perennial argument over shock therapy emerged again, but the Kaminskis' responded that their remarks on this only referred to policies not to institutions. International and peer pressures were seen as having an important influence in reducing corruption: these, together with membership of international organizations, could help to offset the relative weakness of civil society in many countries. It was clear that there were no easy answers to the creation of effective institutions able, *inter alia*, to combat corruption and promote economic development. A great deal depended on informal and intangible factors which only developed slowly. As one participant put it, "you can change the name and the rules of the game quickly, but to train the players takes time".

There are thus no simple answers or policy proposals to meet the problem of corruption, and although many specific suggestions have been made there are no neat conclusions to this Seminar. One issue, however, on which there was wide agreement, is the importance of government in creating and supporting the institutional structure of a market economy. A number of participants also see a crucial development role for the state in creating the conditions for a lift-off into sustained growth. All this represents a marked change from the early 1990s when the general

---

[124] The history of parliamentary democracy in the twentieth century is one of considerable fluctuation and frequent failure. M. Mazower, *Dark Continent. Europe's Twentieth Century* (London, Penguin Books, 1998).

cry was to minimize the role of government and to downplay the idea that it could play any active role in the development process. Although these attitudes were perhaps psychologically understandable in countries emerging from the grip of central planning and one-party rule, they overlooked the strategic necessity of *reforming* the institutions of the state so that it could perform its crucial new role in democratic, market-based polities. This neglect has proven very costly for many transition economies, not least in weakening the defences against corruption and state capture by sectoral interests. For most of the participants at the Seminar, overcoming state failure or weakness, which includes the need to create professional and meritocratic civil services imbued with a public servant ethos, is now a top priority for the transition economies. Most see a useful role for international organizations and other sources of outside pressure in promoting such reform, but it must be asked whether any of them are likely to be effectual without a firm commitment on the part of the ruling elites and with the full support of civil society within individual countries. Ultimately this Seminar must be seen as raising questions of values – values which guide the behaviour of individuals, families and local communities as well as governments, and which shape and underpin the workings of all institutions.[125] As one great scholar put it, "Every social order is one of the possible solutions to a problem that is not scientific but human, the problem of community life".[126]

Corruption has to be seen together with the wider problems of economic development, as suggested by the structure of this Seminar. But Myrdal long ago united these issues in his concept of the "soft state" which comprises "all the various types of social indiscipline which manifest themselves by: deficiencies in legislation and in particular low observance and enforcement, a widespread disobedience by public officials on various levels to rules and directives handed down to them, and often their collusion with powerful persons and groups of persons whose conduct they should regulate. Within the soft state belongs also corruption ... These several patterns of behaviour are interrelated in the sense that they permit or even provoke each other in circular causation having cumulative effects".[127] That corruption is just one manifestation of the "soft state" only underlines the necessity for any programme to combat it to be societal in scope. But this will require committed action by elites with high degrees of personal integrity and competence, as suggested in the Kaminski paper, but in many countries these are unlikely to emerge without a change in unequal power structures. Myrdal saw clearly that "if a large part of the upper class both among politicians and officials and among businessmen and others are out to make the short-term gains from corruption, nothing substantial can be done to stamp it out, however angry the protests of a somewhat lower level of the educated and articulate. Before the power structure has been changed by evolution or revolution it will be difficult to decrease corruption or even hinder its continual increase".[128] This Spring Seminar at least demonstrates a much greater awareness of the problem of corruption than was the case when Myrdal was writing some 30 years ago and this awareness is also reflected in the attention given to it by the leading international economic organizations. The complex societal nature of corruption was also brought out clearly and there was no suggestion that there were any easy solutions. Perhaps UNECE's fourth Spring Seminar should be dedicated to the memory of its first Executive Secretary.

---

[125] Adam Smith, unlike his neo-classical successors, allocated a crucial role to a mature civic ethics in underpinning a cohesive and constructive liberal, market-based polity. J. Evensky, "Adam Smith's lost legacy", *Southern Economic Journal,* Vol. 67, No. 3, January 2001, pp. 497-517.

[126] R. Aron, *The Century of Total War* (London, Praeger, 1954), quoted by Mazower, op. cit.

[127] G. Myrdal, *The Challenge of World Poverty*, op. cit., p. 211.

[128] Ibid., p. 247.

# CHAPTER 2

# PUBLIC AND CORPORATE GOVERNANCE: THE INSTITUTIONAL FOUNDATIONS OF THE MARKET ECONOMY

*William Lazonick*[129]

## 2.1 The triumph of the "market economy"?

In historical perspective, the major debate in political economy in the twentieth century was whether it was the "market" or the "plan" that should constitute the institutional foundations for governing a modern economy. By the end of the century that debate had become old-fashioned as, both politically and economically, capitalism had clearly won out over socialism. The collapse of the Soviet model led nations within the former Soviet bloc to try to transform their modes of economic governance to emulate what they understood to be the institutions of a market economy. In China, beginning with the post-Mao reforms of the late 1970s, a prolonged transition from the plan to the market was set in motion – a transition marked by social disruption and political conflict as well as high rates of economic growth.

Even then, the unquestioned triumph of the market as the dominant mode of economic governance was only secured in the last half of the 1990s with the dynamic growth of the "new economy" in the United States. Building its industrial base on the internet-driven revolution in information technology, the visible features of the new economy that differentiated it from the "old economy" were the heightened mobility of people and money via labour and capital markets as well as rapid changes in product markets.[130]

The emergence of the new economy was widely viewed as the successful culmination of the market-oriented "shareholder-value" movement that had, in the 1980s and 1990s, come to dominate discussions of corporate governance in the United States. Adopting "maximization of shareholder value" as their corporate goal, by the mid-1990s many long-established United States corporations had undergone substantial, and often dramatic, restructuring in attempts to remain competitive in the markets for which they produced. The prime characteristics of this restructuring were the downsizing of corporate labour forces and the increased distribution of corporate revenues to shareholders in the forms of dividends and stock repurchases.[131] The economic theory that rationalized such restructuring was one that posited that, through the market for corporate control, labour and capital resources being misused by corporations could be reallocated by the labour and capital markets to their most efficient uses.[132]

In the mid-1990s such United States-style corporate restructuring did not go down well in most other wealthy capitalist economies. During the 1990s, the Japanese corporate economy maintained the institutions of lifetime employment and cross-shareholding on which it had built its extraordinary economic success during the post-Second World War decades. The ideology of maximizing shareholder value also had little broad appeal in most of western Europe so long as the United States corporate governance regime was perceived as generating employment insecurity and income inequality. Indeed, until the late 1990s in countries such as France, Germany, Italy and Sweden that sought to maintain the integrity of their "social market economies", the issue of corporate governance was hardly discussed and the ideology of shareholder value little known. The western European exception, of course, was Britain, where the Thatcher revolution of the 1980s had sought to give a new lease on

---

[129] Many of the ideas in this paper reflect joint work with Mary O'Sullivan on a project at INSEAD on corporate governance, innovation, and economic performance (http://www.insead.edu/cgep) with funding from The European Commission DGXII (Contract no.: SOE1-CT98-1114; Project no: 053).

[130] W. Lazonick, *Organizational Integration and Innovative Capability in the 'New Economy'*, INSEAD Working Paper (Fontainebleau, France), March 2001; M. Carpenter and W. Lazonick, *The Stock Market and Innovative Capability in the 'New Economy': The Optical Networking Industry*, INSEAD Working Paper (Fontainebleau, France), September 2001.

[131] W. Lazonick and M. O'Sullivan, "Maximizing shareholder value: a new ideology of corporate governance", *Economy and Society*, Vol. 29, No. 1, 2000.

[132] For outlines of this market-oriented perspective as well as historical and theoretical critiques, see W. Lazonick, "Controlling the market for corporate control: the historical significance of managerial capitalism", *Industrial and Corporate Change*, Vol. 1, No. 3, 1992; M. O'Sullivan, "The innovative enterprise and corporate governance", *Cambridge Journal of Economics*, Vol. 24, No. 4, 2000.

life to both the City of London financial elite and masses of pensioners with their savings invested in the stock market. But British industry had entered the 1980s in a much weaker condition than United States industry; as the shareholder-value corporate governance regime took hold in Britain during the 1980s and 1990s, British industrial corporations that downsized their labour forces had much less cash to distribute to shareholders.

In the late 1990s, however, the rise of the United States new economy, based on the internet revolution made the shareholder-value movement more attractive to the countries of continental western Europe. No longer was the pursuit of shareholder value tainted with the charge that it was merely a means of laying off workers to benefit financial interests. Now the shareholder-value movement became associated with economic growth that generated innovative products that people used every day at work and at home. The corporate goal of maximizing shareholder value also became associated with a huge growth in new employment opportunities that called for a workforce with high levels of education and knowledge. Rather than complain that the stodgy corporations of the old economy could no longer provide corporate employees with the employment security that they thought they deserved, it could now be argued that employees had to equip themselves with the transferable productive capabilities to participate in the new economy.

By the late 1990s, as United States stock markets boomed as they never had before, the new ideology of a stock market-driven economy began to erode resistance to the shareholder-value movement in the social market economies of Germany, France, Italy and Sweden. At the microeconomic level, European corporate executives began to see maximizing shareholder value as a recipe for turning staid old-line corporations into nimble innovators, while also perhaps bringing their personal levels of remuneration closer to the extraordinarily high levels that had become the American "norm". At the macroeconomic level, European government officials began to see maximizing shareholder value as a way to generate, on the one side, huge sums of money for state treasuries through the privatization of state owned companies and, on the other side, high returns on retirement savings that United States stock markets were producing and that the public social security systems of western Europe could not hope to provide.

Although, in the years of negative growth of 1997-1998, some Japanese business executives and politicians flirted with the notion that Japan should open its door to the shareholder-value movement, that country maintained its resistance to the ideology throughout the 1990s, as it kept intact its systems of lifetime employment and cross-shareholding, the key "non-market" institutions that had been put in place in its extraordinary ascent from poor to rich nation from the 1950s on.[133] Nevertheless, for the Japanese, the prolonged stagnation of the 1990s, which many came to call the "lost decade", substantially undermined their own confidence in the Japanese model, while in the western economies, market-oriented economists could now dismiss the employment, financial, and regulatory institutions that had underpinned the Japanese "miracle" as obsolete relics of an old economy. The time had come, the western economists argued, for Japan to make the transition to the institutions of a new market economy.

Yet, now in the first half of 2001, the United States stock market is booming no more. With the speculative excesses of the late 1990s becoming more apparent, a "soft landing" for the United States new economy is by no means assured. Even during the new economy euphoria in the United States, the collapse of financial markets in Asia in 1997 and Russia in 1998 had shown the havoc that could be wreaked when masses of money flitted from place to place around the globe in search of higher returns. If, at the beginning of the twenty-first century, the market economy has indeed triumphed as the only possible institutional basis for resource allocation, we now live in a world characterized by profound financial instability and growing income inequality. If one measures superior economic performance in terms of stable and equitable economic growth, then, with the triumph of the market economy, superior economic performance has become ever more difficult to achieve.

This situation poses a conundrum for social reformers and economic policy makers in both richer and poorer nations for whom the achievement of stable and equitable growth is a major objective. The market – and particularly the financial market – often appears as more a problem than a solution for reaching this goal. Yet, especially with the apparent triumph of the market economy, the only respectable theory of the economy to which social reformers and economic policy makers can look for guidance and validation is what can be called "the theory of the market economy"[134] – a theory that posits that an economy in which market institutions allocate resources is the best of all possible worlds.

---

[133] W. Lazonick, "The Japanese economy and corporate reform: what path to sustainable prosperity?", *Industrial and Corporate Change*, Vol. 8, No. 4, 1999; R. Dore, *Stock Market Capitalism: Welfare Capitalism* (Oxford, Oxford University Press, 2000), Part II.

[134] What I am calling the theory of the market economy is, in academic discourse, called "neoclassical economic theory", in contrast to "classical economic theory" of the nineteenth century in which the analysis of production of goods and services rather than their exchange constituted the theoretical core. By focusing on production, the classical economists could address the problem of how the process of economic development could *overcome scarcity*, whereas in focusing on exchange the neoclassical economists posed the economic problem as the "optimal" allocation of *scarce resources* among alternative ends. Unfortunately a growing majority of the current generation of United States-trained economists has never read the classical economists of the nineteenth century – a course in the history of economic thought is not required to obtain a Ph.D. in most leading United States economics departments – and hence the substantive meaning of "neo" in neoclassical has become lost. Throughout this paper, when I refer to "the theory of the market economy", I mean neoclassical theory with its focus on market exchange in the allocation of resources, and when I refer to "market economists", I mean neoclassical economists whose thinking, insofar as it is systematic, is guided by a theory of market exchange as the essence of economic theory. For an elaboration on this theme see also W. Lazonick, *Business Organization and the Myth of the Market Economy* (Cambridge, Cambridge University Press, 1991).

I shall argue that the theory of the market economy propounded by western (and especially United States) economists is more a hindrance than a help in understanding the difficult problems that economies, rich and poor, now face. Specifically I shall argue that the way in which economists are trained to think about the role of market institutions in the operation of the successful market economies has very little to do with how these economies actually operate when they are successful, much less when they falter. As a result, even when market economists recognize that there are major problems with the operation of the market economy, they have great difficulty in making a consistent theoretical case for the regulation of highly speculative markets, or even for government programmes for developing the capabilities of the labour force, investing in new technology or bolstering aggregate demand. The fundamental problem, I shall argue, is that western economists who propound the theory of the market economy – including those who recognize that markets often work "imperfectly" or "fail" – lack a theory of economic development that can explain the successful growth of the wealthy economies. As a result they are intellectually ill-positioned – one might even say that they have a trained incapacity – for explaining why wealthy economies experience crises or why the efforts by poorer national economies to join the ranks of the wealthy go astray.

Lacking a theory of economic development, market economists wrongly tend to see developed markets in labour, capital and products as causes rather than consequences of economic development. The market economy is a very real phenomenon with great economic and political advantages if it can be achieved and controlled. But, in reality, well-functioning markets are much more the consequences than the causes of economic development. To reap the advantages of a market economy, a society must first put in place the organizations and institutions that generate well-functioning markets in labour, capital and products. Then, having put these markets in place, a society must control the operation of markets to achieve stable and equitable economic growth.

## 2.2 The theory of the market economy

An economy is a social system for the allocation of resources to alternative productive uses. Specifically, an economy is a social system that allocates labour and capital inputs to the production of goods and services and then allocates the goods and services that the economy produces to participants in the economy. Through the use of money as a store of value as well as a means of exchange, this allocation process can take place over time as well as at a point in time. As thus stated, this definition of an economy is not controversial.

What are the main political and economic advantages of market exchange as part of a social system for the allocation of resources? The market allocation of labour enables individuals to choose how much and what types of work they wish to do and where they wish to do it. The social advantages of a well-functioning labour market provide the most important *political* argument for a market economy. Some (more privileged) people cannot control the allocation of other (less privileged) people's labour; people are free to choose how and where to seek out a living.

The market for the allocation of capital means that individuals can potentially choose whether they want to work for themselves or for others since, with access to capital markets, they can purchase means of production. Moreover, the existence of a capital market holds out the possibility that they can get a positive return on their savings, and thus can serve as an incentive to save out of their current income. Indeed, whether they work for themselves or for others, the market for capital makes it possible for people to invest, through education and training programs, in the improvement of their own productive capabilities, and thus can enhance their mobility via the labour market and the value of their own productive contributions to the economy. The social advantages of a well-functioning capital market provide the most important *economic* argument for a market economy.

The market allocation of products not only creates consumer choice, but, more importantly, permits access to the purchase of goods and services that are means of production and to the possibility of selling the goods and services that one produces. The existence of product markets thus allows people to combine their labour with access to capital to choose the types of productive activities in which they want to engage. The social advantages of well-functioning product markets are that they can provide people with a greater variety of choices as both consumers and producers, and the existence of well-functioning labour and capital markets can enhance these choices. It is possible, however, for well-functioning product markets to exist in a society that has neither well-developed markets for labour or capital, both of which are more fundamental for the political and economic freedom associated with a market economy.[135]

The advantages of markets, first put forth in a coherent way by Adam Smith more than two centuries ago, have since that time been elaborated by economists into a theory of the market economy – a theory that argues that the more "perfect" the market in terms of the allocation of resources, the better the "performance" of the economy. A "perfect market" is one in which there are no impediments to the mobility of resources from one use to another; labour and capital flow freely and instantaneously to the production of alternative goods and services according to market incentives. Superior

---

[135] The slave economy of the southern United States, catering as it did to world tobacco, sugar and cotton markets, is a dramatic example of such an economy.

performance derives from the ability of individuals to make the best possible use of the allocative mechanisms of labour, capital and product markets to maximize their satisfaction, or utility. The more perfect the market, the more it permits individual utility maximization, and hence (assuming away the thorny problem of interpersonal comparisons of utility), the better the performance of the economy as measured by the satisfaction of its participants.

In this theory, which is readily found in any major economics textbook and which is taken for granted by most of today's professional economists, the key social unit is the household. Although the family household is a small organization that allocates resources internally (and some economists have applied the theory of the market economy to the allocation of resources within the household unit itself),[136] most economists treat the "household" as if it were a utility-maximizing individual. The use of the term household is in effect a concession to the reality that individuals are not able in the early parts of their lives to be active participants in the market economy, and hence are dependent on the allocation decisions of older people who are. The household/individual allocates labour to alternative productive pursuits, income to alternative goods and services, and savings to alternative financial instruments. The important point is that in the theory of the market economy, it is the utility-maximizing decisions of households/individuals acting as atomistic decision-making units that determine the allocation of resources in the economy as a whole. In doing so, they maximize their individual economic performance (i.e. as measured by their utility) in a social system in which, given the pervasiveness of market mechanisms, no individual exercises any power over anyone else.

The individual freedom of choice that is the essence of resource allocation in the presence of markets has led market economists to argue that the perfect market economy is an ideal mode of allocating resources. But many, if not most, market economists also contend that, because of "market imperfections" and "market failures", the perfect market ideal is not always, or even normally, achieved. Market imperfections restrict the free flow of labour and capital to alternative productive activities and household incomes to alternative goods and services, and hence result in less than optimal performance of the economic system as a whole. Market failures occur when there exists a good or a service that society needs, but which is not made available through market resource allocation. Hence the state must step in to supply the good or service directly, or alternatively to influence private allocative decisions so that the market now finds it worthwhile to undertake its supply.

The identification of such market imperfections and market failures then provides market economists with operational concepts with which to focus on reality, and also creates endless possibilities for debate among themselves over whether such imperfections or failures exist, and if so, what to do about them. Some market economists of a more "conservative" bent (for example, Oliver Williamson)[137] argue that market imperfections are inherent in "human nature as we know it", and hence that the organizations and institutions that characterize a market economy are optimal adaptations to these "imperfect" conditions. Other market economists of a more "liberal" bent (for example, Joseph Stiglitz)[138] argue that market imperfections can be reduced through public policy interventions that enhance the free flow of economic resources. Although market economists of different political stripes differ sharply over the efficacy of state intervention, both types agree that the theory of the perfect market economy is the ideal benchmark against which the reality of resource allocation should be compared.

Market failure exists when labour, capital or product markets simply do not generate alternative sources of employment, finance, and goods and services that (according to some normative criteria, which can vary dramatically among economists depending on their political perspectives) are deemed to be necessary or desirable in a well-functioning economy. Examples of market failure are chronic unemployment and poverty-level incomes, an absence of credit facilities for lower income people or smaller firms, and a lack of necessary "public goods" such as primary education and law enforcement that are deemed to be public because the market allocation of resources based on individual (i.e. private) incentives will not generate the demand for these services. Liberal market economists tend to believe in market failure, whereas conservative market economists often argue that state interventions that respond to the existence of purported market failures actually subvert the abilities of markets to allocate resources to achieve the same outcomes. For example, liberal market economists often argue that poverty is market failure, while conservative market economists tend to argue that, if there is indeed a failure, the blame must be laid at the door of the individual for not working hard enough or having sufficient foresight to earn a higher income. What is more, the conservatives would contend, social welfare programmes that treat the problem as if it were a market failure rather than an "individual failure" (given the opportunities of earning incomes that the market provides) only exacerbate the problem by creating incentives for poor people to work less hard and with less foresight.

---

[136] G. Becker, *A Treatise on the Family* (Cambridge, MA, Harvard University Press, 1981).

[137] O. Williamson, *The Economic Institutions of Capitalism* (New York, Free Press, 1985) and idem, *The Mechanisms of Governance* (New York, Free Press, 1996).

[138] For a survey of his work on imperfect markets see R. Bausor, "Joseph E. Stiglitz", in W. Samuels (ed.), *American Economists of the Late Twentieth Century* (Aldershot, Edward Elgar, 1996).

If one accepts these basic terms of the debate, one might be led to believe that the basic explanation for the success of the wealthy market economies is a progressive eradication of market imperfections and market failures that brings the allocation of resources in these economies closer to the perfect market ideal. The clear policy implication of such a perspective for societies that have not achieved such economic success is that their economic future depends on their ability to rely as quickly and as fully as possible on the introduction of markets for labour, capital and products to allocate resources in their economies. Those national economies that want to join the ranks of the wealthy, so the argument goes, should make the transition to the market economy as quickly and as fully as possible, not only within their own political boundaries but also by becoming integrated into the international market economy. And, the policy prescription continues, if these economies continue to experience problems of economic growth, income inequality or financial instability as they make the transition to the market economy, the sources of their problems reside in the persistence of market imperfections and market failures, while the possible solutions will be found in the continued transition to a more perfect market economy.

The theory of the market economy sounds convincing, in large part because of the very real political and economic advantages for the individual of living in an economy in which one can freely allocate one's labour, borrow capital and decide what to consume. The theory of the market economy also seems to be above ideology because there is in fact vigorous debate among market economists with different political perspectives concerning the virtues and vices of state intervention into the resource allocation system. Indeed, in the wealthy economies over the course of the twentieth century, the theory of the market economy attained such a high degree of academic respectability (enhanced immensely by the 30-year-old practice of awarding Nobel Prizes in market economics) that, especially with the collapse of the planned economies, there exists a powerful system of belief that cannot countenance that the theory is fundamentally flawed.

## 2.3 Innovation and development in a market economy

The fundamental flaw in the theory of the market economy begins to become apparent when one asks how an economy can generate higher and higher material standards of living over a prolonged period of time, and when one recognizes that "innovation" – precisely defined as the generation of higher quality products at lower unit costs, given prevailing factor prices[139] – has something to do with the answer. The flaw becomes even more evident once one asks what role business enterprises play in the innovation process, and why indeed business enterprises can grow to employ tens of thousands or even hundreds of thousands of people and persist for decades on end. Can the modern business corporation that controls the allocation of vast amounts of labour and capital be understood either as a massive market imperfection that restricts the free flow of resources via the market or as a set of activities that manifests market failure? Given the importance of the business corporation in a modern economy, might it not make much more sense to have a theory of resource allocation that asks how and under what conditions these business enterprises allocate resources in ways that, by generating higher quality, lower cost products than would otherwise be available, can enhance economic performance? If so, economic theory needs a theory of the innovative enterprise.

The theory of the market economy fails to provide a theory of the innovative enterprise. Indeed, in the theory of the perfect market economy there is no inherent reason why the social unit that we call "the firm" – an entity whose purpose is to transform purchased inputs into saleable outputs – should exist; households engaging in trade on intermediate product (i.e. capital good) markets should be able to perform this function. To create a role for the firm as a distinct unit in the theory of the market economy, one has to assume that there are "economies of scale" in the production of goods and services that make it impossible, or at least economically undesirable, to rely solely on market exchanges to transform inputs into outputs.[140] Even then, in the theory of the market economy, firms, as units responsible for "production", play a passive role in supporting the process of exchange. The firm turns inputs into outputs according to the dictates of factor prices and production technologies that are externally imposed on it by market competition for the allocation of resources, and which the firm therefore takes as given constraints in its resource allocation decisions. As a concession to reality, there are firms in the theory of the market economy, but the theory contemplates only market control, not organizational control, over the allocation of the economy's resources.[141]

This theory of the firm represents the major weakness of the theory of the market economy for

---

[139] W. Lazonick, "The theory of innovative enterprise", in W. Lazonick (ed.), *International Encyclopedia of Business and Management Handbook of Economics* (Stamford, CN, International Thomson, 2001).

[140] I should note that in his famous paper, "The nature of the firm", (*Economica*, n.s., 4, 1937), R. Coase did not explain why firms exist in a market economy, but just how, through the principle of substitution at the margin, existing "firms" (which in his argument could have just as well have been "households") would, within the logic of "imperfect" market exchange, decide to include a greater or lesser range of activities within the firm as an economic unit.

[141] Such is the case even in the theory of monopoly in the market economy, which as I have shown, contains a fundamental logical flaw in comparing perfectly competitive firms with monopoly firms while assuming that both types of firms maximize profits subject to the same cost structures. W. Lazonick, "The theory of innovative enterprise", op. cit.

understanding the way in which actual market economies operate. The main problem with the theory of the firm in the theory of the market economy is that it precludes an analysis of how a business enterprise might allocate resources to transform market and technological conditions in ways that generate innovation – that is, to produce a good or a service that, given the wages paid to labour and the rate of return to capital, is higher quality and/or lower cost than the good or service that this or other firms had previously been capable of putting on the market. Given their belief in the ideal of the perfect market economy, conservative market economists would logically view the modern business enterprise as a massive market imperfection – it exercises organizational control over the allocation of massive amounts of labour and capital and often dominates product markets – while liberal market economists would logically view it as a massive market failure – large-scale business organizations exist because, for some reason, the market "failed" to allocate resources to the particular activities in which the business enterprise is engaged.

It is the introduction of a theory of innovative enterprise into a theory of resource allocation that transforms the market from an explanation to an outcome of economic development, and that, as a result, transforms our understanding of the roles of organizations and institutions, as well as markets, in determining economic performance. Given the importance to the wealthy economies of business organizations of considerable size – and even one that employs 500 people is generally considered large – the adherence to the theory of the market economy leads market economists to ignore systematically the roles of organizations rather than markets in allocating resources to generate superior economic performance. I should stress that markets are important in facilitating the reallocation of resources in the wealthy economies, and the existence of markets for the allocation of labour, capital and products can offer individuals profound political and economic freedom that, once acquired, is to be highly cherished and protected. The problem is that the existence of such socially desirable market opportunities is much more an outcome of the process of economic development than its cause.

A historical and comparative analysis of "the nature and causes of the wealth of nations", not just a century ago but even over the immediate past, shows that in the wealthy economies it has been organizations rather than markets that have been primarily responsible for the allocation of resources that generates economic development. As a result the theory of the market economy provides the wrong benchmark for the "ideal" mode of allocating resources. It is economic development that makes the improvement of markets in labour, capital and products possible, with all the advantages that this improvement brings for political and economic freedom. And it is organizations, not markets, that allocate resources to the production processes that generate economic development. If one wants to learn from the experiences of the wealthy economies, and indeed if a wealthy economy wants to learn critical lessons for the future from its own past, what is needed is a theory of how resource allocation by organizations – both business organizations and government organizations – generates economic development.

The argument that the theory of the market economy lacks a theory of innovation, and hence economic development, is by no means new in the history of economic thought. It is now some 90 years since Joseph Schumpeter, one of the most erudite and creative economists of the twentieth century, made such an argument in a book called *The Theory of Economic Development*. Schumpeter then spent the next four decades of a highly productive career seeking to understand how, why and when innovation contributed to the economic development of the advanced economies. In the 1950s, in an era in which the United States had emerged as by far the most dominant economy in the world, two American economists, Moses Abramovitz and Robert Solow, effectively launched distinguished academic careers by showing, using somewhat different analytical models, that, in the case of the United States from the last decades of the nineteenth century through the first half of the twentieth century, the rate of growth of factor inputs (weighted by their market prices) accounted for only about 10 per cent of the rate of growth in per capita output.[142] Put differently, a growth model based on the theory of the market economy failed to explain most of the economic growth that the United States had experienced during the period when it became the world's richest economy.

In his original article Solow called this unexplained residual "technical change", even though his work, then or subsequently, did not actually demonstrate empirically that technical change, as conceptualized in his growth models, was the source of unexplained growth. In contrast, Abramovitz recognized in a 1962 review of the literature on growth accounting that the index of productivity in this work "has been dubbed by some a 'measure of ignorance,' and it is often referred to simply as the Residual."[143] Indeed, in an article that Abramovitz wrote in 1993, he argued that, over the decades since he and Solow had estimated their first growth models, the empirical work by growth economists had raised more new questions about the causes of growth than they had been able to answer, and hence, relative to what we now know that we ought to know about the growth process, the "measure of ignorance" had if anything increased.[144]

---

[142] M. Abramovitz, "Resource and output trends in the United States since 1870", *American Economic Review*, Vol. 46, No. 2, 1956; R. Solow, "Technical change and the aggregate production function", *Review of Economics and Statistics*, Vol. 39, No. 3, 1957.

[143] M. Abramovitz, "Economic growth in the United States: a review article", *American Economic Review*, Vol. 52, No. 4, 1962, p. 764.

[144] M. Abramovitz, "The search for the sources of growth: areas of ignorance, old and new", *Journal of Economic History*, Vol. 53, No. 2,

Over the past few decades many scholars, including economists who are intent on going beyond the theory of the market economy in understanding how the economy actually operates and performs, have done a considerable amount of empirical research on the process of innovation. Some of this work has been focused on "national systems of innovation", while other work has focused on case studies of particular technological transformations in particular industries.[145]

While different scholars have emphasized different key characteristics of the innovation process, taken as a whole these studies suggest that the innovation process is collective, cumulative and uncertain; the learning process that is the essence of innovation cannot be done all alone, all at once or with any degree of certainty that what needs to be learned will in fact be learned (productive uncertainty) or that, even if it is learned, competitors will not learn to do it better (competitive uncertainty). The innovation process is collective because the transformation of technological and market conditions to generate higher quality, lower cost products requires the organizational integration of the specialized knowledge, skills and efforts of large numbers of people with different functional capabilities and hierarchical responsibilities. When the innovation process is collective, there is a need for organizational, rather than market, control over resource allocation. The innovation process is cumulative because the possibilities for transforming technological and market conditions today and tomorrow depend on the development of those conditions in the past. When innovation is cumulative some or all of the collectivity that engages in learning must remain intact over time. The innovation process is uncertain because the collective and cumulative processes that can transform technological and market conditions to generate higher quality, lower cost products are unknown at the time at which commitments of resources to these processes are made. Hence, an innovative enterprise must be strategic in how it engages in collective and cumulative learning. When innovation is uncertain, investment in organization that is both collective and cumulative can enable an innovative enterprise to transform technological and market conditions that other, less powerful, enterprises might have to accept as binding constraints.[146]

What are the implications of the collective, cumulative and uncertain characteristics of innovation for the mode of resource allocation that defines the institutional foundations of the economy? As Mary O'Sullivan has shown, the theory of the market economy in effect sees resource allocation as individual, reversible and optimal.[147] That the resource allocation is individual means that people make allocation decisions in isolation from one another (as households); that it is reversible means that the decisions that they made yesterday have no bearing on the decisions that they make today; and that it is optimal means that, as individuals who can (but for market imperfections) change their allocation decisions, they make these decisions accepting all of the constraints that the economic (and political) system imposes on them. In the theory of the market economy, participants in the economy have no possibility of strategically changing the technological and market conditions that they face.

Yet the strategic transformation of technological and market conditions is what innovation is all about. Indeed, O'Sullivan then goes on to show that, when the innovation process is collective, cumulative and uncertain, resource allocation to the innovation process cannot be individual, reversible and optimal as in the theory of the market economy. Rather, to allocate resources to a process of transforming inputs into outputs that is collective, cumulative and uncertain, the mode of resource allocation must be organizational, developmental and strategic. Markets cannot engage in resource allocation that is organizational, developmental and strategic; organizations can.

## 2.4 The social conditions of innovative enterprise

There are two main types of organizations that are central in the allocation decisions that result in economic development: the innovative enterprise and the developmental state. Enterprises and states exercise control over the allocation of vast amounts of labour and capital but they differ fundamentally in the ways in which they gain and maintain access to the financial resources that give them control over the allocation of productive resources. Enterprises gain access to what can be called "foundational finance" on an ongoing basis through the revenues that they generate from the sale of goods and services, whereas states gain access to foundational finance through taxation. Both enterprises and states can leverage this finance through debt issues, servicing this debt with the flows of foundational finance. It is, however, the different modes of accessing foundational finance that create the fundamental difference between

---

1993 and idem, *Thinking About Growth* (Cambridge, Cambridge University Press, 1989). It perhaps explains something about the economics profession that when in 1987, Solow won the Nobel Prize in Economics for his work on economic growth, his early work on the measurement of technical change was specifically cited as his seminal contribution. It would appear that Abramovitz's reward for recognizing that the unexplained residual was in fact a measure of ignorance was to be ignored by the Nobel Prize committee, despite an illustrious career in which he contributed numerous penetrating insights into the process of economic growth.

[145] A summary of much of this work on innovation, as well as the conceptualization of its implications for the allocation of resources in the economy, can be found in M. O'Sullivan, "The innovative enterprise and corporate governance", loc. cit., on which the following draws.

[146] For a formal analysis of the implications of innovation that is collective, cumulative and uncertain for the "theory of the firm" and the social conditions in which it is embedded see W. Lazonick, "The theory of innovative enterprise", op. cit.

[147] M. O'Sullivan "The innovative enterprise and corporate governance", loc. cit., and idem, *Contests for Corporate Control: Corporate Governance and Economic Performance in the United States and Germany* (Oxford, Oxford University Press, 2000), chaps. 1 and 2.

the governance of resource allocation in an enterprise and a state. In particular, enterprises are under a compulsion to deliver products that buyers want at prices they can afford. These buyers may be households, other enterprises or even states.

The generation of revenues through the sale of products enables an enterprise to govern "itself". Hence the notion that enterprises operate in the "private" sphere, although the identification of itself – that is, those interests who are deemed to be participants in the enterprise – is a central issue in debates on corporate governance.[148] When these revenues are more than sufficient to allocate expected returns to all parties who have financial claims on the enterprise, the surplus can provide a foundation for financing new productive investments. Hence, the importance of profits for the viability of an enterprise as an ongoing organizational concern. In contrast, the reliance of the state on taxation for foundational finance opens it to societal governance, and hence the notion that the state operates in the "public" sphere. Subject to these very different governance regimes, both the enterprise and the state can choose to allocate resources to organizational learning, with the critical difference that, unlike the enterprise, the state is not normally expected to ensure that the productive resources that are thereby developed are utilized in ways that generate financial returns. The development *and* utilization of productive resources to generate financial returns is the distinctive role of the innovative enterprise, and constitutes the most fundamental reason why the enterprise is more central than the state in affecting the economic performance of an advanced economy.

Both the innovative enterprise and the developmental state can allocate resources to organizational learning processes. In general, the developmental state will undertake investments in technologies that are deemed to be of strategic (e.g. military or medical) importance when the collective, cumulative and uncertain character of the learning process renders the expected scale of the commitment of financial resources so large and the expected duration of time before the generation of financial returns so long (with prospective product markets often non-existent at the outset) that existing enterprises are unwilling to make the investments. Nevertheless, the state typically induces enterprises to participate in these developmental efforts, either through investment subsidies or procurement contracts that, for the activities in which the enterprise invests, make the scale and duration of the commitment of financial resources acceptable from a business point of view.

Although space constraints do not permit an adequate elaboration of the evolving relations between innovative enterprises and developmental states in the growth of advanced economies, it is worth noting that, contrary to what has become conventional wisdom, during the twentieth century the developmental state was of much more direct importance in the growth of the United States economy than in the growth of the Japanese economy. The Japanese state was of critical importance in mobilizing bank finance to help fund the innovative efforts of business enterprises. But the United States government was much more directly involved in the strategic direction of organizational learning processes that spanned state and enterprise organizations in agriculture and health sciences (including biotechnology), aircraft and engines, computers (including semiconductors) and the internet.[149] Indeed, it is safe to say that the internet revolution that has provided the technological foundations for the new economy would not have occurred in the United States, but for decades of United States government support for the development of information and computer technology.

An analysis of how enterprises, with or without the support of the state, develop the productive resources that are ultimately sold on markets to generate returns requires the identification of the "the social conditions of innovative enterprise". From a characterization of the innovation process as collective, cumulative and uncertain, combined with a comparative-historical analysis of successful economic development in the twentieth century, we can identify three social conditions of innovative enterprise: organizational integration, financial commitment and strategic control. The form and content of these social conditions of innovative enterprise depend on the relation between prevailing institutional (financial, employment and regulatory) conditions and organizational (cognitive, behavioural, and strategic) conditions in the economy (see chart 2.4.1). These three social conditions of innovative enterprise all reflect the importance of organizational control rather than market control over the allocation of resources in the economy.

Organizational integration means that it is the organization rather than the market that creates incentives that affect how people allocate their labour. Financial commitment means that it is the organization rather than the market that controls the allocation of money to

---

[148] Ibid.

[149] See, for example, L. Ferleger and W. Lazonick, "The managerial revolution and the developmental state: the case of United States agriculture", *Business and Economic History*, Vol. 22, No. 2, 1993; D. Kash, *Perpetual Innovation: The New World of Competition* (New York, Basic Books, 1989); T. Hughes, *American Genesis: A Century of Invention and Technological Enthusiasm, 1870-1970* (New York, Viking, 1989), chap. 3; T. Heppenheimer, *Turbulent Skies: The History of Commercial Aviation* (New York, Wiley, 1995); E. Constant III, *The Origins of The Turbojet Revolution* (Baltimore, MD, Johns Hopkins University Press, 1980); T. Hughes, *Rescuing Prometheus* (New York, Pantheon, 1998); J. Tilton, *The International Diffusion of Technology: The Case of Transistors*, (Washington, D.C., The Brookings Institution, 1971); D. Mowery and N. Rosenberg, *Paths of Innovation: Technological Change in 20th Century America* (Cambridge, Cambridge University Press, 1998); J. Abbate, *Inventing the Internet* (Cambridge, MA, MIT Press, 1999); M. O'Sullivan, *Contests for Corporate Control ...*, op. cit., chap. 6.

**CHART 2.4.1**

**Industrial, organizational and institutional conditions in the innovation process**

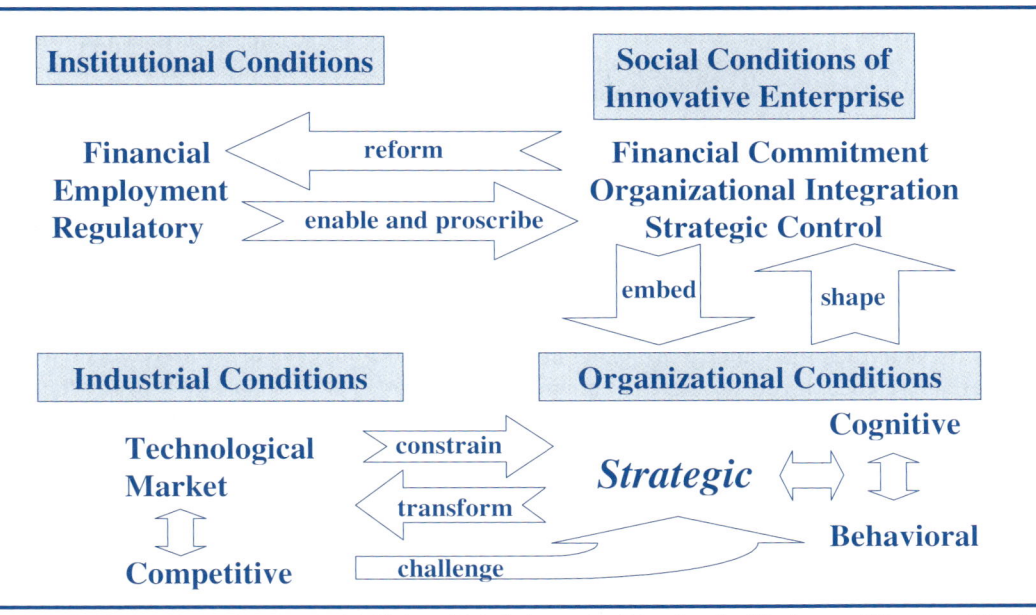

```
Institutional Conditions              Social Conditions of
                                      Innovative Enterprise
  Financial      ← reform ←          Financial Commitment
  Employment                          Organizational Integration
  Regulatory   → enable and proscribe → Strategic Control
                                              ↓ embed    ↑ shape

  Industrial Conditions                 Organizational Conditions
                                                        Cognitive
  Technological  → constrain →          Strategic  ↔
  Market         ← transform ←                            Behavioral
      ↕            ← challenge ←
  Competitive
```

*Source:* W. Lazonick, "The theory of innovative enterprise", in W. Lazonick (ed.), *International Encyclopedia of Business and Management Handbook of Economics* (Stamford, CN, International Thomson, 2001).

alternative uses. Strategic control means that it is the organization rather than the market that determines the types of investment in productive capabilities that the economy makes. Hence, in analysing the process of innovation and economic development, economics needs a theory of the organizational economy rather than a theory of the market economy to understand when, how and whether these social conditions of innovative enterprise are put in place.

Organizational integration is the social condition that creates incentives for participants in the hierarchical and functional division of labour to apply their skills and efforts to engage in interactive learning in pursuit of organizational goals. As a social condition for innovative enterprise, the need for organizational integration derives directly from the collective character of the innovation process. Hence, a theory of innovative enterprise must show how, given the collective character of the transformation of technology and markets in particular, industrial activities, institutions and organizations combine to create the necessary incentives for those who are expected to engage in interactive learning.

Across the wealthiest economies for over a century, the main mode of organizational integration has been the internal career path which has offered employees the expectation that, subject to certain performance criteria, they would find opportunities of stable, remunerative, and, perhaps, creative employment with their existing employer over a long period of time. On these career paths, such employees typically develop skill, knowledge and experience – that is, productive capabilities – that are relevant to the organizations for which they work. Since innovation depends on organizational learning, the enterprise typically has substantial interests in both making investments in "human capital" that enhance the productive capabilities of their employees and ensuring that it can utilize these capabilities by securing the long-term attachment of these employees to the organization.

Although such "organization" men and women possess the right to quit their employment at any time, and by virtue of their accumulated skill, knowledge, and experience are often well-positioned to make use of the labour market, they generally choose to remain with their current employer because as insiders they tend to receive higher pay, greater employment security and more financial stability than people who are outsiders to the organization. Indeed, outsiders to established business organizations who are compelled to look constantly to the market to allocate their labour may be fortunate to live in a society in which they have the political freedom to do so, but, within that society, will tend to be those with the least employment security and lowest remuneration.

Financial commitment is the social condition that allocates financial resources to sustain the process that develops and utilizes productive resources until the resultant products can generate financial returns. As a social condition for innovative enterprise, the need for financial commitment derives directly from the cumulative character of the innovation process – that is from the need for learning. For an enterprise or economy that has accumulated capabilities, financial claims can take on an existence that, for a time at least, are independent of the need to reproduce or augment those capabilities. In effect, financial returns to groups such as employees, creditors, and shareholders may be based on the revenues generated by productive capabilities accumulated in the past without a commitment of financial resources for the regeneration of these returns in the future. But, for innovation to occur within an enterprise or economy, a basic social condition is financial commitment from some source for a sufficient period of time to generate returns. A theory of innovative enterprise must show how, given the financial requirements of the transformation of technology and markets in particular industrial activities, institutions and organizations combine to provide the requisite financial commitment.

To analyse the sources of financial commitment, we must distinguish between new ventures and going concerns. A new venture cannot typically go to capital markets to fund its activities; its ability to generate returns is too uncertain. Hence, for most of the twentieth century in the wealthy economies, the finance for new ventures came from personal savings, friends, business associates and special financial facilities set up by governments. In the post-Second World War decades in the United States, specialized venture capital firms emerged, at first to take advantage of the commercialization opportunities made possible by United States government spending on military research and development during and after the Second World War. The United States venture capital industry expanded rapidly from the end of the 1970s after federal government legislation enabled pension funds to make portfolio investments in risky assets such as corporate stocks and venture capital funds. But, even in the United States, the allocation of resources by venture capitalists to new ventures is by no means a market process; in committing funds until such time that through an initial public offering or a private sale to an established company the venture capitalists can reap returns, venture capital firms that support innovation recognize that eventual success depends critically on their relations with entrepreneurs who are making strategic decisions and on the organizational integration of key personnel.

A new venture becomes a going concern when, through the sale of its products, it can generate sufficient revenues to establish the foundation for ongoing financial commitment. These revenues can be used to enhance the financial commitment that it can make to its personnel in the forms of employment stability and increased remuneration as well as to fund the expansion of its organization in terms of both human and physical capabilities. Growing revenues that are retained within the organization, therefore, can enhance organizational integration. The dependence of a going concern on revenues as an ongoing source of financial commitment means that it places great importance on maintaining its existing customers (households, businesses, governments) by generating higher quality, lower cost products; indeed, its relations with these customers often provide the company with the crucial knowledge of how it can improve its products to serve their needs.

Revenues retained within the business organization can also be used to leverage the access of the enterprise to finance that it can use for expansion. When such debt is secured through market relations, as has been particularly the case in the United States, it takes the form of long-term bonds, so that the enterprise does not have to keep going back to the market to fund investments that require financial commitment. With bonded debt, creditors whose only relation to a company is via the market can force an enterprise into bankruptcy, and hence corporations that use bond finance have historically tended to have low debt-equity ratios to ensure that they will not run into financial difficulty. In general, the use of high debt-equity ratios, with its advantages for funding rapid growth but potential disadvantages for exposing the enterprise to debt-service problems, requires organizational relations with the banking system, as for example, in the case of the Japanese main bank system. These relations support the enterprise by allocating finance based on a company's long-run prospects for sales revenues rather than short-run cash-flow problems. In addition, as part of regulated national banking systems, such bank finance is usually provided at rates well below those that they could actually obtain on the market.

For market economists, organizational control over revenues and relational bank finance is a recipe for the misallocation of resources. Yet the fact is that these forms of financial commitment fuelled the post-Second World War recoveries of Japan and Germany, enabling them to emerge as the second and third largest economies in the world. Moreover, the market economist's favourite form of corporate finance, stock issues, have been relatively unimportant, and even insignificant, as sources of funds for productive investments, not only in Japan and Germany but even in the bastions of shareholder value, the United Kingdom and the United States. In the United States over the past half century net stock issues as a per cent of all net sources of funds of non-financial corporations was at its peak at about 13 per cent in the period 1966-1971, when retentions (undistributed profits and capital consumption allowances) accounted for about 66 per cent and net debt issues 22 per cent of net sources of funds. Since the mid-1980s, net stock issues as a net source of funds has often been negative, mainly because of the growing importance of corporate stock repurchase programs.[150]

But even when a company issues stock, it is generally not to raise money for investments in new productive resources. When a company sells stock on the market, it can use it: (a) to transfer share ownership, so that, as is typically the case in an initial public offering (IPO), the funds raised go to the owner-entrepreneurs who built up the company but not to the company itself; (b) to restructure balance sheets by paying off debt or

---

[150] W. Lazonick and M. O'Sullivan, "Maximizing shareholder value: ...", loc. cit.

building the corporate treasury, a use of stock issues that is particularly attractive to companies during a speculative stock market boom; (c) to acquire other companies, a use of stock issues that has been of particular importance in financing the unprecedented new economy mergers and acquisitions (M&A) activity of recent years; and (d) to compensate employees, a major feature of the new economy that, primarily through stock options, has been designed to recruit and retain employees who have been highly mobile on the labour market.

The use of stock issues on public markets to transfer share ownership has been important in the history of twentieth-century capitalism not only for permitting original owner-entrepreneurs to cash out of the enterprises that they have built up but also as a way for the publicly-listed company to separate share ownership from managerial control. Indeed, the evolution of financial commitment and corporate control in the United States industrial corporation from the last decades of the nineteenth century through the post-Second World War decades, and for most corporations even to the present, cannot be understood without a recognition that a fundamental role of the stock market in the United States has been to separate ownership and control. Public stockholders hold shares in a company because a liquid stock market enables them to participate as "owners" by making only portfolio investments, not direct investments. As direct investors, managers of United States industrial corporations in contrast have been able to exercise control over the allocation of corporate resources in the form of retentions precisely because the so-called owners of their companies are (and want to be) portfolio investors, not direct investors.

The use of stock to restructure corporate balance sheets played a very important role in the evolution of the Japanese corporate economy in the 1970s and 1980s, and indeed is part of the explanation of the origins of Japan's unsustainable "bubble economy" in the late 1980s and the resultant problems of Japanese financial institutions in the last decade. By the late 1980s, the industrial corporations that had driven Japan's economic development, and that since the 1950s had been reliant on high levels of bank finance, no longer needed bank loans on the same scale. From the 1970s, these companies restructured their balance sheets by using retained earnings to retire bank debt. In the late 1980s, with the stock market booming – itself a reflection of the phenomenal growth of the major industrial corporations, especially as they captured foreign markets – these companies sold huge amounts of stock to the public (while at the same time distributing shares to other businesses to maintain existing levels of cross-shareholding) and floated bonds at extremely low rates on foreign capital markets. They then used the proceeds to further reduce their reliance on bank debt or to build up their corporate treasuries (thus, as it turned out, making themselves less vulnerable to financial pressures in the stagnation of the 1990s). This at the same time that, as a result of Japan's successful economic development, the Japanese banks were awash with cash. Having lost much of the business of their best industrial customers, the banks, pressured to find markets for this cash, lent money to real estate and stock market speculators, thus fuelling the rise in land and stock prices. When in 1990 the bubble burst, the banks were left with massive amounts of bad loans that they are still in the process of trying to write off. In other words, the success of Japan's industrial economy, based on its powerful business organizations, led Japan's financial markets to spin out of control, as financial resources were allocated to speculative markets rather than to productive investments.[151]

The use of corporate stock to acquire other companies and to recruit and retain employees through stock-based compensation played a central role in the United States new economy. Innovative high-technology companies such as Microsoft (with about 39,000 employees in 2000), Intel (about 70,000 employees) and Cisco Systems (about 35,000 employees) were able to use their highly valued corporate stock as a private "currency" to accumulate human capabilities, extending stock-based compensation to virtually all of their employees as part of a systematic strategy for recruiting and retaining human resources. It was not because they lacked public currency, i.e. dollars, that these companies chose to use stock as a private currency. On the contrary, their success as innovative enterprises meant that they had huge cash surpluses, and paid little or no dividends to shareholders. Rather they used stock as a currency for accumulating capabilities because the stock market placed an enormously high value on it, and because, by virtue of the alternative employment opportunities created by the expanding numbers of new ventures in the information and computer industries, their highly educated, highly mobile employees demanded this type of remuneration as a condition of employment. The key to the success of these new economy companies was and remains the organizational integration of these employees, which in turn puts pressure on these companies to maintain high stock prices to keep their employees from seeking even better stock-based compensation packages elsewhere. For the older new economy companies such as Microsoft and Intel, the use of

---

[151] It is worth noting that much the same thing happened in the United States in the stock market boom of the late 1920s that resulted in the Great Crash and then the Great Depression. A major difference with Japan in the 1990s was that by the early 1930s United States corporations were laying off hundreds of thousands of workers, thus deepening the depression; the unemployment rate reached 25 per cent in 1933 and did not fall below 15 per cent until 1941, when United States entry into the Second World War ended the Great Depression. Compared with the United States experience of the 1930s, the Japanese experience of the 1990s is one of remarkable economic stability, largely because of the responses to adverse conditions of its business and government organizations. In Japan in the 1990s the descent from recession to depression was avoided, on the one side, by the resolve of Japanese business enterprises to keep people employed and, on the other side, by the willingness and ability of the Japanese state, supported by a very high household savings rate despite extremely low returns on savings, to incur unprecedented levels of debt in order to finance massive government spending. See W. Lazonick, "The Japanese economy and corporate reform: ...", loc. cit.

stock-based compensation meant that, by the late 1990s, they were compelled to repurchase their own stock on the market in order to support its price; in the period, 1997-2000, for example, Intel spent $11 billion on R&D, $1 billion on dividends and $18 billion on stock repurchases.

The current downturn in the stock market is putting these new economy compensation systems to the test. When the stock prices of these companies were generally rising, it was possible to use stock as a major form of compensation covering a broad base of employees that could enhance organizational integration. But a falling stock price can foster general dissension among employees, while a volatile market can erode organizational cohesion as, for example, newly hired employees, who received their options at a lower exercise price, find themselves in a better position than long-time employees who find their options "under water". It remains to be seen over the coming years, whether, how and to what extent organizational learning at United States high-technology companies will be affected by the use of a currency – their company's stock – to provide incentives to large numbers of employees when the value of that currency is so subject to market speculation.

Strategic control is the social condition that enables people within an enterprise who have access to financial commitment, and who influence organizational integration, to allocate resources in ways that can transform technologies and markets to generate innovation. As a social condition for innovative enterprise, the need for strategic control derives directly from the uncertain character of the innovation process. Hence, a theory of innovative enterprise must show how, given the uncertain character of the transformation of technology and markets in particular industrial activities, control over financial commitment and organizational integration rests with those people within the enterprise who, as strategic decision makers, have a willingness and ability to use that control to attempt innovative transformations of technologies and markets.

Strategic control is "insider" control, i.e. the exercise of control over resource allocation within the organization, as distinct from "outsider" control, i.e. the exercise of control over resource allocation from the market. The innovation process is always uncertain and, hence, other than leaving the outcome of resource allocation to pure luck, the only basis for making investments that might result in innovation is to vest control over the allocation of resources and returns with people who are both able and willing to invest in collective, cumulative, and uncertain learning processes. They will be able to do so when they have a broad and deep understanding of the industries and organizations in which they are investing. They will be willing to do so when their own individual success is bound up with the success of the organization as a whole. Put differently, investments in innovation that can confront the inherent uncertainty of the innovation process require the organizational integration of strategic decision makers with the processes of collective and cumulative learning.

Such insiders tend to be career managers, and are rarely people whose main participation in the enterprise is as public shareholders. Hence, once again, the importance of the separation of share ownership and managerial control in the history of successful industrial development. As outsiders to the corporate allocation of resources, the vast majority of shareholders would not hold shares in a company, but for the liquidity that the stock market provides. Through the power of collective shareholding, i.e. institutional investing, shareholders can place pressure on corporate managers to increase the allocation of returns, either in the form of higher dividends or, increasingly, stock repurchases. But public shareholders, the major institutional investors included, generally have neither the ability nor the incentive to participate in the process of strategic control that allocates corporate resources to innovative investments.

Indeed, especially in the United States, but increasingly elsewhere, the demands of corporate shareholders for higher returns have been supported by a theory of corporate governance that claims that superior economic performance in the economy as a whole depends on maximizing shareholder value. Yet, as O'Sullivan has shown, rooted as it is in the theory of the market economy, the shareholder perspective on corporate governance lacks a theory of innovation.[152] The shareholder perspective has nothing to say about the roles of organizational integration, financial commitment and strategic control as social conditions of innovative enterprise. Rather, the shareholder perspective inserts itself into a social environment in which innovative enterprises and development states have for decades allocated resources that have created value, and makes the ideological claim that, as the "principals" in the modern corporation, shareholders should now have the predominant, if not the only, voice in determining the distribution of value that has already been created.

## 2.5 Economic theory and economic development

The social conditions of innovative enterprise make organizational control rather than market control over resource allocation central to the development of the economy. Organizational integration requires the management of labour mobility, financial commitment, the management of capital mobility, and strategic control and management of the transformation of the resources, human and physical, that the enterprise has accumulated into high-quality, low-cost goods and services. The economist should not view these social conditions of innovative enterprise as market imperfections but as the institutional foundations for economic development.

---

[152] M. O'Sullivan, "The innovative enterprise and corporate governance", loc. cit.

The identification of organizational integration, financial commitment and strategic control as the key social conditions of innovative enterprise derives from comparative-historical analyses of the development of the world's wealthiest economies over the past two centuries.[153] While the social conditions of innovative enterprise that we have identified are common to the development of national economies such as those of Germany, Japan, the United Kingdom and the United States, the particular configurations of institutional and organizational conditions that created these social conditions vary markedly among these economies, even to the present, and, within a given national economy, have undergone significant transformation over time. Moreover, different social conditions of innovative enterprise, including distinct differences in the functional and hierarchical divisions of labour that characterize innovative organizations, vary both across different industrial activities and, for given industrial activities, over time, with different outcomes in terms of product quality and cost.[154] The theoretical perspective on the social conditions of innovative enterprise and economic development that I have proposed is not in itself an explanation of successful economic performance, but should be seen as a tool for systematic study of the comparative and historical realities of the development of the wealthy economies. Innovation and economic development are processes of change that are highly dependent on the particular institutional, organizational and industrial conditions under which they occur. A theory of economic development that fails to comprehend how, when and to what effect specific institutional and organizational arrangements have yielded superior economic performance will soon lose touch with reality. For a social scientist to comprehend these processes of change requires the integration of theory and history.[155]

As for social reformers and economic policy makers intent on contributing to the wealth of their particular countries, the main implication of the perspective that I have set out is that they have to combine a relevant theoretical analysis of the development process with a deep understanding of the social context in which they expect reforms and policies to have their effects. Specifically, they must understand the ways in which particular institutional, organizational and industrial conditions will promote or impede resource allocation that is organizational, developmental and strategic. To do more good than harm, social reformers and economic policy makers must be both astute observers of the social environments in which they work and insightful analysts of the development processes that they are trying to influence.

Adherence to the logic of the theory of the market economy will not help them in this task. By portraying resource allocation as individual, reversible and optimal, the theory of the market economy reduces such social phenomena as organizational structures, development paths and strategic choices to imperfections or failures of the market mechanism that it would be best to eliminate. The result is that the basic policy recommendations that one derives, quite logically, from the theory of the market economy will, in all probability, erode rather than support the social conditions of innovative enterprise.

The market economist would recommend policies that increase labour mobility. But from the perspective of economic development labour mobility is only beneficial if it enables people to choose more attractive employment opportunities than the ones they already have. The process of economic development is not promoted when people are uprooted from their traditional employments and pushed into the "modern" sector. Rather, economic development generally depends on enhancing the capabilities of people in their traditional employments, and then forcing the modern sector, itself often the result of the direct or indirect result of strategic policy choices, to compete for these productive capabilities. A well-functioning labour market will be the result rather than the cause of economic development.

The market economist would recommend the creation of financial markets, and particularly stock markets, to encourage the mobility of capital. But, in and of themselves, financial markets simply create opportunities for those with financial assets to engage in portfolio investment which, unless highly regulated, tends to evolve into speculative investment. From the perspective of economic development, what is needed is financial commitment, not financial liquidity, which means that funds have to end up in the hands of direct investors who are able and willing to exercise strategic control over the particular investments that they undertake. Financial markets can be useful for mobilizing savings, but the allocation of financial resources to the process of economic development requires organizations and institutions that are designed to protect the innovative enterprise and the developmental state from the "individual, reversible and optimal" decisions of portfolio investors. The emergence of innovative enterprises that can generate financial returns on the financial commitments that have been made expands the portfolio opportunities for those who, as outsiders to the enterprise, seek to reap financial returns on financial markets. Moreover, the success of innovative enterprises places more disposable income in the hands of employees who can accumulate financial

---

[153] See e.g. W. Lazonick and M. O'Sullivan, "Finance and industrial development", *Financial History Review*, Vol. 4, Nos. 1 and 2, 1997; and idem, "Big business and skill formation in the wealthiest nations: the organizational revolution in the twentieth century", in A. Chandler, Jr., F. Amatori and T. Hikino (eds.), *Big Business and the Wealth of Nations* (Cambridge, Cambridge University Press, 1997).

[154] W. Lazonick and M. O'Sullivan, "Organization, finance and international competition", *Industrial and Corporate Change*, Vol. 5, No. 1, 1996.

[155] W. Lazonick, "Understanding innovative enterprise: toward the integration of economic theory and business history", in F. Amatori and G. Jones (eds.), *Business History Around the World* (Cambridge, Cambridge University Press, 2001).

assets, join the ranks of portfolio investors, and thereby increase the liquidity of financial markets. Well-functioning capital markets are the result rather than the cause of economic development.

The market economist would recommend the creation of markets in goods and services that would expand the consumption choices of households. But what is the source of the incomes that enable households to consume beyond their basic needs? And for any particular good or service, what determines the quality and cost of the products that consumers find available on product markets? The theory of the market economy cannot provide answers to these questions because it contains no theory of innovative enterprise – an enterprise that can generate higher quality, lower cost products that, depending on the distribution of enterprise revenues, can simultaneously result in higher returns to labour, higher returns to capital and lower prices to consumers, even while providing consumers with a higher quality product than previously existed and leaving surplus revenues in the enterprises to make further investments in innovative processes and products. There is nothing inevitable about either the success of innovative enterprises or the ways in which they distribute the gains from successful innovation. But the evolution of higher standards of living, reflected in both the incomes of the population and the quality and cost of the products that they can consume, cannot be understood without a theory of innovative enterprise. A wide variety of consumer choice of goods and services, including the relation between quality and cost, is the result, not the cause, of economic development.

Innovation and economic development do not just happen. Just as the allocation of resources by the state must be governed, so too must the allocation of resources by the enterprise. In ignoring an analysis of the innovation process and economic development, theories of corporate governance based on the theory of the market economy cannot address the difficult organizational and institutional questions concerning the governance of innovative enterprise. Nor can they learn from the varied experiences in innovative corporate governance that can be found within and across the wealthy economies. Moreover, without a theory of innovation and economic development, debates on the role of the state in the allocation of resources will be limited to the extent to which its activities are predatory (as most conservative market economists would argue) or regulatory (as most liberal market economists would argue). To be sure, the state often plays both these roles. But at times it also plays a developmental role that, as I have suggested, has been critical to the success of all of the wealthy economies, not the least to the "new economy" of the United States.

Innovation and economic development are not easy processes. They require hard thinking and hard choices. They are social processes that can generate stable and equitable economic growth if the people who participate in them as workers, managers, investors and consumers understand and accept the organizational, developmental and strategic challenges involved. The ideology of the market economy does not further such widespread understanding and acceptance. In a real economy, the widespread and engrained belief in the theory of the market economy tends to render ungovernable those corporate executives and political elites who wield power over the allocation of resources while it tends to leave vulnerable the vast majority of the population who depend on the strategic decisions of the enterprise and the state to create economic opportunity.

# DISCUSSANTS' COMMENTS ON PROFESSOR LAZONICK'S PAPER

## 2.A Victor Polterovich

In his very insightful and stimulating paper, Professor Lazonick contrasts the neoclassical economic theory, which he calls the theory of the market economy, with the theory of the innovative enterprise.

The benchmark of neoclassical economics is a model of the perfect market so that all economic activities of the state and even the modern business corporation have to be explained as market imperfections or market failures. Within this theory, "... the basic explanation for the success of the wealthy "market economies" is a progressive eradication of market imperfections ..." (section 2.2). This theory, however, has a fundamental flaw: it is unable to explain the innovation process. Not the market, but the innovative enterprise and the developmental state, Professor Lazonick says, are the two main organizations that generate economic development. In the neoclassical model an enterprise maximizes its profit under given market and technology constraints, and therefore cannot be innovative. Therefore, an alternative theory, a theory of the innovative enterprise, has to be developed. This theory takes into account that the innovation process is collective, cumulative and uncertain, and includes the learning process. The allocation of resources to the innovation process is organizational, developmental and strategic, so that three social conditions of the innovative enterprise can be identified: organizational integration, financial commitment and strategic control (section 2.3).

I have four comments on Professor Lazonick's paper.

First, I would prefer a more positive evaluation of the neoclassical economics. In the neoclassical framework there has been developed a macroeconomic theory of endogenous economic growth.[156] This is a macroeconomic theory of innovation. In addition, a microeconomic theory of innovation has arisen as a chapter in a new theory of industrial organization.[157] It exploits the fundamental ideas of Josef Shumpeter and Kenneth Arrow, and includes a rich literature devoted to the modelling of R&D rivalry, technology diffusion and the interaction of innovation and imitation processes. Macro- and microeconomic models both describe technology that is changing due to learning-by-doing, public good externalities and human capital improvements. Some of the models also take into account that R&D efforts give every enterprise a chance to find a profitable innovation and to become a monopolist for a short time until other enterprises learn how to imitate it. I don't think that changing technology is an obstacle to employing the neoclassical maximization approach. I would add also that the modern theory recognizes that the innovation process is collective, cumulative and uncertain, and that there are many models that take into account at least some of these features.[158] Thus neoclassical economic theory already contains important elements of the theory of the innovative enterprise.

What I would agree with, however (and this is my second comment) is that the modern micro-theory of industrial organization and the modern macro-theory of endogenous economic growth need to be much better integrated. In growth theory we admit changing production technology, but assume that the technology of changing the production technology is fixed. However, the theory of industrial organization tells us that the goals and performance of the modern corporation depend on strategic interaction among owners, managers and employees. Moreover, power and the strategies of different agents, on the one hand, and market conditions, on the other, are interconnected, and this makes the structure of innovative behaviour even more complicated. We need a synthesis of the micro- and macro- theories to explain market structure and economic growth through institutional dynamics. Professor Lazonick writes that "the existence ... of desirable market opportunities is much more an outcome of the process of the economic development than its cause". This statement seems to me to be very profound and important.

The lack of an institutional theory of economic development and a dogmatic interpretation of the

---

[156] P. Romer, "Increasing returns and long-run growth", *Journal of Political Economy*, Vol. 94, 1986, pp. 1002-1037; R. Barro, "Government spending in a simple model of endogenous growth", *Journal of Political Economy*, Vol. 98, 1990, pp. 103-125; R. Barro and X. Sala-I-Martin, "Public finance in models of economic growth", *Review of Economic Studies*, Vol. 59, 1992, pp. 645-666; R. Lucas, "Making a miracle", *Econometrica*, Vol. 61, No. 2, 1993, pp. 251-272.

[157] J. Tirole, *The Theory of Industrial Organization* (Cambridge, MA and London, MIT Press, 1988).

[158] The collective nature of the innovative process is taken into account in the models that consider knowledge as a factor of production, or other public good externalities or imitation behaviour. Learning-by-doing and human capital improvement are cumulative processes. In many macroeconomic models a representative producer takes into account uncertainty in production and in market conditions.

neoclassical models have led to serious mistakes in economic policy. A striking example of such mistakes was the proposition that privatization of state property, in itself, would improve the performance of the transition economies. Empirical research does not find any essential difference between the performance of privatized and non-privatized enterprises in Russia or in most of the other newly independent states.[159] It is not privatization but restructuring that is important. Restructuring will not occur if the market infrastructure is not developed properly and if property rights are not protected.[160]

Here we come to another problem that is discussed by Professor Lazonick in his paper and that is the subject of my third comment: the role of institutions and the state in the promotion of economic development. Again, the experience of the transition economies turns out to be very helpful in understanding this problem more deeply. Ten years ago when more than 20 socialist countries started on their way to the market economy, many experts believed that efficient market institutions would arise spontaneously as soon as state control of prices, foreign trade and property were eliminated and market legislation introduced. This belief in Hayek's "spontaneous order" proved to be misplaced. Inside any legislative framework there exists room for the development of different institutions or behavioral norms. Economic agents may find themselves in a stable equilibrium which is the most profitable for everyone but inefficient for society as a whole. Such an equilibrium is called a "lock-in" or an "institutional trap".[161] Institutional traps usually arise in periods of drastic social and economic change and continue to prevail due to a lack of coordination among economic agents. Mutual arrears, large-scale tax evasion, and endemic corruption are examples of institutional traps. The more agents are involved in these activities the lower is the probability that anyone will be punished and the larger will be the costs of changing the norms of behaviour for all individuals. However everyone would be ready to change the norm if everyone else does the same. Institutional traps slow down the rate of economic growth and may lead to an underdevelopment trap so that a country's economic backwardness may turn out to be permanent.

The role of the state is particularly important for preventing a system from falling into an institutional trap and for helping it to get out when it has fallen into one. To be effective the government must be of high quality. A number of recent empirical investigations have confirmed that the quality of government is an important factor of economic growth.[162] The state is responsible for creating the main preconditions of efficient economic interactions – property rights protection, law enforcement and regulation. And it is the state which must promote growth through balanced and wise social and industrial policies.

My last comment concerns the main elements of an efficient economic structure. Innovative enterprise, the developmental state and an efficient market can be considered as three main elements of an efficient economic organization, but mention must also be made of a fourth, namely, developmental culture.[163]

Turning again to the causes of the reform failure in Russia, we have to explain why the quality of governance was so low and why the state was not a helping hand. I think that substantial features of governance in Russia may be at least partially explained by the cultural legacy of the Soviet period, first of all by paternalism, social passivity and what I call "habitual deviationism" – ordinary and routine deviation from official rules and laws.[164] Paternalism, passivity and habitual deviationism determine the system of people's attitudes towards the state, the law, property and liberal values. In the Russian case this system entails a strong resistance to reforms, an adversarial (using Stiglitz's term) style of governance, and opportunism and corruptibility of the ruling elite.

Culture is much more inertial than the political and economic organization of a society. We know that prices may be liberalized in a day, that parliament can be elected in a few months. But nobody believes that modern civic culture, a necessary element for the success of a market economy, can arise in a short time. A realistic theory of economic institutions must take into account the prevailing culture in a society.

---

[159] D. Jones, "The economic effects of privatization: evidence from a Russian panel", *Comparative Economic Studies*, Vol. XL, No. 2, Summer 1998, pp. 75-102; S. Djankov, "Ownership structure and enterprise restructuring in six newly independent states", *Comparative Economic Studies*, Vol. XLI, No. 1, Spring 1999, pp. 75-95.

[160] O. Blanchard, *The Economics of Post-Communist Transition* (Oxford, Clarendon Press, 1997); J. Stiglitz, "More instruments and broader goals: moving toward the post-Washington consensus", *1998 Wider Annual Lectures 2*, The United Nations University, January 1998 and idem, "Distinguished lecture on economics in government: the private uses of public interests: incentives and institutions", *Journal of Economic Perspectives*, Vol. 12, No. 2, Spring 1998, pp. 3-22.

[161] D. North, *Institutions, Institutional Change and Economic Performance* (Cambridge, Cambridge University Press, 1990); V. Polterovich, "Institutional traps", in L. Klein and M. Pomer (eds.), *The New Russia: Transition Gone Awry* (Stanford, Stanford University Press, 2000), pp. 93-116.

[162] M. Olson, Jr., N. Sarna and A. Swamy, "Governance and growth: a simple hypothesis explaining cross-country differences in productivity growth", *Public Choice*, Vol. 102, No. 3/4, March 2000, pp. 341-364; J. Aron, "Growth and institutions: a review of the evidence", *The World Bank Research Observer*, Vol. 15, No. 1, February 2000, pp. 99-135.

[163] A. Hillman and H. Ursprung, "Political culture, political liberalization and economic decline", Bar-Ilan University, mimeo, 1998.

[164] J. Kornai "Efficiency and the principles of socialist ethics", *Contradictions and Dilemmas* (Budapest, Corvina, 1985), pp. 124-138; V. Polterovich, "Civic culture and economic transition in Russia", paper presented to the 15th Annual Congress of The European Economic Association (Bozen-Bolzano), 30 August-2 September 2000.

## 2.B Eugen Jurzyca

The paper by Professor Lazonick attacks the basic postulates of the current mainstream in economics. I will not repeat the theoretical foundations of that school as it is taught in the majority of economic universities all over the world. Trying to create added value, I first try to summarize what the paper claims, and then underline what I think is illuminating in the paper and deserves more attention from scholars and practitioners. At the end, I will raise some questions and express some doubts about some of the non-traditional ideas of Professor Lazonick.

To summarize Professor Lazonick's paper I would mention the following ideas:

The market is clearly the winner in the competition against the plan (both practically and theoretically), but the corporate goal of "maximization of shareholders' value", valid for the United States, has not been very successful in many other economies until recently (Japan, western Europe). This goal is broadly perceived as necessary to generate innovations and economic growth, and, in turn, employment. However, a "soft landing" of the United States new economy is not clear. If this version of the market wins, the world will be marked by inequality and financial instability. Stable and equitable economic performance will tend to be undermined by the market. Mainstream market economics claims that the allocation of resources via the market is the best possible solution: the more perfect is the market, the better will be the performance of the economy. This theory, however, does not help us to understand the problems faced by the world's economy.

Market economists have problems in making a theoretical case for the regulation of speculative markets, government programmes for developing the capabilities of the labour force, investing in the new technology, or bolstering aggregate demand. They fail to explain economic growth, and they wrongly see developed markets in labour, capital and products as causes rather than consequences of economic development.

Society must first create the institutions which generate well-functioning markets and then control the market.

Both conservative and liberal market economists believe that the theory of the market economy is the ideal benchmark, against which the reality of resource allocation should be compared: making markets more perfect should improve economic performance, and thus (and this is the wrong lesson also for transition economies) *problems are caused by a slow transition to the perfect market.*

Markets are important for facilitating the reallocation of resources in wealthy economies, but the innovation process is collective, cumulative and uncertain.

Economists should see the social conditions for innovative enterprise not as a market imperfection, but as an institutional foundation of economic development.

Economic policy makers need to combine a relevant theoretical analysis (e.g. market institutions) of the development process with a deep understanding of the social context in which the reforms are to take place. A well-functioning labour (or capital) market will be the result, rather than the cause of economic development. In the real economy, the widespread belief in the market economy model tends to render ungovernable those corporate executives and political elites who wield power over the allocation of resources, while it tends to leave the vast majority of the population vulnerable to those who make the strategic decisions.

Should the enterprise then be understood as a massive market imperfection? A theory of innovative enterprise should be developed (explaining how a business enterprise allocates resources to generate innovations, how it transforms inputs into outputs to generate products that are of higher quality and lower cost than their predecessors).

It is organizations, not markets, that allocate resources to the production process that generates economic development (we need a theory how resource allocation by organizations is carried out).

If the innovation process is collective, cumulative and uncertain, the allocation of resources should be organizational, developmental and strategic, instead of individual, reversible and optimal, as the market economists suppose. Thus, there are three conditions of the innovative enterprise: organizational integration (derived from the collective character of the innovation process); financial commitment (derived from the cumulative character of the need to provide financial support until the product generates returns); and strategic control (derived from the uncertain character of the innovation process).

The key to success of the new economy is organizational integration. The shareholder perspective has nothing to say about the roles of organizational integration, financial commitment and strategic control.

The state is often predatory and regulatory, but it is also developmental and Lazonick argues that the developmental state has played a larger role in the United States than in Japan.

To summarize: rules among economic subjects are less important for the creation of economic wealth, than rules within them. The economic subjects, not their relations, generate growth.

One can probably *agree* with many elements of this argument. The claim that, the more perfect the market the better the performance of an economy, *is* misleading (markets cannot solve the problems in small societies, for example), and problems are not always caused by the slow transition to a perfect market. (However, one should probably mention that the globalization process has its own dynamism and not complying with it may cause problems, thus a relatively speedy transition may be necessary.)

Policy makers should certainly combine theoretical analysis with an understanding of the social context in which reforms are to take place, and the enterprise should not be understood as a massive imperfection of the market. But equally one should not call the market a massive imperfection of organization. Although the development of communications has brought the costs of organizing close to zero, organization is still a prevailing arrangement of the society. I can agree that the theory of the innovative enterprise needs to be developed further; and I also agree that although the state is often predatory and too regulatory, it can also play a developmental role.

However, there are a number of factual questions I would like to raise:

Criticism of the mainstream economic theory requires some supporting data: does the empirical evidence really show that countries with higher economic freedom are marked by higher inequality? There are a number of studies showing the contrary. Are we sure that the market economy leaves more of the population vulnerable than in a planned economy? The inequality in planned economies was also high, although it was difficult to measure.

At the level of ideas I would question the following:

Is the market really the clear winner of the twentieth century? There is an enormous and growing body of literature on corporate governance studying the internal aspects of organizations. Are market economics and the theory of innovative enterprise in contradiction, or are they complementary? The market should serve as a tool for cooperation among "small societies" based on hierarchies. They should exist together. Let me illustrate this.

Human beings living as nomads or living in small communities did not need a market for cooperation. They were well informed of what each individual produced and what that individual should receive in exchange for it. Once they left their original group and started to cooperate, they needed some tool for cooperation. For example, in 450 BC, Hanno, a sailor from Carthage, reached for the first time the region around Guinea-Bissau. The Carthaginians soon developed a so-called quiet barter trade. The (hierarchically organized) crew landed on the coast, put their goods close to the sea, made a fire and returned to their ship. The local people, attracted by the fire, came to inspect the goods and deposited gold beside them. The sailors then evaluated the gold and, if satisfied, took it and left. If they were not satisfied, they went back to the ship and waited. If the other party to the contract gave more gold, the exchange was concluded.[165] So, the market developed naturally, not because of a mistaken economic theory.

Nevertheless, imperfect markets also developed historically and naturally as well. For example, silk was imported to Rome in 105 BC from China through the Silk Route (9,600 kilometres). The monopolistic position of the producer, as well as similar positions of traders along the route, led to dramatically high prices in Rome. The reactions of two Roman Emperors were different: Marcus Aurelius opted for regulation via education and refused to wear silk himself, while Justinian ordered the prices of silk to be fixed and regulated in the year 540.[166]

Are developed markets the consequences (or causes) of economic development? Do they not develop simultaneously, as the chicken and the egg? Consider the examples above. Are markets really important only in developed economies? Haven't we seen them developing thousands of years ago? The market was simply developed as a tool for cooperation in a globalizing world.

Are we sure that development is taking place mainly because of the internal activities of organizations, instead of improvements in the relations among them? We should admit that at least some (if not the majority) of the motives for increasing efficiency come from the outside environment (the literature on soft versus hard budget constraints provides an example).

Are the main engines of innovation really located within organizations instead of their external environment? The empirical evidence does not confirm that. Isn't there a mutual interrelation? Weren't the innovation processes in history often organized beyond the organization despite their collective character?[167] Isn't the Silicon Valley model an example of how venture capitalists may enter the innovation process?[168]

Professor Lazonick's paper is provocative and thus, also refreshing. Surprisingly, this and some other similar papers have not been discussed more broadly in the corporate governance literature.

---

[165] F. Barker in F. Barker, M Ross-Macdonald and D. Castereagh (eds.), *The Search Begins* (London, Aldus Books and Jupiter Books, 1973).

[166] Ibid.

[167] James Watt invented his steam engine in 1769, on the basis of general knowledge. However, he was inspired by repairing another steam engine that had been constructed in 1712 by Thomas Newcomen, who in turn had improved an engine patented in 1698 by Thomas Savery. The Englishman Savery was inspired by the Frenchman Denis Papin who proposed the idea around 1680. J. Diamond, *Guns, Germs and Steel: The Fates of Human Societies* (New York and London, W.W. Norton & Company, 1997).

[168] M. Aoki, "Information and governance in Silicon Valley model", in X. Vives (ed.), *Corporate Governance, Theoretical & Empirical Perspectives* (Cambridge, Cambridge University Press, 2000), pp. 169-196.

# CHAPTER 3

# INSTITUTIONAL CHANGE AND ECONOMIC PERFORMANCE IN THE TRANSITION ECONOMIES

Paul G. Hare[169]

## 3.1 Introduction

From the very beginning of transition, institutional reforms formed a key component of the policy package that was recommended to the economies of central and eastern Europe and the former Soviet Union. The overall package was designed to transform them from centrally planned economies operating under the socialist system, into market-type economies operating under the capitalist system in a democratic political framework. However, for understandable reasons rather more emphasis was placed initially upon the urgent need to achieve macroeconomic stabilization, accompanied by extensive price and trade liberalization, soon giving way to privatization and enterprise restructuring. This basic agenda was already a huge one, not only for the established states but also – and even more so – for the new ones that emerged from the ruins of communism.

For all the states concerned, the basic transition agenda already entailed substantial institutional reform in such areas as central banking (especially, but not only, for the new states), taxation and fiscal policy, industrial and trade policy, property and commercial law, and so on. Further, as failing enterprises shed labour and unemployment became a reality in the region for the first time in several decades, it became necessary to develop a raft of labour market institutions and policies to deal with retraining, unemployment benefit and other forms of income support, and the collection of suitable statistics. These considerations help to explain why institutional change was not at the forefront of the reform programmes undertaken by the transition economies in the early 1990s, except as a by-product of other policies initially perceived as of higher priority.

A second factor is the simple observation that the economic theory of the time seriously underrated the role and importance of the institutional framework needed to operate a well functioning market-type economy. Most textbooks, even relatively recent and advanced ones, had little to say about key institutions such as private property, business contracts or the role of trust in conducting business successfully. Implicitly, regarding the market system as a form of economic mechanism on a par with central planning, it tended to be assumed that the "operating costs" of the system were close to zero – indeed this line of thinking was often used (erroneously) to support the claim that a market economy must be inherently more efficient than a centrally planned one. In any event, once central planning was swept away, it was taken for granted that the opening of markets would bring with it – rather quickly and painlessly – the needed institutional structures to make the new market system work properly. By now, it is well understood that arguments like this are fundamentally wrong, and that active efforts to create and support new institutions are vital. For instance, McMillan puts it very nicely when discussing markets in transition economies: "A market is an institution, which needs rules and customs in order to operate. Given the uneven distribution of information among them, the rules of exchange must be cleverly structured for a market to work smoothly. Institutions and organizations must evolve, to transmit information and to provide appropriate incentives".[170]

Third, some mistakes or misjudgements in early transition influenced the direction of institutional development even where the need for new or reformed institutions was acknowledged. Here I have in mind the widespread views that privatization was the best way to create a new private sector rapidly, rather than new business formation; and that banking reform could sensibly be delayed until later in the transition.

Last, from the mid-1990s onwards, for the 10 countries in the region that have applied to join the EU, a factor operating in the opposite direction and promoting institutional change across a broad spectrum of policy domains has been the requirement that these countries must fully conform to the Community's *acquis communautaire* prior to their accession to the EU. Commitment to this programme entails a huge amount of institutional modernization, renewal and, in some

---

[169] I am grateful to the discussants and to other participants in the UNECE Spring Seminar for helpful comments on the paper, which have enabled me to improve it. Remaining errors are my own.

[170] J. McMillan, "Markets in transition", in D. Kreps and K. Wallis (eds.), *Advances in Economics and Econometrics: Theory and Applications*, Seventh World Congress of the Econometric Society, Vol. 2 (Cambridge, Cambridge University Press, 1997), chap. 6, p. 222.

instances, creation. While confirming quite properly the key importance of the institutional framework, this approach raises some important questions of its own, notably whether the particular institutional framework that has evolved across the EU is the one most suitable for a group of transition economies still in the throes of major economic restructuring. I return to this question later on.

The above remarks tacitly assumed a certain uniformity across the transition economies in order to justify discussing them together. Specifically, it was assumed that the political entities that existed or came into being at the start of transition were the "correct" units for analysis, that the states concerned had the capacity to bring forward and implement deep reforms, and that all these political units shared the common objective of transforming themselves into "normal" market-type economies. Unfortunately, as the experience of the past decade makes abundantly evident, none of these assumptions is wholly accurate.

### (i) Political configuration

Concerning the political point, several states that existed in 1990 or 1991 have since split into components or merged with others. First, the former German Democratic Republic became part of the existing Federal Republic of Germany following unification in 1990, whereupon the new eastern Länder found themselves both part of the EU, and undergoing comprehensive institutional reforms to align their structures and administrative practices with those already established in Germany; although traumatic in various ways – especially for the east Germans – this change occurred peacefully.

Then the former SFR of Yugoslavia (as it was in 1990) started to disintegrate, accompanied by bitter fighting whose legacy is still with us. The new states that have emerged to date are: Bosnia and Herzegovina (this state still comprising two "entities", one Serb dominated, the other Muslim dominated and comprising the Muslim and Croat parts of the country), Croatia, Slovenia and The former Yugoslav Republic of Macedonia. What remains of Yugoslavia, still officially named the Yugoslav Federal Republic, consists of Montenegro and Serbia, and the Federation took some important steps towards democratization and political renewal in late 2000. At the time of writing, though, it is unclear whether the two republics will remain together; it is also unclear what the future holds for the Serbian province of Kosovo, with its overwhelmingly Albanian population. Not surprisingly, as a result of these political upheavals, economic reforms have been seriously delayed in much of the former SFR of Yugoslavia.

Disputes over the nature and pace of economic reforms led to the disintegration of Czechoslovakia from the beginning of 1993, the Czech Republic and Slovakia being the two successor states. Since this was a peaceful and orderly separation, the course of economic reforms was not greatly disturbed, although the two states pursued divergent approaches to reform for some years after the break. Ironically, in some respects their policies are now converging, stimulated by their shared goal of joining the EU.

From the former Soviet Union, 15 states emerged, the three Baltic states (Estonia, Latvia and Lithuania), and the 12 states that formed the Commonwealth of Independent States (CIS). Of the latter, the states based in the Caucasus region – Armenia, Azerbaijan and Georgia – have experienced civil or international conflict, some of which is still not settled. In central Asia, Tajikistan has been afflicted by intense civil war. Even Russia has not been free from separatist tendencies, with some regions openly discussing separation for a time, and sustained armed intervention occurring in Chechnya. For the time being, under the leadership of President Putin, the country appears to be politically somewhat more stable for the first time since the dissolution of the Soviet Union. Interestingly, Belarus currently seeks a political union with Russia, though the commitment to this idea from the Russian side appears to be rather limited.

Overall, the picture that emerges is of a region whose political contours are far from finally settled. While in what follows I shall not further discuss the issue of possible further changes in the region's political contours, the possibility should nevertheless be borne in mind.

### (ii) State capacity

There is a tendency for economists to assume that where economic policy advice is to be proffered, the recipient state will possess the ability to take the advice on board and implement it effectively. However, as the World Bank[171] has made clear, states vary enormously in their capacities, for many different reasons. The transition economies are no exception to this general observation. Across the region, one can find examples of states with competent, reasonably well-functioning and largely corruption free administrations, while others lie at the opposite pole – corrupt, inefficient, largely incapable of delivering anything but the simplest of policies. In the discussion below about institutional reforms, it will sometimes be necessary to qualify what is said in the light of the nature and functioning of the different state structures that one can observe.

### (iii) State objectives

Last, some of the transition economies may not wish to transform themselves into market-type economies in the sense we generally think of, and for such countries the whole project of institutional change under discussion in this paper is immediately much less applicable. States that have not yet evinced a notable commitment to market-oriented reforms include Belarus, Turkmenistan and Uzbekistan among others. This is not to say that at some point these countries might not change tack, but for the time being, although there have been some limited

---

[171] World Bank, *World Development Report 1997: The State in a Changing World* (Washington, D.C.), 1997.

market reforms reflecting the ending of some of the state controls that characterized the communist period, serious talk of transition is probably premature.

### (iv) Outline of paper

In the following sections of this paper I proceed as follows. Section 3.2 offers some definitions and explores what sorts of institutional structure are needed for a market-type economy to function well. It also examines some recently developed theoretical approaches to questions of institutional change. On this basis, section 3.3 surveys the transition economies to assess how far they have undertaken market-oriented institutional change up to the year 1999 or 2000 (the cut-off year depending on data availability). Next, section 3.4 investigates the links between alternative indicators of institutional change and various dimensions of economic performance. Section 3.5 concludes and draws out the most significant policy implications from the foregoing analysis.

## 3.2 Institutional change – concepts and definitions

What sorts of institution are helpful in supporting a well functioning market-type economy? To answer such a question, one approach is to appeal to the findings of various historical studies that seek to identify the factors that have led certain countries, or groups of countries, to perform well over long periods, as compared with other parts of the world. Thus Landes[172] points to "culture" as key, while others have highlighted "the Protestant ethic", the existence of a strong "civil society", or the environment that nurtures "small-scale business enterprise".[173] Roberts[174] places the emphasis on numerous institutional features of the "western model of civilization", often copied even in countries no longer, or never, dominated by western powers. More firmly grounded in the underlying economic reality, Kennedy[175] argues for a link between "economic success and military might". His thesis is that economically successful countries apply resources to developing their military power, then later, strive to maintain their prowess through military means, eventually at the cost of productive investment and further development. Finally, Fukuyama[176] acknowledges a tendency towards convergence across countries in many basic institutions of government and economic regulation, accompanied by continuing divergence in economic performance. This line of argument leads him to the view that deep-seated cultural differences between societies explain the observed differences in performance. Some of these differences he sums up in the notion of "trust".[177] Clearly, there are as many views as authors, all of some interest and relevance to the present study, but none providing a sufficiently complete or compelling analysis to serve our purposes.

An alternative approach to these broad, conceptual overviews is more pragmatic and empirical, simply examining the institutions we find in modern, market-type economies in various policy domains, and arguing for their necessity in transition economies. This approach quickly encounters three types of problem: (i) it is descriptive and all encompassing, offering no obvious means of judging which institutions are more or less vital, which are desirable but not indispensable; (ii) it does not provide a definition of what we mean – or ought to mean – by an institution, so is open to multiple interpretations; and (iii) it gives us no theory of the market economy to explain the roles and significance of the various institutions observed therein.

Accordingly, this section is organized in the following way. First, I attempt – albeit briefly – a definition of what is meant by an economic institution. Then I list the typical institutions that characterize a "normal" market-type economy, while acknowledging along the way the huge diversity of practice and structure that can be found for any given institutional form. Third, since economies function even in the absence of what might otherwise be considered as key institutions, we consider some examples of what can happen in such situations. Fourth, we examine some theories of institutional development and change in market economies in order to throw light on what is more or less important, and on the processes whereby institutional change comes about.

It should be emphasized, however, that "we are still very ignorant about institutions",[178] so that a good deal of what follows must be regarded as rather tentative or even speculative. There is still much to be learned.

### (i) Definitions

Economic institutions are social arrangements possessing a number of special features: (a) they *regulate economic behaviour* in ways which, in the short run, often conflict with individual preferences; (b) they are based on *shared expectations*, derived from custom, trust, legal provisions, etc.; (c) they make most sense if the economy is thought of as a *"repeated game"* in which

---

[172] D. Landes, *The Wealth and Poverty of Nations* (London, Little, Brown and Company, 1998).

[173] On the last of these, see F. Braudel, *Civilization and Capitalism, 15th-18th Century, Volume 3 – The Perspective of the World* (London, Fontana Press, 1985) (originally published in French, 1979).

[174] J. Roberts, *The Triumph of the West* (London, BBC, 1985).

[175] P. Kennedy, *The Rise and Fall of the Great Powers* (London, Unwin Hyman, 1988).

[176] F. Fukuyama, *Trust: The Social Virtues and the Creation of Prosperity* (London, Hamish Hamilton, 1995).

[177] Zak and Knack formulate an interesting growth model in which the level of trust in society is part of its equilibrium. In this model, high trust equilibria are associated with high rates of investment and faster growth, a conclusion that is supported empirically. P. Zak and S. Knack, "Trust and growth", *Economic Journal*, Vol. 111, April 2001, pp. 295-321.

[178] O. Williamson, "The new institutional economics: taking stock, looking ahead", *Journal of Economic Literature*, Vol. 38, No. 3, September 2000, pp. 595-613.

most types of transaction occur many times;[179] and (d) *anonymity*, in the sense that the functioning of a given institution should not be dependent upon the identity of the economic agents seeking to conduct the types of transaction to which this institution relates.

Given such characteristics, many institutions are likely to have the character of *public goods*. Among other things, this implies that the "supply of institutions" generated by the market mechanism left to itself, is unlikely to correspond to the socially efficient level. Under these conditions, there is evidently a role for the state both in creating institutions that the market does not provide and regulating in the public interest those that it does. What this means in practice we shall see through various examples in the subsequent discussion.

### (ii) Typical institutions

Well functioning market-type economies are generally found to contain institutions or institutional arrangements to provide for the following key economic functions:

- private property rights and contracts;
- banks and other financial markets: existence, functioning and regulation;
  - reliable access to credit on reasonable terms;
  - bankruptcy/liquidation policy in place to facilitate orderly exit;
- labour market institutions: social policy and the social safety net;
- clear fiscal environment for firms, perceived as fair, predictable and enforced (this means, for instance, that in a multi-level country such as Russia it should not be possible for the regions to set taxes that conflict with national policies, and taxes should not be changed frequently);
- institutions dealing with competition policy, industrial policy and trade policy;
- trust between economic agents, trust and honesty in public institutions (lack of corruption, reliable law enforcement, including as regards business taxation).

Especially in economies undergoing rapid change, such as the transition economies, it is not uncommon for them to possess institutions in several of the above categories, but institutions that do not function very well.

Hence the problem is often not so much to create institutions *ab initio*, but to improve the functioning of those already in place. This suggests a possible role for the state in fostering the conditions for such improvements to occur. It also raises an interesting question, namely whether references to the state should always imply the usual, national-level entity, or whether it might sometimes refer to subnational levels of government.

### (iii) Economic behaviour with missing institutions

In the transition economies, it has sometimes been the case that important institutions have not been created at an early stage of transition, or that the relevant laws are incomplete, imperfectly enforced or still subject to serious political controversy. In such unsettled institutional environments, several outcomes are possible, all of which can be found in one or another transition country.[180]

First, the private sector can step in to create a missing institution. For instance, in parts of the former Soviet Union, notably Russia itself, where business contracts and private property rights have not been reliably secured through adequate legislation, private means of contract enforcement have developed. Sometimes these private institutions act in the interests of particular firms or groups of firms, and can entail the use of violent methods to compel payment where necessary. This approach can prove effective for the firms concerned but it would not generally be regarded as desirable since it fails to offer a universal service and violates important principles to do with the "rule of law". It does, however, have the merit of filling a clear "gap" in the institutional space.

In contrast, where an existing institution is weak or the legal provisions supporting it are poorly enforced, more predatory private sector "solutions" can be observed, commonly associated in the public mind (and perhaps in reality – the evidence is seriously incomplete) with mafia-like criminal structures.[181] These include the widespread practice of demanding "protection money" and the like from many firms, as a condition for them to continue in business. Needless to say, such practices are wholly undesirable. They are likely to inhibit or delay the expansion of existing firms, and seriously discourage new business formation. Nevertheless, practices of this sort are common across the CIS, less common in central and eastern Europe.[182]

---

[179] This is not the place for an exposition of game theory. Suffice it to say that a repeated game is one in which the players make a series of moves, and their choices in later moves can be influenced by what happens in the early moves. This situation can provide incentives for good behaviour that could not be explained in a one period or one move game. For a thorough analysis of repeated games in the context of social institutions see A. Schotter, *The Economic Theory of Social Institutions* (Cambridge, Cambridge University Press, 1981). For a wider economic analysis of the institutional structure of a market economy see T. Eggertsson, *Economic Behavior and Institutions*, Cambridge Surveys of Economic Literature (Cambridge, Cambridge University Press, 1990).

[180] Some aspects of this issue are studied in J. McMillan, op. cit., chap. 6.

[181] In some countries, the state itself can act in a predatory manner. This idea is discussed by Evans, who distinguishes between predatory and developmental states in an interesting way. P. Evans, *Embedded Autonomy: States and Industrial Transformation* (Princeton, NJ, Princeton University Press, 1995). For an interesting case study of Russia see W. Buiter, "Predation to accumulation? The second decade of transition in Russia", *Economics of Transition*, Vol. 8, No. 3, November 2000, pp. 603-622.

[182] Unfortunately, no economy is totally free from such criminal practices. However, from the standpoint of economic policy what matters is the general expectations that firms hold. If most firms expect to have to

Second, the state itself can step in to create a missing institution. A good deal of the EU's aid to the transition economies under the PHARE and TACIS programmes serves this purpose, and some institutional development also accompanies World Bank and EBRD projects in the region. Provided that the private sector has not already rushed in to pre-empt a state solution, and the aid is not diverted to finance those who strongly resist institutional innovations, this can be highly effective. Good examples of successful institutional creation through this sort of route are numerous, and only a couple of examples are cited here: (a) the development of local and regional development agencies in Hungary, as part of the country's evolving industrial policy; and (b) the design of new tax systems in many transition economies (although there are often residual problems of tax administration, income definition, coverage, etc.).

Third, a given institution might not exist and it may be impossible to create it due to political, legal or other obstacles. Thus in Russia and some other CIS countries, it is still the case that there is no legal private market in agricultural land, a circumstance resulting from vociferous political opposition to such an institution. The result, however, is that private farming in Russia is severely inhibited, while the state lacks the resources to adequately fund the existing state farms and other non-private organizational forms (e.g. the remaining cooperatives). This is not the only problem in Russian agriculture, of course, but it is an important one, and remains unresolved despite several (unsuccessful) attempts to get suitable legislation through the state's Duma.

Last, sometimes countries can lack an apparently important institution, such as private property, and yet find ways around the missing institution – possibly by accident – in order to enable successful development to take place. The most spectacular instance of such a serendipitous process can be found in China, in terms of the unexpectedly rapid growth and spread of township and village enterprises since the late 1970s. Neither state owned in the old sense, nor strictly private, and unprotected by clear laws on private property and commercial contracts, these firms have nevertheless thrived. They are established at a very local level, but serve both local and wider markets – provincial, national, international – and are obliged to operate competitively. There is no protection for those that fail commercially. These firms operate as they do, not quite in the institutional vacuum that one might imagine, but in a secure political framework and strong local networks of trust that take the place of the missing institutions. Local authorities support "their" firms, because a share of the resulting profits is what finances the development of local infrastructure. Everyone therefore has an interest in encouraging highly profitable, fast growing firms.[183] An interesting question, however, is whether such firms might soon reach the limits of their possible development in the absence of more fundamental institutional reforms.

### (iv) Theories of institutional change

Theoretical approaches to institutional change in transition economies are already quite diverse, ranging from broad theories of the reform process as a whole to very specific models of particular aspects of institutional reform. For space reasons, it is only possible here to cite some of the more interesting of the available studies, without going into great detail.

Probably the most ambitious, and most general approach is that of Roland.[184] Roland considers gradual and so-called big bang approaches to reform, and investigates the formation of various types of coalition for or against different stages of reform. The approach is general in that no specific reform measure is characterized in the analysis, but the approach is virtually the only one that starts to get to grips with the complex political configurations that can inhibit or favour reforms.

Although not directly focusing on our institutional concerns, Stiglitz draws attention to the weaknesses of the standard, neoclassical model of a market economy as a basis for advising transition governments on appropriate reform strategies.[185] He draws attention to numerous informational and incentive issues that arise in many markets, the resolution of which entails various forms of state intervention and regulation. In other words, Stiglitz's book can be regarded as providing a conceptual foundation for many of the institutional reforms now widely acknowledged as essential for economies in transition. Sometimes the outcome of such analysis, drawing on the economics of information, can lead to remarkably powerful conclusions, as for instance in the analysis of credit markets and banks, the role of competition, and privatization and property rights.

Contrary to the definition of institutions given above, several transition economies – notably in the CIS – have developed practices that seriously conflict with the desirable features I listed. In particular, the "anonymity property" is frequently violated through forms of state capture by large enterprises and other economic interest groups whereby these agents are able to influence state policy in their favour, and hence undermine the proper functioning of economic and political institutions. Such practices have been analysed in some depth for Russia by Ericson,[186] and more widely, based on extensive

---

pay protection money that is a far more serious situation than that where only a small proportion encounters these difficulties.

[183] This raises the obvious question as to why such firms are not to be found everywhere, why only in China? Elsewhere there must be other obstacles such as a more centralized state structure (with less local autonomy than in China); or political elites could continue their old association with traditional state owned enterprises rather than promoting new businesses (this option would not be available in most Chinese counties); or there could be insufficient trust and confidence at the local level to serve as a substitute for the missing private property rights.

[184] This has been set out in numerous papers but is nicely summarized in G. Roland, *Transition and Economics: Politics, Markets and Firms* (Cambridge, MA, MIT Press, 2000).

[185] J. Stiglitz, *Whither Socialism?* (Cambridge, MA, MIT Press, 1994).

[186] R. Ericson, "The Russian economy: market in form but 'feudal' in content", Colombia University (New York), mimeo, 2000.

enterprise-level empirical data through 1999, by Hellman, Jones and Kaufman.[187] The former sees a continuation in Russia of the traditional inter-penetration of political and economic structures, and characterizes what market structures there are as "fragmentary". The latter conclude that "improved property rights protection and civil liberties can significantly reduce the capture economy".

An alternative route for firms seeking to operate in a poorly defined institutional environment is the second, or shadow economy. While some firms operate outside the official system in all countries, the scale of such activities can be regarded as a measure of how badly the formal structures are functioning. The nature and diversity of second economy activities are explored by Schneider and Enste,[188] who also present estimates of the size of the shadow economy for many developing countries and for transition economies. In the latter, the shadow economy is estimated to account for 35 per cent of GDP on average in the former Soviet Union, and about 21 per cent in central and eastern Europe (taking, in each case, the lower of two estimates given in the paper). The corresponding average for OECD countries is estimated (on the same basis) as about 11 per cent of GDP. The authors conclude along lines highly material to the present study that, "Most studies of the shadow economy focus on the influence on the allocation of resources and the loss of revenue for the state. But the impact on official institutions, norms and rules is even more important. The shadow economy can be seen as an indicator of a deficit of legitimacy of the present social order and the existing rules of official economic activities".[189] Set against this remark, the above-cited figures speak for themselves.

We expect economic performance – including overall economic growth and sustained enhancements in productivity – to depend fundamentally on what happens at the level of individual enterprises across any given economy. Key aspects of enterprise sector behaviour in this connection are: entry, exit, and the restructuring of incumbent firms. Hence institutional measures that foster these processes are highly desirable. Aghion and Schankerman[190] provide a valuable formal analysis of such measures, under the heading of "market-enhancing infrastructure". Their model captures three aspects of infrastructural investment (often interpreted as measures that cut transport costs or reduce other forms of transactions costs), namely: direct market selection (more high-cost firms are forced out of business), restructuring (stronger incentives for firms to engage in cost-reducing activities), and entry (less incentive for new, high-cost firms to enter, stronger incentives for low-cost firms to enter). In a dynamic framework, these processes generate productivity improvements across a sector, region or entire economy, although the relative importance of the different effects depends on initial conditions, the initial level of infrastructure and the costs of restructuring and entry. These are likely to vary across countries and over time. The authors conclude that an interesting extension of their work would be to examine the political economy of infrastructural investment (since existing high-cost firms are likely to oppose it, low-cost firms will support it), and to study the way in which infrastructural investment affects firms' learning processes both through experimentation and via demonstration effects.

In informal discussions of transition institutions, the legal environment is often highlighted as critical for successful private sector development. Analytically, this can be approached at a general level, in terms of the economics of law enforcement, and at a more concrete level one can investigate specific areas of the legal regulation of market institutions, drawing conclusions about good practice from cross-country comparisons. Roland and Verdier[191] provide an example of the first approach, their general approach being illustrated nicely in the context of property rights, by Rapaczynski.[192] Black presents an interesting case study of the second, in the context of securities market regulation and investor protection.[193]

Roland and Verdier draw attention to a (social) coordination problem associated with law enforcement in that (a) people have to agree to levy and collect taxes in order to pay for law enforcement – but in jurisdictions where law enforcement is weak, tax collection also tends to be weak; and (b) for given expenditure on law enforcement, if people choose to be mostly law abiding then enforcement is effective, while if they do not so choose then the enforcement effort can be largely ineffective. In models with such coordination problems, it is well known that there can be multiple equilibria, with generally law abiding outcomes in some ("good" equilibria), serious disregard for the law in others ("bad"). While these models are rarely very specific about exactly which laws are under discussion, there is a presumption that law abiding behaviour is good for private sector development, business confidence and the like; and conversely. Since there is little theory to help us determine which equilibrium will occur in a given situation, the authors consider what concrete institutional mechanisms might exist in transition economies to help eliminate the bad equilibrium. They identify two

---

[187] J. Hellman, G. Jones and D. Kaufman, *"Seize the State, Seize the Day" – State Capture, Corruption and Influence in Transition*, World Bank Policy Research Working Paper, No. 2444 (Washington, D.C.), September 2000.

[188] F. Schneider and D. Enste, "Shadow economies: size, causes and consequences", *Journal of Economic Literature*, Vol. 38, No. 1, March 2000, pp. 77-114.

[189] Ibid, p. 108.

[190] P. Aghion and M. Schankerman, *A Model of Market Enhancing Infrastructure*, CEPR Discussion Paper, No. 2462 (London), May 2000.

[191] G. Roland and T. Verdier, *Law Enforcement and Transition*, CEPR Discussion Paper, No. 2501 (London), July 2000.

[192] A. Rapaczynski, "The roles of the state and the market in establishing property rights", *Journal of Economic Perspectives*, Vol. 10, No. 2, 1996, pp. 87-103.

[193] B. Black, *The Legal and Institutional Preconditions for Strong Securities Markets*, Olin Program in Law and Economics Working Paper, No. 179, Stanford Law School (forthcoming in *UCLA Law Review*), 2000.

possible mechanisms: Chinese style "dualism", and the prospect of EU accession (the argument here is that the prospect of effective law enforcement in the future provides incentives to be law abiding in the present). Since neither of these arguments applies to Russia or other CIS countries, it is perhaps not so surprising that they continue to perform so poorly.

In regard to the securities markets, Black draws attention to informational asymmetries and reputational issues in such markets. While formal, legal regulation is important, he also notes that securities markets cannot function well without a network of intermediaries prepared to invest in a reputation for honest dealing. Minority shareholders, in particular, need good information about company values, and need to have reasonable confidence that managers or majority shareholders will not cheat them. Ensuring both these conditions in an open and transparent manner is surprisingly difficult, and few countries can claim to be completely successful. Sometimes a (small) country can manage by issuing shares through the stock exchange of another country with already well-established institutions, but mostly this option is not available. However, failure to develop a well functioning securities market compels firms to resort to self-financing or bank financing. While clearly making important points about the complex and subtle structure of institutions necessary to support effective securities markets, it seems to me that Black overgeneralizes his case. For many countries have experienced successful and sustained economic growth with very poor or virtually absent securities markets. We should therefore be careful, especially in the context of transition economies where many new institutions are being created *ab initio*, not to insist that all must adopt a rather uniform model that might only suit some of them.

A wholly different approach to institutional change is through the provision of an external model. Such a model both provides something to be copied – hence reducing considerably the costs of learning that would otherwise arise, and avoiding many mistakes – and if it is embodied in some form of international agreement, it can strengthen the hand of reforming politicians and assist them in resisting interest groups opposing reforms. The simplest and most widespread instance of such a model is provided by membership of the World Trade Organization (WTO), since the disciplines it entails make it harder for domestic industrial lobbies to secure special protection, and the trading practices supported by the WTO are generally conducive to efficiency. Without such an external frame of reference, though, it must be doubtful whether many countries would be willing to adopt the open and liberal trade policies that they do follow. Most transition economies are either already WTO members, or are applicants for membership.

In the economic sphere, the EU provides the most comprehensive external model, not only for the applicant states but also for other transition economies. Through its network of Association Agreements and Trade and Partnership Agreements, and by means of the PHARE and TACIS aid programmes to the transition economies, the EU promotes institutional change and development across the region. Adaptation to Community norms, by implementing various provisions of the *acquis communautaire* in their domestic economies, has proved a powerful tool in fostering rapid institutional change. Indeed, although I question later whether the EU model is necessarily the most suitable one for the transition economies, its existence has almost certainly accelerated a great deal of very important institutional reconstruction. As with the WTO, EU agreements have often provided domestic politicians with useful levers to help them resist pressures from sectional interests.

## 3.3 Institutional change in transition economies

What exactly have the transition economies done as regards institutional transformation and the development of market institutions? Their progress in economic reforms has been tracked by the EBRD in successive issues of its annual *Transition Report*, which reports a set of qualitative transition indicators for 26 transition countries. The EBRD also tabulates summary indicators of macroeconomic stabilization (using GDP growth and the rate of inflation as key indicators), and of the extent and effectiveness of legal reforms, especially those to do with business contracts and property rights. The IMF[194] also reports a number of measures of institutional conditions: (a) an indicator of initial conditions; (b) a liberalization index; and (c) a measure of institutional quality. Luckily, among the various indicators there is an extremely high positive correlation,[195] implying that although they purport to measure somewhat different dimensions of institutional change, they are actually picking up essentially equivalent indicators of institutional development. This remark does not, of course, apply to the IMF's index (a) since that does not measure change at all. Instead, it measures various aspects of the extent of distortions and institutional deficiencies in the transition economies around 1990. In other words, it is an attempt to measure the distance to be travelled once reforms get underway.

---

[194] IMF, *World Economic Outlook: Focus on Transition Economies* (Washington, D.C.), October 2000.

[195] From a methodological standpoint, it should be pointed out that, strictly speaking, the calculation of simple correlation coefficients across sets of indicators based on rankings and qualitative scoring is not valid. Instead, Spearman rank correlation coefficients should be employed. However, performing the calculations correctly does not materially affect the finding reported in the text. This issue, and the related one of performing ordinary least squares regressions with qualitative/ordered variables where some form of ordered probit would be a more appropriate technique, crops up frequently in the empirical work reported in this paper.

## TABLE 3.3.1

**Economic growth and inflation in selected transition economies, 1989-2000**
*(Percentages)*

| | GDP in 1999 (relative to 1989) | GDP growth 1999 | Forecast GDP growth in 2000 | Highest annual inflation rate (1989-1999) | Inflation in 1999 | Estimated inflation in 2000 |
|---|---|---|---|---|---|---|
| **EU accession countries** | | | | | | |
| Bulgaria | 67 | 2.4 | 4.0 | 1 082 | 0.7 | 7.0 |
| Czech Republic | 95 | -0.2 | 2.0 | 52.0 | 2.1 | 3.9 |
| Hungary | 99 | 4.5 | 6.0 | 35.0 | 10.1 | 9.5 |
| Poland | 122 | 4.1 | 5.0 | 586 | 7.3 | 9.9 |
| Romania | 76 | -3.2 | 1.5 | 256 | 45.8 | 45.0 |
| Slovakia | 100 | 1.9 | 2.0 | 61.2 | 10.6 | 11.9 |
| Slovenia | 109 | 4.9 | 5.1 | 1 306 | 6.1 | 8.6 |
| Estonia | 77 | -1.1 | 5.0 | 1 076 | 3.3 | 3.8 |
| Latvia | 60 | 0.1 | 4.5 | 951 | 2.4 | 2.9 |
| Lithuania | 62 | -4.2 | 2.2 | 1 021 | 0.8 | 1.0 |
| **Other south-east European countries** | | | | | | |
| Albania | 95 | 7.3 | 7.0 | 226 | 0.4 | 0.4 |
| Bosnia and Herzegovina | – | – | – | – | – | – |
| Croatia | 78 | -0.3 | 3.5 | 1 518 | 4.2 | 6.5 |
| The former Yugoslav Republic of Macedonia | 74 | 2.7 | 5.0 | 1 664 | -1.3 | 8.5 |
| **CIS** | | | | | | |
| Armenia | 42 | 3.3 | 3.5 | 5 273 | 0.6 | -0.5 |
| Azerbaijan | 47 | 7.4 | 7.5 | 1 664 | -8.5 | 1.5 |
| Belarus | 80 | 3.4 | 2.0 | 2 221 | 294 | 168 |
| Georgia | 34 | 3.0 | 3.0 | 15 607 | 19.3 | 4.4 |
| Kazakhstan | 63 | 1.7 | 8.2 | 1 892 | 8.4 | 13.2 |
| Kyrgyzstan | 63 | 3.6 | 5.0 | 855 | 36.8 | 18.6 |
| Republic of Moldova | 31 | -4.4 | -3.0 | 1 276 | 39.3 | 32.0 |
| Russian Federation | 57 | 3.2 | 6.5 | 1 526 | 86.1 | 20.7 |
| Tajikistan | 44 | 3.7 | 5.0 | 2 195 | 27.5 | 24.2 |
| Turkmenistan | 64 | 16.0 | 20.0 | 3 102 | 24.2 | 10.0 |
| Ukraine | 36 | -0.4 | 3.0 | 4 735 | 22.7 | 28.0 |
| Uzbekistan | 94 | 4.1 | 1.0 | 1 568 | 29.1 | 30.0 |

*Source:* EBRD, *Transition Report 2000* (London), November 2000.

Tables 3.3.1 to 3.3.3 summarize selected indicators of economic performance and reform progress across the transition economies, principally drawing on the above sources. Thus, table 3.3.1 presents data on growth of GDP and on the rate of inflation experienced by the transition economies. Table 3.3.2 shows indicators concerning the initial conditions for transition, a liberalization index, an indicator of institutional quality, and the EBRD's average transition indicator for each country in 1995 and 1999. Table 3.3.3 gives more EBRD transition indicators, this time to do with the extent and effectiveness of legal reforms in the areas of commercial law and the regulation of financial markets.

From the tables it is abundantly clear that the experience of the transition economies over the past decade, both in terms of growth and inflation, and in terms of the extent and nature of institutional reforms, is extremely diverse. Their starting points, measured by the indicator of initial conditions shown in 3.3.2, were also very diverse. These points make it hard to discuss the countries as a group, since virtually their sole common feature is the fact that they had communist governments in place prior to 1989. At the same time, the sheer diversity revealed in the tables helps to explain why it is nevertheless interesting to try to understand better the linkages that can be found between the economic performance of the transition economies and other indicators of their status and progress: initial conditions, effectiveness of macroeconomic stabilization, extent and nature of (market-oriented) institutional reforms, and so on. These linkages are explored in the next section, both at the macroeconomic level, and at a more microeconomic level by referring to studies based on the analysis of enterprise survey data.

To a large extent, the CEE countries that are candidates for EU accession stand out as having relatively favourable initial conditions, and most have also substantially liberalized, achieved high levels of institutional quality and performed well on the EU transition indicators. They suffered relatively modest post-communist recessions and their GDPs are already close to pre-transition income levels (Poland is already well ahead of this level). Mostly, these countries are growing at respectable rates now and there is every indication that their achieved rates are sustainable.

TABLE 3.3.2

**Progress with institutional reform in selected transition economies**

| | Initial conditions index[a] | Liberalization index[b] 1989 | Liberalization index[b] 1997 | Institutional quality[c] 1997-1998 | EBRD transition indicators[d] 1995 | EBRD transition indicators[d] 1999 |
|---|---|---|---|---|---|---|
| **EU accession countries** | | | | | | |
| Bulgaria | 2.1 | 0.13 | 0.79 | 0.1 | 2.5 | 2.9 |
| Czech Republic | 3.5 | – | 0.93 | 6.8 | 3.5 | 3.4 |
| Hungary | 3.3 | 0.34 | 0.93 | 8.7 | 3.5 | 3.7 |
| Poland | 1.9 | 0.24 | 0.89 | 7.0 | 3.3 | 3.5 |
| Romania | 1.7 | – | 0.75 | -0.8 | 2.5 | 2.8 |
| Slovakia | 2.9 | – | 0.86 | 2.8 | 3.3 | 3.3 |
| Slovenia | 3.2 | 0.41 | 0.89 | 8.5 | 3.2 | 3.3 |
| Estonia | -0.4 | 0.07 | 0.93 | 6.1 | 3.2 | 3.5 |
| Latvia | -0.2 | 0.04 | 0.89 | 2.6 | 2.8 | 3.1 |
| Lithuania | – | 0.04 | 0.89 | 2.6 | 2.9 | 3.1 |
| **Other south-east European countries** | | | | | | |
| Albania | 2.1 | – | 0.78 | -7.1 | 2.4 | 2.5 |
| Bosnia and Herzegovina | – | – | – | -9.9 | – | 1.8 |
| Croatia | 2.5 | 0.41 | 0.85 | 0.3 | 2.8 | 3.0 |
| The former Yugoslav Republic of Macedonia | 2.5 | 0.41 | 0.82 | -3.3 | 2.5 | 2.8 |
| **CIS** | | | | | | |
| Armenia | -1.1 | 0.04 | 0.72 | -4.4 | 2.1 | 2.7 |
| Azerbaijan | -3.2 | 0.04 | 0.62 | -7.8 | 1.6 | 2.2 |
| Belarus | -1.1 | 0.04 | 0.51 | -7.6 | 2.1 | 1.5 |
| Georgia | -2.2 | 0.04 | 0.72 | -6.1 | 2.0 | 2.5 |
| Kazakhstan | -2.5 | 0.04 | 0.86 | -5.3 | 2.1 | 2.7 |
| Kyrgyzstan | -2.3 | 0.04 | 0.75 | -4.2 | 2.9 | 2.8 |
| Republic of Moldova | -1.1 | 0.04 | 0.75 | -2.0 | 2.6 | 2.8 |
| Russian Federation | -1.1 | 0.04 | 0.83 | -5.4 | 2.6 | 2.5 |
| Tajikistan | -2.9 | 0.04 | 0.45 | -15.0 | 1.6 | 2.0 |
| Turkmenistan | -3.4 | 0.04 | 0.36 | -11.5 | 1.1 | 1.4 |
| Ukraine | -1.4 | 0.04 | 0.65 | -5.8 | 2.2 | 2.4 |
| Uzbekistan | -2.8 | 0.04 | 0.57 | -10.4 | 2.4 | 2.1 |

*Source:* IMF, *World Economic Outlook: Focus on Transition Economies* (Washington, D.C.), October 2000; EBRD, *Transition Report* (London), various issues.

*Note:* Summing and averaging qualitative/ordinal indicators is, strictly speaking, methodologically rather suspect, but there is no obvious alternative and with careful judgement the method is probably not too bad. In this connection it is reassuring that indicators, scores and rankings based on different sets of underlying measures seem to yield broadly similar results.

[a] **Initial Conditions Index** - a weighted average of indicators for level of development, trade with CMEA, macroeconomic disequilibria, distance to the EU, natural resource endowments, market memory (measured by number of years of communist rule), state capacity.

[b] **Liberalization Index** - a weighted average of three components: domestic market liberalization (weight 0.3), foreign trade liberalization (weight 0.3), enterprise privatization and banking reform (weight 0.4). Each component is scored in the range [0, 1].

[c] **Institutional Quality** - index based on five components, namely, extent of democracy, government effectiveness, extent of regulation, rule of law and extent of graft/corruption. Each indicator is scored in the range -25 to +25 and so the average lies in the same range. For developed market economies the average score is 12.6.

[d] **Transition Indicators** - based on a simple average of eight indicators each scored in the range of 1 (no market reforms) to 4 (conditions as in a developed market economy) or 4* (exceptionally strong development of market-based institutions). Scores with an asterisk are treated as adding one third of a point to the aggregate.

In contrast, many of the CIS countries had poor initial conditions, have not (yet) liberalized so decisively, still have poor institutional quality and their reform progress as measured by the EBRD transition indicators is modest at best. Many CIS countries have resumed economic growth, but their post-communist recessions were so deep and prolonged that they still have an enormous amount of catching up to do to get back to pre-transition income levels. Within the CIS group, Belarus and Uzbekistan are of interest as a pair of countries that have experienced post-communist recessions that were quite minor, while undertaking virtually nothing in terms of serious market-oriented reforms. This might be interpreted as an argument for institutional stability except that the longer-term viability of these economies, based on their current policies, is questionable. At the opposite pole are Georgia, the Republic of Moldova and Ukraine, whose (officially measured) GDPs fell by two thirds during the 1990s. This truly shocking economic decline, virtually unprecedented except in wartime (both Georgia and the Republic of Moldova saw serious civil conflict in the early 1990s, although Ukraine remained totally peaceful during the period), was accompanied by limited and inconsistent reform progress, albeit far from the worst in the transition economy region. Hence there remains a good deal to be explained in the exceptionally poor performance of certain countries.

### TABLE 3.3.3
**Indicators of legal reform and effectiveness in selected transition economies**

|  | Commercial law | | Financial regulations | |
|---|---|---|---|---|
|  | Extent | Effectiveness | Extent | Effectiveness |
| **EU accession countries** | | | | |
| Bulgaria | 4 | -4 | 3 | 2 |
| Czech Republic | 3 | 3 | 4 | -3 |
| Hungary | 4 | -4 | 4 | 4 |
| Poland | -4 | 4 | 4 | 4 |
| Romania | 3 | -4 | 4 | 3 |
| Slovakia | 3 | 3 | 3 | -3 |
| Slovenia | 4 | -4 | 4 | 4 |
| Estonia | -4 | 3 | 4 | -3 |
| Latvia | 4 | -4 | 3 | 3 |
| Lithuania | 4 | 3 | 4 | -4 |
| **Other south-east European countries** | | | | |
| Albania | 3 | -2 | 2 | 1 |
| Bosnia and Herzegovina | 3 | 1 | 1 | 1 |
| Croatia | 4 | 3 | 3 | -3 |
| The former Yugoslav Republic of Macedonia | 3 | 2 | 3 | -2 |
| **CIS** | | | | |
| Armenia | -4 | 2 | 2 | 3 |
| Azerbaijan | 3 | 2 | 2 | -2 |
| Belarus | 1 | 2 | 2 | 2 |
| Georgia | 3 | 2 | 3 | 2 |
| Kazakhstan | 4 | -4 | 3 | -3 |
| Kyrgyzstan | 3 | 3 | 3 | -3 |
| Republic of Moldova | 3 | 2 | -3 | -2 |
| Russian Federation | -4 | 3 | 3 | -3 |
| Tajikistan | 2 | -2 | 2 | 1 |
| Turkmenistan | .. | .. | .. | .. |
| Ukraine | 3 | 2 | 3 | 2 |
| Uzbekistan | 3 | 2 | 2 | -2 |

*Source:* EBRD, *Transition Report 2000* (London), November 2000.

*Note:* Each indicator is scored on a basic scale from 1 to 4, with 1 signifying little or no development of the relevant legal provisions and/or poor implementation due to badly formulated laws, poor administration, defective judicial procedures, etc. A score of 4 indicates that the legal rules are quite comprehensive in relation to what one would expect in a well ordered market economy, or that the implementation of legal provisions attains a high standard. Indicators + or - indicate that the achieved level is just above or just below the given score, but not enough to move to the next level.

## 3.4 Links between institutional change and economic performance

By now, numerous studies have sought to understand the connections between various dimensions of institutional reform in transitional economies, and their economic performance. Some of these studies have focused on macroeconomic performance, looking at broad features of the institutional environment and their impact upon aggregate economic growth. Others have focused more narrowly, studying the ways in which institutional change has supported or inhibited restructuring at the level of individual enterprises. Both levels of investigation are important, and examples of both are reported in this section. In principle, one might also wish to understand the forces explaining institutional change itself in various countries, so that one could see why some countries are able to rebuild new institutions rather quickly while others seem to get "stuck" or "diverted". However, this causal direction has not yet been much studied, though a useful start was made by Raiser et al.[196]

### (i) Macroeconomic studies

Already by the mid-1990s it was perceived that economic performance among the transition countries was diverging, with many of the countries of central and eastern Europe already recovering from post-communist recessions and resuming growth, sometimes – as in the notable case of Poland – at quite rapid rates. In contrast, all 12 CIS states were still plunged in deep recession, with output in most cases still falling rapidly or at best, starting to bottom out well below the levels that had been achieved around 1990. These observations, rough and ready as they were, made it natural to investigate whether and to what extent the earlier and more sustained recovery in central and eastern Europe could be explained by (a) earlier and more effective macroeconomic stabilization; and (b) greater commitment to market-oriented economic reforms, including the institutional reforms that form the focus of the present paper.

As background to such studies, it is worth pausing to note that there are now many investigations of economic growth and performance across a much wider range of countries than the transition economies alone, which are seeking to explain why some countries grow rapidly while others do not. Sachs and Warner[197] provide a particularly interesting example of such studies, focusing on the question of economic convergence. Rather than pointing to increasing returns to scale as a reason for persistent divergence between rich and poor countries, they argue instead for the importance of sound policies. They find, by performing appropriate regressions across a large number of countries, that open trading policies and effective protection of private property rights are usually sufficient to enable poorer countries to achieve higher than average growth rates. In a later study, Sachs and Warner[198] found that increased global integration (i.e. greater openness), higher government saving and better quality institutions all increase steady state income and hence boost transitional growth rates. Interestingly, countries with abundant natural resources can generally expect lower growth rates, other conditions being the same.[199]

---

[196] M. Raiser, M. Di Tommaso and M. Weeks, *The Measurement and Determinants of Institutional Change: Evidence from Transitional Economies*, EBRD Working Paper, No. 60 (London), 2000.

[197] J. Sachs and A. Warner, *Economic Convergence and Economic Policies*, NBER Working Paper, No. 5039 (Cambridge, MA), September 1995.

[198] J. Sachs and A. Warner, "Fundamental sources of long-run growth", *American Economic Review*, Vol. 87, No. 2, Papers and Proceedings, 1997, pp. 184-188.

[199] The reasons for this, at first sight puzzling, result are manifold. The most obvious points to make, however, are: (a) large natural resource exports can push up the equilibrium exchange rate, making much of the

The first serious attempts to investigate econometrically the performance of transition economies are associated with the names of Fischer, Sahay and Végh. They perform a regression on a set of 20 transition economies using data for the years 1992-1994 (some countries could not be included due to data limitations), which examines the impact of macroeconomic stabilization and early reforms on growth.[200] A second equation, estimated in other studies but quoted by these authors, looks at more conventional determinants of growth that are postulated to come into play once a country is well down the transition path. The equations are as follows (figures in brackets are $t$-statistics):

$$GR = 0.10\ S\ 2.73\ LINF + 0.24\ FISCAL + 0.26\ OFAST + 12.97\ LIP \quad (1)$$
$$(0.02)\ (-5.27)\quad\quad (1.90)\quad\quad (2.32)\quad\quad (2.86)$$

$$GR = -0.83\ S\ 0.35\ Y_0 + 0.62\ GN + 3.17\ SEC + 17.5\ INV \quad (2)$$
$$(0.98)\ (2.50)\quad (2.82)\quad (2.46)\quad (6.53)$$

The variables in these equations are defined thus: GR is the growth rate of aggregate income (i.e. GDP growth rate), LINF is the *log* of the inflation rate, FISCAL is the budget surplus as a percentage of GDP, OFAST is official aid as a percentage of GDP, LIP is an indicator measuring the extent of banking and enterprise sector reforms. $Y_0$ is initial per capita income at world prices, GN is the population growth rate, SEC is the secondary school enrolment rate (used as a measure of human capital), and INV is the share of investment in GDP.

Equation (1) shows the decisive role of high inflation and fiscal deficits in producing deep recession at the start of transition (second and third terms on the right hand side). Conversely, in countries that stabilized rapidly, the last term (LIP) shows the importance of early market-oriented – and largely institutional – reforms in kick-starting the growth process. After a time, equation (2) should be the more relevant one. No longer are institutional changes highlighted. Rather, aside from variables that cannot be changed quickly or are not amenable to policy intervention at all, the key variable in (2) is INV. Thus, in the medium and longer term the message for transition economies wishing to grow rapidly is utterly clear, namely, that they must devise policies to raise their rates of investment and then maintain investment at high levels. As Fischer, Sahay and Végh show, with an investment ratio of 20 per cent it will take transition economies an average of 45 years to catch up to current average OECD per capita income levels, while if investment can be raised to 30 per cent of GDP, the average catch up time falls to 30 years, just one generation.[201] In this regard, we note that those transition economies that have successfully stabilized and introduced many market reforms are already experiencing increases in investment, although not yet to levels much over 20 per cent. In most cases the levels (expressed as a share of GDP) are still not very high in international comparison. I return to this issue towards the end of the paper when discussing industrial policy.

Fischer, Sahay and Végh perform a somewhat similar analysis for 26 transition countries, strongly emphasizing the stabilization variables in their conclusions.[202] This stress on stabilization led the authors to predict that even the relatively laggard CIS countries, once they got inflation rates down to tolerable levels (taken to mean under 4 per cent per month) and gained control of their fiscal positions and external balances, would quickly start to grow again. However, this did not happen to any marked extent, with Russia in particular only just starting to grow a little in 1997 before being knocked off course by the August 1998 financial crisis; Russian growth only resumed in late 1999. This experience suggests that while stabilization might be necessary for growth, it is not sufficient. As I argue below, some of the institutional reforms discussed in this paper appear to be critical for economic growth.

Perhaps the most comprehensive study of the links between institutional change and economic performance is to be found in Havrylyshyn and van Rooden.[203] This paper starts by explaining the GDP growth of transition economies in terms of three sets of variables: (i) current inflation, INFL, taken to indicate the effectiveness of stabilization; (ii) current and lagged indicators of structural reforms; and (iii) two indicators of initial conditions, one reflecting macroeconomic distortions, the other reflecting the distortions associated with socialism before 1989/1991. The empirical work was based on data for 25 countries over the period 1991-1998 (with a few missing observations due to incomplete data). The basic regressions found a highly significant impact of the stabilization variable as well as the structural reform indicator, with initial conditions significant but rather less important in explaining GDP growth. The typical format of their regressions was:

$$GR = a_1\ INFL + b_1\ RI + b_2\ RI_{-1} + c_1\ IC1 + c_2\ IC2 + [d_1\ IR1 + d_2\ IR2] \quad (3)$$

The variables in (3) are as follows: INFL is the current rate of inflation, RI is an indicator of structural reforms (the suffix, -1, simply denotes a lagged value),

---

rest of the economy internationally uncompetitive ("Dutch disease"); (b) incomes earned in the resource sectors may not be taxed adequately to spread the benefits across the whole economy (firms in the resources business may establish strong lobbies, for instance), so domestic demand for goods and services can then be very weak due to the low incomes; and (c) the resulting incentives for new firms to enter non-resource sectors can be weak at best.

[200] S. Fischer, R. Sahay and C. Végh, "Economies in transition: the beginnings of growth", *American Economic Review*, Vol. 86, No. 2, Papers and Proceedings, 1996, pp. 229-233.

[201] S. Fischer, R. Sahay and C. Végh, "Stabilization and growth in transition economies: the early experience", *Journal of Economic Perspectives*, Vol. 10, No. 2, 1996, pp. 45-66.

[202] S. Fischer, R. Sahay and C. Végh, "From transition to market: evidence and growth prospects", in S. Zecchini (ed.), *Lessons from the Economic Transition: Central and Eastern Europe in the 1990s* (OECD and Dordrecht, Kluwer Academic Publishers, 1997), Part I, chap. 4, pp. 79-101.

[203] O. Havrylyshyn and R. van Rooden, *Institutions Matter in Transition, But So Do Policies*, IMF Working Paper WP/00/70 (Washington, D.C.), March 2000.

IC1 and IC2 are the two measures of initial conditions, and the terms in square brackets, IR1 and IR2, are measures of institutional change in the narrow sense employed by Havrylyshyn and van Rooden.

Havrylyshyn and van Rooden distinguish between what they call *structural reforms*, which can be measured by some of the EBRD's transition indicators to do with price liberalization, trade and exchange rates, banking and financial market reforms, private sector growth and the like; and measures of *institutional reform*, which they take to be about the basic legal and political framework of the society concerned. The latter they measure using the EBRD's indicators of legal reforms (coverage and effectiveness of legal aspects of business), and measures of the extent of political liberalization. Including such indicators in their original regressions simply as additional variables confirms that institutional reform as defined by Havrylyshyn and van Rooden is important for growth, but is overwhelmingly dominated by what they call structural reforms. For purposes of the present paper, however, this distinction does not appear terribly useful, since both structural reforms and institutional reforms as defined here are included in the rather broad definition of economic institutions given in section 3.2 above. Accordingly, our conclusion must be that institutional change in the widest sense is absolutely vital for sustained growth and recovery of the transition economies. This does not, of course, deny the importance of sound government policies, both to maintain macroeconomic stabilization and to promote through deliberate state action the very institutional reforms that must be in place for growth.

Grigorian and Martinez[204] adopt a somewhat narrower focus, in that they study the impact of various institutional quality indicators on industrial growth in 27 Asian and Latin American countries. Five qualitative indicators are used to assess institutional effectiveness in these countries, covering: government repudiation of contracts, risk of expropriation (especially important for foreign investors), corruption, rule of law and bureaucratic quality. These dimensions each reflect different aspects of the interface between the state and private sector business, and taken together they tell us a great deal about how easy it is, and what sort of barriers there are, to doing business in a given country. Although not explicit in this study, there tends to be a very high correlation between these types of indicator and the creation of specific institutions such as well functioning banks and financial markets, liberal and open trade, bodies to enforce competition policy, and so on.

The study found that their measures of institutional quality had a strongly positive effect on industrial growth rates, with indicators to do with the legal and regulatory framework apparently working through two channels: via encouraging higher levels of investment; and via improvements in general resource allocation. The authors conclude that rather than developing full blown industrial policies which states may be unable to manage effectively, for many developing countries it is likely to be better to pay more attention to basic steps to reduce corruption, to remove bureaucratic barriers to business formation and growth, and to strengthen the legal environment. For transition economies, such steps, it is suggested, should complement other policies to promote investment, education, R&D and large-scale privatization.

Finally, Kushnirsky[205] estimates country-level production functions for 10 transition countries (Tajikistan and Turkmenistan being omitted due to incomplete data), incorporating indicators of initial conditions and institutional change into the formulation of his estimating equations. The role of initial conditions was found to be significant, while the impact of institutional factors – measured using EBRD indicators, including the private sector share in production – was more mixed, turning out to be clearly significant in only seven out of ten cases.

### (ii) Enterprise level studies

What really brings about changes in economic performance in any economy is changes in the behaviour of enterprises. Hence to understand fully the role of institutions in influencing economic performance, we should think in terms of a two-stage process: (a) the institutional environment affects enterprise behaviour in various ways; and (b) the resulting changes in enterprise behaviour lead to different, and one hopes improved, economic outcomes and performance. As far as the first stage is concerned, one can look for three types of change, all of which are potentially important:

(i) the entry of new firms or the relative growth of high productivity firms;

(ii) the restructuring and reorganization of existing firms;

(iii) exit from the market by failing firms, or the decline of low productivity firms.

This is the dynamic process whereby economies adapt, modernize, develop and grow, spurred by competitive pressures, market opportunities and different dimensions of the policy and institutional environment. Under different conditions, one would expect the balances between changes of types (i), (ii) or (iii), and hence their relative contributions to overall economic performance, to vary a good deal, for there is little in economic theory to tell us what these contributions ought to be although, as noted above, Aghion and Schankerman[206] set out a possible framework for analysing these matters. However, the importance of careful empirical work cannot be underestimated.

---

[204] D. Grigorian and A. Martinez, *Industrial Growth and Quality of Institutions: What Do (Transition) Economies Have to Gain from the Rule of Law?*, World Bank Policy Research Working Paper, No. 2475, (Washington, D.C.), November 2000.

[205] F. Kushnirsky, "A modification of the production function for transition economies reflecting the role of institutional factors", *Comparative Economic Studies*, Vol. XLIII, No. 1, 2001, pp. 1-30.

[206] P. Aghion and M. Schankerman, op. cit.

The extreme instance of this dynamic process could be found under socialism, when, in the era of central planning, there was generally no entry except by new firms provided for in the prevailing plan; there was effectively no exit as enterprises were not permitted to go bankrupt (an aspect of the familiar *soft budget constraint* problem);[207] and the extent of restructuring of existing firms was very limited indeed. Under these conditions, the distinctly lacklustre performance of the socialist economies by the 1980s should not have been very surprising. At the very microlevel of individual firms, stability and rigidity are disastrous for economic performance, and in all successful economies, success is built upon the ruins of many thousands of failed businesses. This elementary (but not very popular) observation has implications for policy that are taken up in the concluding section of the paper.

As Carlin et al.[208] elucidate, the above dynamic process can be thought of as a combination of two parallel processes falling under the headings of *incentives* and *selection*. Incentives basically drive the type (ii) change listed above, while selection drives types (i) and (iii). In order for these processes to work well, it is important that market signals should be tolerably reliable, i.e. they should reveal real market opportunities. Also, as a form of positive feedback, it is important that there are plentiful examples of economic success for others to emulate.

These processes are studied first for the United Kingdom, using a large sample of establishments in United Kingdom manufacturing, over the period 1980-1992. For each year, the dataset contained around 140,000 establishments. The findings for the United Kingdom are quite striking, with about one half of productivity growth occurring in surviving establishments, and a further 30 per cent due to net entry (exiting units having lower productivity than entrants, on average). Productivity gains were especially strong in multi-plant firms, perhaps due to the operation of some form of internal capital market. Some evidence was found that in booms, entry is less selective in that some low productivity firms are able to enter the market. More surprisingly, recessions do not necessarily speed up the exit of low productivity firms/establishments, and appear not to encourage innovation. Last, stronger product market competition raises productivity, both levels and growth.

For the transition economies, such detailed data over many years does not exist. But Carlin et al. draw on the more detailed enterprise-level results reported in a previous study,[209] which made use of a survey of over 3,500 firms in 25 transition countries carried out jointly by EBRD and the World Bank. From this survey, some important conclusions emerged, most notably that: "Large-scale shifts of resources from inefficient to efficient sectors of the economy have played no significant part in productivity growth in transition countries".[210] Further, although most firms – both old and new – have engaged in a great deal of restructuring, and many new firms have expanded rapidly to achieve large market shares, the results in terms of overall productivity growth have not been as spectacular as might have been expected. Outside the CIS, where the institutional conditions are both stronger and more settled, outcomes have been fairly good; but in the CIS countries themselves, "turbulent" restructuring is at best only weakly associated with productivity gains.

Policy lessons for the transition economies are consequently not straightforward, but it is clear that weak property rights and poor market signals are major parts of the story. It is not sufficient for many firms to be undertaking restructuring, since for productivity improvements it is crucial that restructuring should be in appropriate directions – not merely frenetic activity for the sake of it, but purposeful, guided by tolerably reliable market signals. Similarly, enterprise owners, and the managers acting for them, need to be reasonably secure in their property rights in order to be willing to invest[211] in product or process innovations. This is why economic environments that are too turbulent are not conducive to useful economic advance. There can be too much, or the wrong sorts of competition. For as Carlin et al.[212] note, while monopoly power unambiguously gives rise to poor performance, excessive competition can both restrain firms' ability to finance needed investments (an important consideration in environments where the market for bank credit is likely to be severely limiting) and confuse the market signals that guide them.

An extremely comprehensive survey of the research on enterprise restructuring in transition economies has been made by Djankov and Murrell.[213] By reviewing over 125 studies of restructuring, the aim of the authors is to identify what general findings there might be, and to elicit some general lessons, including possible implications for

---

[207] J. Kornai, *The Socialist System: Political Economy of Communism* (Princeton, Princeton University Press, 1992); P. Hare, "Shortage and the soft budget constraint", in E. Maskin and A. Simonovits (eds.), *Planning, Shortage and Transformation* (Cambridge, MA, MIT Press, 2000), chap. 11; and M. Schaffer, "Do firms in transition economies have soft budget constraints? A reconsideration of concepts and evidence", *Journal of Comparative Economics*, Vol. 26, No. 1, March 1998, pp. 80-103.

[208] W. Carlin, J. Haskel and P. Seabright, "Understanding 'the essential fact about capitalism': markets, competition and creative destruction", *National Institute Economic Review*, No. 175, January 2001, pp. 67-84.

[209] W. Carlin, S. Fries, M. Schaffer and P. Seabright, "Competition and enterprise performance in transition economies: evidence from a cross-country survey", University College London, mimeo, June 2000.

[210] Ibid., p. 81.

[211] There is also the question of firms' *ability* to invest, since their access to bank credit or their ability to secure funding through new share flotations are also likely to depend on secure property rights.

[212] W. Carlin, S. Fries, M. Schaffer and P. Seabright, op. cit.

[213] S. Djankov and P. Murrell, "Enterprise restructuring in transition: evidence from a quantitative survey", World Bank (Washington, D.C.), mimeo, April 2000.

economic policy. The (mostly implicit) presumption in this survey is that enterprise restructuring is desirable, associated with long-term improvements in economic performance, and that in transition economies, therefore, we should expect to see a good deal of restructuring going on. Further, insofar as there are policy lessons, they concern how best to create the conditions likely to foster restructuring. However, from the study we have just discussed, it is apparent that at best restructuring can only form part of the full story, and that the factors leading to the exit of unproductive/unprofitable firms and the entry of better ones are at least as important for long-term performance.

Nevertheless, some of the broad findings of Djankov and Murrell are of interest for our present exercise. In particular, they highlight the key importance of privatization, hardened enterprise budget constraints and a moderate degree of product market competition in promoting restructuring. At the same time they note that policies in these areas have proved far more effective in most of central and eastern Europe than they have in the CIS region, when evaluated in terms of improved economic efficiency and enterprise competitiveness. This raises questions about the wider institutional environment supporting market processes, and its links with enterprise ownership/corporate governance. Unfortunately, at this point the research is rather lacking in the detail that would be needed to reach very precise conclusions.

Some of the institutional features that influence how far hardening budget constraints are likely to be effective in compelling improved enterprise performance are investigated by Frydman et al.[214] What they found was that efforts to harden budget constraints usually led to improvements in the performance of already privatized firms, since these firms lacked the political connections that might have enabled them to bypass credit constraints and seek financial assistance through tax arrears, delayed payments to state creditors and the like: hence such firms had no choice but to restructure or exit. But for firms still state owned, these quasi-political links were still in play, and tended to be exploited. On the one hand, such firms then survived where they might not otherwise have done so; on the other, essential restructuring to facilitate longer-term viability was often deferred. The implication drawn in the paper is that privatization should not be delayed, since only then will the remaining state owned firms start to behave "properly". In this way, the paper does correctly pinpoint the key role of achieving an effective organizational separation between enterprises and the state: the state sets the "rules of the game" – in accordance with my definition of institutions set out at the start of section 3.2 – and enterprises then decide how they will play it. However, this rather stark, and perhaps oversimplistic approach to economic institutions requires further discussion in the concluding section of this paper.

Last, Commander et al.[215] focus on regulatory aspects of the nexus between enterprises and the state, noting enormous disparities across the transition economies. Broadly speaking, regulation becomes less effective, less satisfactory and in many respects more bureaucratic as one moves across the region from west to east. The authors focus on two principal areas where good policies and institutions can make a big difference to overall economic performance, namely: (a) supporting new entry and the growth of firms; and (b) dealing with firms in distress.

On new entry, this is partly a question of having in place the proper legislation (with enforcement) to protect shareholders, so that outsider control as well as new investment can both be encouraged. It is also important to have an effective approach to competition policy, not merely to stop the abuse of monopoly power by large firms, but far more importantly, to eliminate barriers to entry and limitations on competition that are frequently arbitrary, politically motivated and economically wholly unjustified. For instance, in both Russia and Ukraine, regional authorities have frequently restricted trade with other regions, supposedly in the interests of protecting supplies to local markets, but the main effect of such regulation is anti-competitive, and it tends to discourage new entry. Likewise, the prevalence of non-market (barter) transactions has a similar effect. Far from positively promoting new entry and competitive behaviour, many transition economies – especially in the CIS – still operate complex arrays of regulations involving numerous permits and often licences to set up almost any kind of new business. Not only are such regulations economically inefficient, but they facilitate corruption and often encourage new firms to operate in the informal sector (hence avoiding taxation and regulation altogether) or result in extremely low rates of new business formation. Thus rates of business formation in Russia are scarcely a tenth of the rates found in central and eastern Europe. As Commander et al. point out, this is not surprising when: "... in Russia on average it took shopkeepers almost four months of permission-seeking to set up in business."[216]

Yet although the connection is not immediately apparent, high rates of new business formation are actually crucial for the successful treatment of firms in distress. In regions with many firms and substantial new entry, for instance, the best way of dealing with firms in difficulties is simply to allow them to close as rapidly and in as orderly a manner as possible, with all remaining subsidies, credit concessions and tax favours withdrawn forthwith. It is far more effective, in the medium and longer term, for available public resources to be devoted to new firms, mostly for the provision of business-related infrastructure. But in areas with few firms, or dominated

---

[214] R. Frydman, C. Gray, M. Hessel and A. Rapaczynski, "Ownership and hard budget constraints in transition economies", *Economics of Transition*, Vol. 8, No. 3, November 2000, pp. 577-601.

[215] S. Commander, M. Dutz and N. Stern, "Restructuring in transition economies: ownership, competition and regulation", paper presented at the World Bank's ABCDE (Washington, D.C.), mimeo, April 1999.

[216] Ibid., p. 30.

by a single large firm, and with little new business entry, such a "hard line" approach is likely to prove politically unacceptable, regardless of its economic merits. The question then is where or how to start the process. Partly, it becomes important to reduce the resistance to change by lowering its perceived costs, e.g. through provision of a generous social safety net for displaced workers, including retraining and relocation assistance. This needs to be combined with measures to foster new firm creation in relatively depressed areas, a task that we know to be immensely hard from decades of experience in developed western economies. Realistically, given the costs and the difficulties, it is probably essential to adopt a targeted approach, either restructuring or shutting down these firms over a defined period. In the end, the resources thereby released can be used to stimulate a virtuous circle of productivity improvement and income growth. But designing an effective industrial policy to achieve such a delicate balance and get this process started is not easy.

## 3.5 Conclusions and policy implications

This paper has already covered a lot of ground and the purpose of this final section is to take stock, both by summarizing the main findings and conclusions and outlining the associated institutional basis, and then by drawing out what seem to me the most pertinent implications for policy – both for the transition economies themselves and perhaps more widely.

### (i) Findings and conclusions

In accord with the broad theme of the paper, I take it that our principal aim is to identify those aspects of the institutional framework of a market-type economy that are conducive to strong economic performance. The empirical work reported above enables us to highlight a few exceptionally important indicators of a flourishing economy – the first is essentially a macroeconomic condition while the next three are its microeconomic counterparts:

(i) high rates of domestic savings and investment;

(ii) high rates of new business formation;

(iii) sufficient, but not excessive competition;

(iv) effective arrangements for orderly exit by failing firms.

On item (i), while there are virtually no cases of economies growing rapidly while only saving at a low level in relation to GDP, there are examples of economies with high savings ratios growing slowly. Hence high rates of savings and investment are necessary but not sufficient conditions for growth. In addition, it is important that the investment that is undertaken should be efficiently selected, with regard to its economic effectiveness (i.e. in most cases, its profitability). Further, no economy can simply take the view that it will rely upon foreign direct investment (FDI) in order to fund its investment programme, and hence avoid the need to mobilize domestic savings. The notion that FDI can be brought in on a large scale where domestic savings are insufficient is a widely held myth, but a myth nonetheless. For economies to grow, and sustain growth, they must save, regardless of how poor they are at the start of the process.

To give an indication of orders of magnitude for the transition economies, assuming they wish to advance at rates that enable them gradually to catch up with the per capita income levels of the more developed market economies, they need to grow by 4-5 per cent per year at least (for several decades). To achieve this, savings rates of 25-30 per cent of GDP will be needed. For most of the region, especially in the CIS countries, this condition is a long way from being satisfied. The institutions and policies that can give rise to such high rates of savings, and transform them into efficient investment, are sketched below.

Taking items (ii), (iii) and (iv) together, the key point to re-emphasize is that overall economic performance has to relate back to what happens at the enterprise level. All economies are undergoing constant change, evolution and development, and how this translates into broad productivity improvements and higher income levels through adaptation at enterprise level – entry, exit and improved performance by surviving firms – is critical. Hence the institutions and policies that influence these processes are crucial for economic development; and as far as the economy is concerned, nothing else greatly matters.

### (ii) Institutions

In order to discuss the institutional underpinnings of the transition to a market economy reasonably compactly, I need to put on one side the whole issue of macroeconomic stabilization since its complex ramifications lie beyond the scope of this paper. Suffice to say that stabilization is undoubtedly a necessary condition for sustained growth, and that it requires reasonably sound monetary policy to keep inflation down, and a state capable of managing the public finances while avoiding large deficits. Among other desiderata, stable, predictable and collected taxes are important.

Besides stabilization, though, the macroeconomic condition (i) requires extensive development of banks, financial markets, and financial intermediation, both to mobilize savings and to direct it towards profitable investment opportunities. This simple statement entails a long list of requirements, from which I summarize only a few key ones (mostly without being too prescriptive, since in most cases there are many ways of achieving a given objective):

- a sound network of savings institutions/banks to hold and manage individual and small business deposits;

- competent, competitive commercial banks able to select projects and monitor loans for investment;

- a wide range of financial institutions and/or financial "products", e.g. various forms of insurance, pension funds and the like;

- a wide range of financial assets for individuals and firms to hold – government paper, commercial paper (both bonds and equity shares), unit trusts, etc.;
- effective regulation of the financial sector;
- appropriate investor and saver protection (striking a proper balance between the need to encourage investment and savings, and the need to ensure that those involved bear a reasonable share of the accompanying risks).

Clearly, the institutional infrastructure required for all this, and the competence/probity required to run it properly, can be quite problematic for a small country with little or no prior experience of operating modern financial markets. In such cases, there can be advantages in a group of countries working together to set up joint markets, or working through the already established markets of third countries.

Turning to the microeconomic/enterprise level aspects of institutions, it is worth starting by asking what the role of the state might be. Fundamentally, the evidence both from transition economies and much more widely, is that the state should be as disengaged as possible from the day-to-day functioning of the microeconomy in almost all sectors. States – including in the transition economies – have an exceptionally poor record of running businesses, and are inclined both to protect the firms that already exist, to restrict new entry, and to delay or block productivity enhancing restructuring. In a sense the problem is that states tend to be too "soft" and responsive to lobbying, and are hence far more tolerant of business failure than are markets. To move away from these conditions, the following points sum up what is required:

- extensive and well protected private property rights;
- effective legal protection for private business contracts;
- effective bankruptcy/liquidation legislation to facilitate the orderly closure of persistently loss-making firms;
- termination of most tax favours, subsidies, directed credits and other means to protect established firms in difficulty (i.e. the ending of "soft budget constraints");
- completion of privatization, possibly including a second round of the process to ensure that firms are mostly held by owners capable of supplying new funding, new technology and competent management;
- effective and credible regulation of pricing policy, customer service and investment in those sectors – whether the producers are privatized or not – of public utility type;
- institutions to facilitate entry by new firms;
- institutions to facilitate labour mobility and lower the costs to workers of essential restructuring.

### (iii) Policy implications

As shown in section 3.3, many of the transition economies have already undertaken a good deal of the institutional restructuring and innovation that they need in order to function as "normal" market economies, although it is clear that many also have a long way to go. In some instances, the steps already taken have created barriers to further institutional reform, and these barriers – political, economic and social – need to be overcome if market-oriented reforms are to advance much further.

In the published discussions of transition policies there is relatively little to be found on what I have identified here as a key issue, namely policies to foster high rates of domestic savings and investment. There is rather more on industrial, trade and competition policy, both describing what exists and making recommendations about what else should be done. Policies for these broad areas are sketched out below. In addition, since it has become so critical for 10 transition economies already, I remark on the EU institutional model being adopted by these countries as they prepare for accession.

### (a) Savings and investment

In addition to the institutional requirements already listed above, what other policies can help to boost savings and investment rates? It probably sounds anodyne and feeble to include "propaganda" here, but it is worth doing because of the prevailing view in many transition economies (which I have personally encountered many times) that they are "too poor to save", and merely need "massive injections of foreign investment" to put them right. As I implied above, this line of thinking is pure fantasy, and repeating it does not make it any more sensible. Thus, a vital educational role for governments in the region is to stress repeatedly the importance – for individuals, for firms and for the whole society – of high savings rates, and to build the institutions to correspond to that. In Asia, the examples of China and Singapore come to mind, but there are many others that one could point to, where high savings have played a critical role in bringing them up from the deepest poverty to medium income levels within a generation or so.

High savings, however, only deliver if most investment turns out to be profitable, and ensuring that is a more complex task than merely mobilizing the savings in the first place. On the other hand, high savings and profitable investment are likely to be mutually reinforcing since individuals and firms will see their savings generating returns that will encourage them to go on saving. But what role can government policies play in creating the conditions for investment to be profitable?

At the macroeconomic level, perhaps the most important requirements are to maintain stability of prices and the exchange rate, and to keep the public finances in order, with current spending on goods and services, plus net transfer payments, plus interest on accumulated government debt, kept in line with government revenues.[217] At the microeconomic level, much more

---

[217] Here I am being deliberately brief for lack of space. But my comment should be taken to include all off-budget government commitments and funds, and all levels of government.

depends on the individual investors, although they can be helped by a combination of measures:

- substantial deregulation of the process of authorizing or giving permission for a given investment project to go ahead; the simpler and quicker this is the better, both to reduce opportunities for corruption, to reduce involvement in the economy by a generally incompetent state and to enhance economic flexibility. I acknowledge, of course, the need to have some controls over business location, health and safety issues, building regulations and the like, but much of this can be handled in a far less bureaucratic and more permissive manner than is usual, especially in the CIS. What regulations there are should be openly published, fairly implemented and open to timely appeal in cases of dispute;

- effective monitoring of investment projects by banks and/or shareholders, to ensure that when investments fail (as many will, as in any normal economy) they are stopped quickly, funding is reallocated to other activities and any equipment already in place is disposed of for others to use as productively as they can. Governments should rarely step in to perform rescue operations; nor should they lean on the banks to do so. On the other hand, activities that turn out to be successful, perhaps unexpectedly so, should quickly have access to additional funds so that they can expand.

*(b) Industrial, trade and competition policy*

When states are weak, incompetent or corrupt, as is the case for many transition economies, especially in the CIS region and in the Balkans, it is pointless to put forward recommendations for complex and sophisticated policies. Instead, whatever policies are advanced should be simple and, where possible, self-enforcing. The state has to set the "rules of the game" which the business community then follows, without providing opportunities for all sorts of *ad hominem* bargaining, deals and corruption.

To illustrate what I mean, I give some examples of very bad policies, and then move on to make some positive recommendations. For instance a trade policy based on a uniform tariff or a small number of standard rates covering broad commodity groups is far better than many of the existing tariffs which have several thousand separate rates, often with little or no economic rationale to justify them. A highly complex tariff is vulnerable to very easy manipulation by importers who can simply bribe customs officials to reclassify goods to obtain a lower tariff rate; this practice is widespread, and extremely inefficient.

Second, a policy to support small businesses that requires them to approach 20 different offices for permissions connected with different aspects of their business, in circumstances where most of the rules are not published and are subject to arbitrary amendment, invites corruption. At least 20 officials might have to be bribed, and even then the person setting up the business may not feel secure in their rights to do so. In these conditions, many businesses will not even get off the ground, let alone flourish.

A third example concerns the inspection of already existing businesses. If, as in Russia and Ukraine among others, many public agencies have the right to inspect businesses with little or no notice, checking their accounts, their tax payments, compliance with health and safety rules, labour regulations, and so on, then again this can prove a fruitful source of bribes for officials (as well as consuming valuable management time), simply to get them to go away and allow normal business to continue. Inspection is not always a bad thing, but it should be openly conducted, based on clear, published criteria, with written reporting of any problems identified, and proper appeal procedures.

Last, the toleration of non-cash settlement either of tax payments or of inter-enterprise sales is almost always very inefficient. Barter generally only arises in environments where governments are too weak to enforce cash payments, or where other institutional deficiencies encourage it. For instance, in several transition economies, taxes from enterprises are collected through the agency of the banks, and this very bad institutional arrangement gives a positive incentive for firms to evade taxes by engaging in non-monetary transactions where they can. Likewise, the toleration of non-payment of energy costs by public agencies (e.g. in Russia) contributes to a climate in which non-payment is considered acceptable (with the damaging implication that some users of energy can treat it as a free good!), and encourages the energy utilities to engage in bargaining with the government over their tax liabilities.

Turning to more positive aspects of industrial, trade and competition policy, there is some useful recent literature that draws out lessons from relevant western experience, notably Elsner and Groenewegen,[218] and Cowling.[219] The former argue that industrial policy should focus on securing benefits for the home country, limiting the extent of siphoning off of incomes by transnational corporations; that it should support local companies that are not "footloose" in preference to multinational firms and foreign investors; and that rather than supporting specific businesses, most attention should go towards building up the business infrastructure and related institutions. Cowling argues against traditional, sectoral policies that would involve governments trying to "pick winners", usually unsuccessfully, but proposes support for what he calls "local selection processes". These might involve a sectoral aspect, but the choice of sector and mode

---

[218] W. Elsner and J. Groenewegen (eds.), *Industrial Policies after 2000* (Dordrecht, Kluwer Academic Publishers, 2000).

[219] K. Cowling (ed.), *Industrial Policy in Europe: Theoretical Perspectives and Practical Proposals* (London, Routledge, 1999).

of support would be determined more locally, more democratically than in the traditional approach to industrial policy. It is not wholly clear to me how this would work in practice, but in spirit it fits in with the analysis of productivity improvement reported in section 3.4(ii) above.

On the trade policy of transition economies, I have earlier made a comprehensive survey[220] of country-by-country experience, with advice on constructing sound trade policies (including WTO membership and the disciplines it imposes), allowing for different stages in the overall reform process and different degrees of competence/honesty in the governments concerned. Further, Brown and Earle[221] argue that reductions in import barriers and geographic market segmentation – the latter being especially significant in the geographically larger transition economies – should encourage more rapid growth of industrial productivity by encouraging the right kinds of competition.

In relation to labour market aspects of industrial and social policy, Boeri offers some interesting suggestions, learning from the mistakes made in certain countries in the early transition years.[222] He points out that there is a high welfare/high tax model that has been followed to a large extent in the candidates for EU accession, and a low welfare/low tax model that has been followed elsewhere, notably in the CIS. It is argued that for all its weaknesses of implementation, the first model has proved itself unambiguously superior, since the low welfare model has seriously discouraged enterprise restructuring in many transition economies. However, there is an issue of feasible choices that Boeri does not directly address, in the sense that it might simply not have been possible for the CIS countries to choose the high welfare model because of their inability to raise sufficient tax revenues. Hence these countries might well have been politically constrained to develop low welfare models, despite their (known) damaging consequences. If this is correct, then a large research agenda is immediately opened up, to explore how countries initially trapped in such a bad model can find ways out into a better equilibrium. Such an investigation extends far beyond the limited scope of the present paper.

### (c) EU institutional model

Boeri also refers to aspects of EU accession and the credibility of the whole process that have an important bearing on the institutional focus of our discussion.[223] For the accession countries themselves, it can be questioned whether the institutions and practices they are taking on as they gradually adhere to more and more of the *acquis*

*communautaire* are good for their development at its present stage. And for non-accession countries, one might ask whether something like the "EU model" provides a conception of the institutional structure for a market-type economy that merits wider emulation – partly because it is a ready-made and working model, partly because adopting it lowers the otherwise high costs of rethinking all aspects of the institutional structure from scratch, partly for the practical reason that adoption might facilitate trading and other economic links with the EU.

These arguments are important for both sets of countries, but for the accession countries the overwhelming argument for adopting the *acquis* must be the political one, namely that doing so helps to lock these countries firmly into the western system of alliances. So far, only the Czech Republic, Hungary and Poland have already joined NATO and are members of the OECD. Thus, for all 10 associated states EU entry will be a very important step. Hence it is regrettable that the entry process is turning out to be so prolonged now, with a few countries ahead of the game in the negotiations process (Czech Republic, Estonia, Hungary, Poland, Slovenia,), the others (Bulgaria, Latvia, Lithuania, Romania, Slovakia,) lagging far behind in terms of their satisfaction of the Copenhagen Criteria for accession, and even more in terms of their detailed compliance with the *acquis*. But if joining the EU is politically important for these countries, then it must surely happen, and be seen to happen. Instead, what we are witnessing at present in Brussels is a laborious series of intricate negotiations on detailed parts of the *acquis*, much of which, in my view, could safely be left for post-accession discussions.

As Boeri rightly stresses, continuing delays over accession reduce the credibility of the EU "project" in the transition economies, and are likely to undermine or delay their willingness to undertake the associated institutional reforms, especially as these are both complex and costly. The EU's institutional structures have evolved to suit a group of relatively developed economies with high incomes, accustomed to steady but quite slow economic growth. For the transition economies, however, seeking to catch up with the more advanced countries within a reasonable period, there must be substantial chunks of the *acquis* that might not be appropriate for them. Most obvious is the environmental component of the *acquis*, because of its high cost, but many of the institutional arrangements and policies operated by the EU are very complex and require well-trained staff, relatively free from corruption, to work effectively.[224] For much of the transition economy region, simpler policies with a better chance of being implemented properly would actually be more effective for the countries concerned.

---

[220] P. Hare, "Trade policy during the transition: lessons from the 1990s", *The World Economy*, Vol. 24, No. 4, April 2001, pp. 483-512.

[221] J. Brown and J. Earle, *Competition and Firm Performance: Lessons from Russia*, CEPR Discussion Paper, No. 2444 (London), May 2000.

[222] T. Boeri, *Structural Change, Welfare Systems and Labour Reallocation* (Oxford, Oxford University Press, 2000).

[223] Ibid.

[224] Some of the EU policies and recommendations concerning the regulation of public utilities, e.g. electricity supply or telecommunications, are nice in theory but rely on strong, credible government. Where that is missing, quite different approaches are likely to be called for.

# DISCUSSANTS' COMMENTS ON PROFESSOR HARE'S PAPER

## 3.A Danica Popovic

My first and final impressions of the paper are the same: Professor Hare has provided us with an ambitious, comprehensive and rather useful guide to the theory, policy and possible outcomes of institutional change in transition economies.

My comments are as follows:

### Concepts and definitions

The question that remains unanswered here is the following: What is the driving force of an institutional reform? There is no single answer to the question, neither in this paper nor in the many others in this field. It seems that the driving force is still a mystery. We know more about the stones on the planet of Mars than how institutions develop.

Apart from that, the notion of institutions as a public good is rather compelling. However, the insufficiency of the market supply of institutions does not seem to fit the theory. Namely, the state itself is effectively a market-supplied institution. Following Buchanan's contractual theory of the state[225] and Hayek's postulates on natural order,[226] the state was created by a collective action of various agents, but essentially in response to market demand. This obviously contradicts Professor Hare's statement concerning the insufficiency of supply of institutions (as a public good).

Coming back to more pragmatic lines, we discover that when the state steps in to create institutions no one can be sure that it will be done in the right way. The crucial question is what is the incentive for the state to create proper institutions. Is it political, based on improving relations with the EU, for example, or is it economic progress? Usually the link between good institutions and good economic performance is not too close: there is a substantial lag, so that progress cannot materialize in one term of the government. Again the issue of the driving force is not clear. Finally, there is enough evidence that bad institutional reform (or the lack of it) can be linked to bad economic performance. The crucial issue is what is the incentive to the government to trigger the institutional reform, taking into account that the results will only appear with a lag. What happened in those countries where things went well and what did *not* happen in the others?

Finally, there is no doubt that economic behaviour with missing institutions can be devastating. There are substantial economies of scale in providing law enforcement so private enforcement, apart from being biased, is very costly. But the mechanism of devastation is not disclosed. The origin of the mafia in Sicily is a much better example of private enforcement of rules than Russia today. Namely, following the line of Anderson,[227] three types of mafia can be recognized. The Sicilian case is a perfect example of the rise of a mafia in a vacuum of power, in a region that turned away from state power to private means of protecting property and ensuring order. In contrast, the Russian mafia originates in excessive bureaucratic power rather than the lack of institutions, where government agencies and people with power exercise it over the distribution of resources and entry into legitimate business, sometimes with the help of gangs. Only ethnic mafias in the autonomous regions of Russia follow the Sicilian pattern.[228] Essentially, this does not go beyond the research in this paper, simply because it implies the restructuring of government bureaucracy and the rebuilding of institutions by enforcing new rules. The case of Serbia might be a good example of mismanagement in fighting an oil-importing mafia (also arising from excessive bureaucratic power) via the introduction of a state monopoly of oil imports. Namely, it was more plausible that this move would create a new mafia (like the Russian GAZPROM) rather than destroy the previous one.

### A "moderate degree of competition"

A "moderate degree of competition" is an interesting concept. The only problem with it is that it is very difficult to define the yardstick. What degree of competition can be labelled as "too much"? There was no attempt to specify that even theoretically, let alone practically. The second problem, more on the normative side, is what the author tells us about the adverse effects of too much competition.

One concept is that "too much competition" is bad for the economic advance of the transition economy (country), and the other is that too much competition is not beneficial for the economic restructuring of the enterprises. They are not the same. It is quite understandable that tough competition is not good for

---

[225] J. Buchnan and G. Tullock, *The Calculus of Consent: Logical Foundations of Constitutional Democracy* (Michigan, Ann Arbor, 1962).

[226] F. von Hayek, *The Road to Serfdom* (London, Routledge, 1944).

[227] A. Anderson, "Theories of the state and the origin of criminal organizations", in G. Fiorentini and S. Peltzman (eds.), *The Economics of Organized Crime* (Cambridge, Cambridge University Press, 1995).

[228] The American mafia is rooted in the management of illegal markets.

enterprise restructuring, because new entrants may force the exit of an incumbent firm rather than allow it enough time to restructure. But development is quite consistent with strong competition – the exit of less efficient firms and the entry of more efficient firms. There is nothing wrong with that. Why too much competition is bad for economic advance remains a mystery. As Professor Hare stated in his paper, Carlin et al.[229] provide some clues, that too much competition can: (i) restrain firms' ability to finance needed investments (pretty unclear, both from the positive and normative aspect); and (ii) can confuse the market signals that guide enterprises (even more unclear). Obviously both clues have nothing to do with Baumol's theoretically sound concept of destructive competition.[230]

### Microeconomics

"High rates of new business formation are actually crucial for the successful treatment of firms in distress". Good point, although the euphemism "successful treatment" should be switched to scrapping. The crucial point is sequencing. If no firm is in distress, or if the government would like to preserve these firms one way or another, there is no incentive for the government to create a good environment for new entrants. The government will focus on various ways to keep the firm going. On the other hand if the firm in distress goes bust with substantial labour shedding, the government (local or central) has incentives to create a good environment for new entrants. Čačak in Serbia is a good case. First the local big company (a blend of military and civilian production) went bust, then it was completely destroyed (both military and civilian production) by NATO air raids. All labour was shedded and the only way to generate new employment was via new entrants. In other words, local government had a very strong incentive to implement institutional reform to foster new entrants.[231] Anyway, the task of the government is to insist on the inevitability of labour shedding. Again, the question of the funding of a generous social safety net without the "myth of FDI" or explicit foreign aid, remains both painful and unanswered.

### The FDI myth and domestic savings

Many authors would oppose Professor Hare's thesis of the FDI myth, but my opinion is that it should be considered very carefully.

My first objection is that this statement is not founded in any data or consideration in the paper. The data might indicate a somewhat different conclusion – a credible path of reforms might have attracted FDI, thus reinforcing each other, raising savings, exports, growth and new investment.

My second objection derives from the key macroeconomic accounting identity $(S-I) = (G-T) + (E-M)$. If the government must run a balanced budget, and a reasonably small foreign trade deficit (say, up to 3-5 per cent GDP), there is no room for domestic savings to foster growth, at least not at the beginning, whereas FDI (rather then international loans) really can do the job. Why? Simply because it is not only money (although this is beneficial for creating a "generous social safety net", pension funds, regional imbalances, etc.), but even more it brings: (i) technology and know-how; (ii) management skills; and (iii) opens export markets which are deficient in post-socialist countries. FDI seems to be both inevitable and welcome in this respect.

Moreover, it remains unclear in the paper as to what specific element of saving should increase. Following the paper, government saving must rise since "stable, predictable and collected taxes are required" for the macroeconomic purposes of balancing the budget. Following also the very good point that the high welfare/high tax model is superior in regard to economic and enterprise restructuring, we come to the following problem. Namely, this implies: (i) a rise, rather then a fall, in government saving; (ii) a crowding out effect; and (iii) a fall, rather than a rise, in domestic investment – inevitably leading to a drop in private savings for the simple reason that one cannot reduce consumption as much as this action would require. If the government seeks to maximize the saving rate at all costs (which the paper obviously suggests), that would unduly punish current consumers. One cannot expect to foster a sustainable growth path by reducing consumption, which turns out to be the only source for increased savings.

One of the problems with raising private savings is that banking reforms are costly and take a long time, and the domestic banking sector lacks both funds and credibility. As Hunya[232] shows, in recent years especially, the banking sector has attracted significant amounts of FDI. Also, the difference in savings rates between the EU and the transition economies, which as Professor Hare states roughly corresponds to the ratio of banking assets to GDP, amounts to 60 per cent in transition economies and 250 per cent in the EU. If FDI is a myth, there may be no way out for transition economies at all. Unfortunately, there is no way to follow the experience of Asia (a continent), China and Singapore (a city) in this domain, primarily due to the need to build EU-type institutions.

It seems to me that the mechanism is as follows: (i) credible reformers do attract FDI and thus escape from the "impossibility theorem" of increasing both government and private savings at all costs. Only then (ii) does investment foster growth, which raises

---

[229] W. Carlin, S. Fries, M. Schaffer and P. Seabright, "Competition and enterprise performance in transition economies: evidence from a cross-country survey", University College London, mimeo, June 2000.

[230] W. Baumol, "Entrepreneurship: productive, unproductive and destructive", *Journal of Political Economy*, Vol. 98, No. 5, 1990, pp. 893-921.

[231] Again, a quite unexpected institutional development – a collateral benefit of NATO raids.

[232] G. Hunya, *Recent FDI Trends, Policies and Challenges in South-East European Countries,* The Vienna Institute for International Economic Studies, Research Reports No. 273 (Vienna), December 2000.

consumption, only then in turn (iii) making room for private savings to emerge and grow. So, the key point for a country is to be a credible reformer, since there is no other way to attract FDI, and hence, there is no other way to effect a successful transition.

This is completely in line with the authors' remark that "institutions and policies that influence these processes are crucial for economic development; and as far as the economy is concerned, nothing else matters". Nor does the dilemma of whether FDI is a myth or not.

In addition, a very good suggestion is that countries where the civil service has limited capabilities should not implement complicated industrial policies or trade protection. A straightforward rule of law is definitely more efficient and easier to handle. So far, this is also in line with the policies enacted by the Serbian (Yugoslav) government, but the results are yet to be seen.

## 3.B László Csaba

### Institutions and social learning

Paul Hare presents a masterful summary view of both the policy issues and outcomes of 12 years of transition and the theoretical implications of these controversial issues for the broader theoretical perspective of economic development. The author takes a middle ground in the international debate on the role of policies versus institutions in bringing about sustainable development in the post-communist world, but also in less developed countries in general. While adherents to the post-Washington consensus tend to emphasize the role and responsibilities of policy makers as the prime explanatory variable, adherents of the path dependency school underline that only those countries where history and social endowments allow for the "right policy mix" to become politically feasible will "graduate" from the state of "vicious circles" to "virtuous circles". While subscribing to Hare's conclusion that policies do not work without institutional foundations, and conversely, institutions are not a substitute for common sense policies and sequencing, it seems overambitious to state that the key role of institutions was acknowledged from the very outset. Initially the debate did centre around the crucial issue, if and when policy should precede, substitute or complement institutional arrangements that were known to require long periods, decades rather than months, for their completion. While some contemporary analysts argued for policy first, others highlighted the decisive role of institution building in rendering good policies sustainable and non-reversible (for contemporary expositions see the contribution of Anders Åslund for the first line of argument, and Grzegorz Kolodko and Gerard Roland for the second).[233]

The debate should not be trivialized, even with hindsight, for a number of reasons. First, as Hare lucidly explains, calls for radical policies were often based on inadequate knowledge of circumstances and disregard of feasibility conditions. On the other hand, it would be hard to deny that references to institutional arrangements and the need to consider historical legacies have frequently been invoked by those opposing the entire rationale of radical transformation and the ingredients of what Hare lists among the components of "solid" policies.[234] The diversity of the transforming countries, an issue correctly highlighted in the paper, has not always been appreciated either by theorists or by policy makers. There may be many reasons for this. But obviously, the quest, inherent in economic theory, for generalizations, and the fear of policy analysts and makers to create new divisions in the newly reintegrating Europe, might have been superimposed on what are inherent features of the subject of our analysis. As a consequence, little or no attention has been paid to the different starting points and the different institutional heritages of the post-communist countries. This has backfired, especially in the less developed and in the disintegrating transition countries, where the crucial issue of state failure, correctly underscored by Hare, has not even been considered in most analytical and policy papers. This is all the more regrettable, as state failure may, and often is, at the root of market failure, which in turn calls for even more state intervention of dubious quality and efficiency – a model well demonstrable in Russia,[235] but probably also in other CIS and south-east European states as well.

One of the most insightful sections of the paper is the one dealing with the state's inability to create institutions at its will, and conversely, the often perverse effects of private initiative in substituting for formal institutions. This is clearly at odds with the naive beliefs of the early transition period, when the euphoria following the collapse of the "evil empire" allowed for a great deal of theorizing on what were the optimal institutional arrangements "to be chosen": some tended to preach the minimal state, implying that private initiative would inevitably produce superior alternatives. The findings cited by Hare reflect both of these widely propagated suggestions and call for a more conservative appraisal of the role of public institutions in bringing about the rule of law, transparency of regulations, and the self-propelling evolution of what he calls "trust" and others call social intangibles, i.e. those elements of capital formation normally not represented among the quantitative variables of econometric models, but still generally accepted (e.g. by the business community) to play a major role in the way established and civilized markets function.

There is still no consensus – and probably will never be – on the precise mechanics of the virtuous circle,

---

[233] L. Csaba (ed.), *Systemic Change and Stabilization in Eastern Europe* (Aldershot, Dartmouth Publishing Co., 1991).

[234] See, for example, K. Poznanski, "Post-communist transition as institutional disintegration: explaining the regional economic recession", *Acta Oeconomica*, Vol. 50, Nos. 1-2, 1999, pp.1-36.

[235] V. Mau, "Russian economic reforms as seen by an insider", Royal Institute of International Affairs/International Economic Programme (London), June 2000; L. Csaba, "A decade of transformation: Russia and Hungary compared", in A. Brzeski and J. Winiecki (eds.), *A Liberating Economic Journey* (London, CRCE, 2000), pp. 99-139.

whereas there is ample evidence, from other parts of the world with poor quality or no institutions, as in Africa as to how the vicious circle works.[236] Lacking fundamental institutions, such as a state monopoly of violence and the existence of private property, no amount of natural resources, land, manpower or physical capital will lead to development if the incentives to predatory behaviour outweigh those for accumulation and investment. This is by and large what we find as well in the story of state failure leading to market failure. But probably it would be excessive to expect the author to offer solutions going beyond the conventional proposition, mostly based (implicitly at least) on the "Daddy State" concept, or on attributing more powers and capabilities to both the administration and polity of new democracies than these may realistically possess. What international business people and transparency watchers deplore as corruption is nothing more (or less) in traditional societies than the only way of sustaining public authorities, and those working in them, by creating and inviting various forms of self-financing. What standard textbooks describe as a bribe may in these cases be equivalent to direct taxes or simply payments for skills and services employed but not fed by the formal state structures. And certainly, a lack of transparency is in the best interest of all those operating under such circumstances (on both sides of the counter), and not only for the fear of prosecution. Such arrangements pave the way for, and even invite, directly unproductive profit-seeking activities, that are estimated in some developing countries at 5-7 per cent of the GDP per annum.[237]

It is less surprising to see these phenomena re-emerging in the post-communist context. The problem arises more in the normative part of the analysis, since for developing countries the literature usually calls for creating a strong and democratic state, and cooperation with international agencies and investors. If we accept that at the heart of the failure to bring about sustainable – or mostly any – economic growth in post-communist countries is a weak state, we may lack the Archimedean point for overcoming the crisis, at least through conventional statist developmental methods. From this angle, steps such as President Putin's legislative initiative to create private property for arable land, approved by the Duma in October 2000, or even his crusade against the oligarchs, seem to make sense, provided the latter serves more than redistributory and disciplinary aims (as many analysts interpret them). But the evidence in many countries suggests that an aggressive state is not necessarily an efficient one, and political activism is frequently used in less developed countries to camouflage the lack of administrative efficiency. Thus, it would probably take a Baron Munchausen to find a logically coherent and practicable solution.

Fuzzy institutions lead not only to state capture but also, as a consequence, to a network-dominated economy in which inherited and newly acquired power positions are used for rent-seeking and the creation of monopolies irrespective of the formal legislation.[238] The theoretical framework referred to above may convincingly explain a large part of the inefficiency of post-privatization enterprise behaviour, as well as the reasons why the traditional mechanisms formulated in the Coase theorem are not dominant in the post-socialist context. Therefore, the resemblance of pre- and post-transition power structures for large parts of the population and the ensuing lack of legitimacy of the new arrangements – both of democratic pluralism and private property – is less than surprising. The wider is the scope for the irregular economy, and the less we can see, even in theory, the dividing line between the official, grey and unofficial sectors, the less is the credibility of dual track policies, such as those promulgated rather than employed in China – a proposition that has been contentious from the very outset.[239] Any dualism, to make sense, would presuppose effectual enforcement mechanisms for separating the various sectors – the lack of which is the empirically observed foundation on which the entire network-based approach is built. Jumping to the policy conclusions for a moment, it is plausible that the Archimedean point for reversing the developmental vicious circle should indeed be seen in fostering the elementary trading and financing activities that emerge spontaneously at the lowest possible level of development, assuming that the public authorities do not forcibly extinguish them.[240] Building on this evolutionary process, and adopting a first-things-first approach also in state building, certain of the elementary conditions of state power and democracy can be built. This, in turn, may start the virtuous circle (more on that later). The question, however, in the post-communist context, is whether societies that commonly see democrats as thieves, and private property as a result of predation, can learn or be "re-educated" in a spirit that would be conducive to the virtuous circle solution, or, rather, whether they are more likely to be lastingly trapped in the vicious circle – a proposition which is more dominant in the social sciences.

Criticizing the overzealous attempts of the early transition years, Hare rightly deplores the disregard of the first-things-first approach in policy advice and policy-making alike. With reference to Stiglitz he seems to belittle the role of capital markets in transition economies – a proposition I tended to share myself earlier. While

---

[236] P. Collier and J. Gunning, "Explaining Africa's economic performance", *Journal of Economic Literature*, Vol. 37, No. 1, March 1999, pp. 64-111.

[237] J. Bhagwati, "Directly unproductive profit-seeking activities", *Writings on International Economics* (Delhi, Oxford University Press, 1998).

[238] L. Bruszt and D. Stark, *Post-socialist Pathways* (Cambridge, Cambridge University Press, 1998).

[239] O. Bogomolov (ed.), *Market Forces in Planned Economies* (Basingstoke, MacMillan, 1989).

[240] P. Bauer (1992), "Subsistence, trade and exchange: understanding developing economies", reprinted in J. Dunn, S. Hanke and A. Walters (eds.), *The Revolution in Development Economics* (Washington, D.C., Cato Institute, 1998), pp. 275-284.

agreeing completely with the need to take a bottom-up approach to institution building and policy measures, moving from simpler to more sophisticated tasks, it is hard to see how else capital can be injected in the banking sector, how regulation can be made credible, and generally, how money, technology and management can be secured for major, pace-setting investments if not by reliance on foreign money, foreign owners, and on readily available, thus tested and credible, pieces of legislation. The more we accept the argument that not only high rates of investment, but also efficient allocation is needed, and that banks and other intermediaries play a vital role in monitoring microlevel restructuring and project selection, the more we appreciate the role of foreign direct investment, foreign ownership and – last but not at all least – the crucial role of the capital market. The latter acts both as a disciplinary mechanism against imprudent expansion and a decentralized mechanism enabling business ventures to raise money irrespective of what the government believes to be the desirable allocation of funds. The more large firms can go international, both as recipients and investors, the higher is the probability that market-based selection will actually materialize. Moreover, the more firms can raise money abroad, the less the backwardness of domestic financial intermediation and all the inherited problems, discussed in the vicious circle model, are constraints on their development. Should the largest firms be able to rid themselves of these concerns, as they seem to have done in Hungary in the 1993-2000 period, the higher is the endogenous growth potential of the economy. Their growing reliance on external funding enables them to grow before all their domestic problems are settled. Their growth may set the pace for others, and their adoption of international business practices may civilize otherwise traditionalist local business practices. Last but not least, if the domestic capital market is from the very outset integrated with the international market, the controlling function of standards, disclosure requirements and many other components of solid business practices is likely to spread and become dominant earlier than would be the case via self-learning and reinvention.

The more we buy the argument of the author on the crucial importance of enterprise level responses and the microeconomic foundations of sustainable growth, the more we have to disagree with his scepticism as to the role of foreign direct investment. He is, of course, right, in calling it a myth insofar as foreigners are assumed to substitute for domestic savers. However, FDI has been acting as a pace-setter, a point of orientation and even as an indicator of the speed and depth of transformation, both at the macrolevel[241] and at the level of microeconomic adjustment.[242] The bottom line is not the quantity of saving that can be substituted for consumption, but the presence of strong owners, new technologies, and joining in inter-firm division of labour on a global scale, a precondition of entry into the markets for more sophisticated goods and services. I share, of course, the author's view that sustaining high savings rates are a must for lasting growth, and overall policies should indeed be geared towards meeting this requirement (although this is easier said than practised, especially around elections). But also the quantity of FDI matters: if one thinks of China attracting $40-$45 billion of FDI annually in a poor economy, it is clearly not irrelevant for financing growth.

On the whole I tend to share the author's quest for the microfoundations of macroeconomic performance and also his emphasis on the entry and exit of firms being crucial to success. I would perhaps add the role of non-creative, standardized international accounting both at the macro- and microeconomic levels, for banks and corporations alike. Hare's finding is all the more relevant, especially with the benefit of hindsight, in that one of the most politicized controversies in Hungary in 1992-1994 was the one blaming the allegedly too harsh arrangements fostering exit and the too ambitious application of international accounting standards, which brought down some of the biggest banks. A decade later it is clear that postponing the process of market clearing, or trying to substitute for this process with grand revitalization programmes managed by governmental departments led nowhere in all countries.

Last but not least I personally would tend to be less Eurosceptic than the British author on taking over the *acquis*. Briefly, much of the substance of the *acquis* is by and large in line with the inherent/endogenous tasks of transforming economies, particularly in central Europe. True, we may perhaps know that a 10 per cent rate or more of inflation is not a very good indicator, or that 5-7 per cent real rates of interests may be too high, or that environmental standards should have been enforced even before the fall of communism. However, the polity tends to be more responsive to common sense when it comes from a mighty and respected partner. Thus, in this respect, I share the author's view of the EU accession being politically advantageous. It would be tempting to reflect upon many other aspects of this truly rich and illuminating paper. The fundamental question is obviously what we should think about the abilities of societies to learn, and what are the grounds on which we may base our judgements. It seems there are more questions than answers. A complete inability to learn would support the vicious circle model. A good ability to learn is implicit in the arguments in favour of the frontrunners for EU accession. But the big question – i.e. what makes one country more able to learn than another, or what makes certain periods in the same country more conducive to efficient reforms than another – remains to a large degree open, despite individuals' unquestionable ability to learn.

---

[241] L. Csaba, "Commonalities and differences in transition", in K. Maenicke (ed.), *Zehn Jahre Wende in Osteuropa* (Frankfurt, Verlag Lang, 2001).

[242] G. Hunya (ed.), *Foreign Direct Investment and Restructuring in Eastern Europe* (Cheltenham, Edward Elgar, 2000).

# CHAPTER 4

# CORRUPTION IN ECONOMIC DEVELOPMENT: GREASE OR SAND?

*Shang-Jin Wei*[243]

"If you look under most banking crises, there's always a degree of fraud and abuse, and there's often a large amount of criminal activity. Corruption threatens growth and stability in many other ways as well: by discouraging business, undermining legal notions of property rights and perpetuating vested interests."

*Lawrence Summers
Speech to the Summit of Eight, Denver, 10 June 1997*

"In terms of economic growth, the only thing worse than a society with a rigid, overcentralized, dishonest bureaucracy is one with a rigid, overcentralized and honest bureaucracy."[244]

*Samuel P. Huntington*

## 4.1 Introduction

Corruption scandals are recurrent news items around the world. Like cockroaches, corruption has been with human society for a long time and is still widespread even beyond the transition economies. Does corruption stifle the process of economic development and transition? Or does it actually provide "grease" that smoothes the path of economic progress? As indicated by the two conflicting quotes above, one can in fact find very different opinions from intelligent people. One can probably find anecdotes to support both (mutually inconsistent) statements. There appear to be examples of value-destroying corruption (as the first quote suggests) as well as value-creating corruption (as the second implies). However, there is a limit to what anecdotes can tell us. What does a careful examination of the facts and data tell us? The first purpose of this paper is to review recent studies of the consequences of corruption on economic development.[245] The second purpose is to reflect on the effectiveness of various anti-corruption reform strategies, including my own proposal of how to start a reform to overcome political risk, financial affordability and other difficulties.

This paper is organized in the following way. Section 4.1 discusses how cross-country difference in corruption may be measured. Section 4.2 reviews the evidence of the consequence of corruption on economic growth. Section 4.3 makes the argument and presents evidence that anti-corruption is becoming more important as the world economy becomes more globalized. Section 4.4 discusses how cultural differences affect the consequences of corruption. Section 4.5 discusses factors that may contribute to the different extent of corruption in different countries, and looks at possible remedies to the problem. Section 4.6 provides some conclusions.

## 4.2 Corruption: which corruption?

This paper focuses on corruption in the economic sphere involving government officials. Corruption here is defined as government officials abusing their power to extract or accept bribes from the private sector for their

---

[243] I thank Ivo Bićanić, Gunnar Eskeland, Daniel Kaufmann, John Montgomery, Denis Osborne, Pasuk Phongpaichit, Susan Rose-Ackerman, Krassen Stanchev, Vito Tanzi and participants at the Spring Seminar of the United Nations Economic Commission on Europe on 7 May 2001 for helpful discussion. However, I am responsible for all the views and errors in the paper.

[244] S. Huntington, *Political Order in Changing Societies* (New Haven, Yale University Press, 1968), p. 386.

[245] Classic work in this area includes A. Krueger, "The political economy of rent-seeking society", *American Economic Review*, Vol. 64, No. 3, June 1974, pp. 291-303; S. Rose-Ackerman, "The economics of corruption", *Journal of Public Economics*, No. 5617, June 1975; idem, *Corruption: A Study in Political Economy* (New York, Academic Press, 1978); F. Lui, "An equilibrium queuing model of bribery", *Journal of Political Economy*, Vol. 93, No. 4, August 1985, pp. 760-781; A. Shleifer and R. Vishny, "Politicians and firms", *Quarterly Journal of Economics*, Vol. 109, No. 4, November 1994. Previous survey papers include J. Andvig, "The economics of corruption: a survey", *Studi Economici*, No. 43, 1991, pp. 57-94; P. Bardhan, "Corruption and development: a review of issues", *Journal of Economic Literature*, Vol. 35, No. 3, September 1997, pp. 1320-1346; D. Kaufmann, "Corruption: some myths and facts" (an earlier version was published in *Foreign Policy*, Summer 1997, pp. 114-131); UNDP, *Corruption and Good Governance*, Discussion Paper No. 3 (prepared by S. Rose-Ackerman), 1997; and V. Tanzi, *Corruption Around the World: Causes, Consequences, Scope and Cures*, IMF Working Paper WP/98/63 (Washington, D.C), December 1998. Several new features of the current paper are worth pointing out. First, it emphasizes data-based studies rather than mere theories. Second, as the research on corruption has expanded rapidly in the last half decade, it reviews more recent studies including those using firm-level observations in addition to cross-country regressions.

personal benefit. This is to be distinguished from political corruption (e.g. vote-buying in an election, legal or illegal campaign contributions by the wealthy and other special interest groups to influence laws and regulations), and bribes among private sector parties. The other types of corruption are undeniably important as well for economic transition and development. But due to the need to stay focused and to the limitation of my expertise, this paper reviews the evidence primarily concerning official corruption as I have first defined it.

Given the very nature of corruption (its secrecy, its illegality, its variation across different economic activities), it is impossible to obtain precise information on its extent in a country: it cannot, for example, be measured like inflation. This difficulty also precludes a precise grading of countries according to their relative degree of corruption.

Moreover, what is called "abuse of power" or "corruption" in one cultural context may not be so labelled in another. However, just because something has a long history, is widespread and is difficult to control, does not justify ignoring it. The cockroach has been with human society for a long time and is widespread; but we still have to have pest control. Of course, the exact boundary of what is considered "corruption" may depend on culture. A survey in Thailand in the early 1990s revealed that the Thai people are willing to accept a range of behaviour by government officials as "permissible" that would have included activities described as "corruption" in the United States or western Europe.[246] Still, there are abuses of power that are considered as "corruption" everywhere in the world. The same survey revealed that the Thai people, despite displaying a higher level of tolerance of such behaviour by their government officials, still considered official corruption to be a major issue in their country and an overwhelming majority wished something could be done to substantially reduce it. This is not just a Thai phenomenon. From Russia to Indonesia, and from China to Venezuela, corruption is denounced as public enemy number one.

Like pornography, corruption is difficult to quantify, but you know it when you see it. In spite of all the difficulties associated with any attempt to quantify corruption, one can still get useful information on the seriousness of corruption in a country by surveying experts or firms in that country. There are essentially four types of "corruption ratings":

**(i) Corruption ratings based on "expert opinions"**

A prominent example is the International Country Risk Guide (ICRG) index, which has been produced every year since 1982 by Political Risk Services, a private international investment risk service. The ICRG corruption index is apparently based on the opinion of the "in-house experts" and is supposed to capture the extent to which "high government officials are likely to demand special payments" and to which "illegal payments are generally expected throughout lower levels of government" in the form of "bribes connected with import and export licenses, exchange controls, tax assessments, police protection or loans."

**(ii) Corruption ratings based on surveys of firms or citizens**

Two prominent examples in this category are the Global Competitiveness Report (GCR) corruption index, and the World Development Report (WDR) corruption index. Unlike the ICRG indices, the GCR index is based on a survey (every year since 1996) of firm managers, rather than experts or consultants. Sponsored by the World Economic Forum, a European-based consortium with a large membership of firms, and designed by the Harvard Institute for International Development, this survey asks the responding firms about various aspects of "competitiveness" in the host countries where they invest. In 1996, 2,381 firms in 58 countries answered the question on corruption which asked the respondent to rate the level of corruption on a one-to-seven scale according to the extent of "irregular, additional payments connected with import and export permits, business licenses, exchange controls, tax assessments, police protection or loan applications." The GCR corruption index for a particular country is the average of all respondents' ratings for that country.

Similar to the GCR index, the WDR index is based on a 1996 survey of firms conducted by the World Bank for its 1997 World Development Report. Every respondent was asked a long list of questions, one of which concerned the perceived level of corruption. The question is essentially identical to the one in the GCR survey. The WDR survey covers over 70 countries (many of which are not in the WDR sample, and the reverse is also true). The WDR survey tends to cover more medium and small firms whereas the GCR surveys include more large firms.

**(iii) Corruption ratings based on "a poll of polls"**

The best-known index in this category is the index produced annually since 1995 by Transparency International (TI), an international non-governmental organization dedicated to fighting corruption worldwide. The TI corruption index is based on a weighted average of approximately 10 surveys of varying coverage.[247] It ranks countries on a one-to-ten scale.

As a survey of surveys, the TI index has advantages and disadvantages. If the measurement errors in the different surveys are independent and identically

---

[246] P. Pasuk and S. Piriyarangsan, *Corruption and Democracy in Thailand* (Thailand, Silkworm Books: Chiang Mai, 1994).

[247] The exact number of constituent surveys varies over the years, but has tended to increase over time.

distributed (*iid*), the averaging process used to produce the TI index may reduce the influence of measurement errors. But the *iid* assumption may not hold. Moreover, since different surveys cover different subsets of countries, the averaging process may introduce new measurement errors when cross-country rankings are produced. It should also be noted that, as the TI indexes in different years are derived from potentially different sets of surveys, they should not be used to measure changes over time in the level of corruption for a particular country.

Recently, Kaufmann, Kray and Zoido-Lobaton[248] applied an unobserved component framework to derive an aggregate indicator of governance (or corruption) that pools the diverse array of individual perception indexes (including the ICRG, WDR and GCR indexes). This approach has the virtue of producing an index that has a larger country coverage than any single index, and is statistically better justified than the Transparency International's method.

### (iv) Corruption ratings based on more "objective" and "harder" data

All of the previous indexes are based on the subjective judgement of the respondents. A more "objective" way to measure corruption might be to quantify the fraction of business transactions in a country that involves bribery to government officials. This is inherently difficult, because very few business people in any country would admit to having paid bribes to domestic officials since this would be a criminal offence in most countries (even if bribery is common). The only example that I have found is what I would label the Neumann Index.[249] This was based on information obtained by Peter Neumann in 1994 for the German publication, *Impulse*, from interviewing people (mainly German exporters), with business experience in foreign countries. He interviewed on average 10 individuals (with a minimum of three) per foreign importing country with a guarantee of strict confidentiality. The measure indicates the proportion of transactions involving the payment of bribes. (The numbers are apparently rounded to the nearest 20 per cent.) It is useful to note that at the time of the interview, it was not a crime in Germany for German businessmen to bribe foreign government officials. Unfortunately, the exercise was not repeated in later years.

As examples, table 4.2.1 shows the corruption ratings from three sources (WDR, TI and Neumann indexes) for the transition economies in Europe. Most other ratings, such as the GCR index, have only a sparse coverage of the transition economies. (In the original WDR and TI indexes, large numbers refer to low levels

TABLE 4.2.1

**Measures of corruption in the transition economies of Europe and the CIS**

|  | WDR 1997 | TI 1999 | Neumann index |
|---|---|---|---|
| Albania | .. | 8.7 | .. |
| Azerbaijan | 4.6 | 9.3 | 6 |
| Belarus | 4.2 | 7.6 | 4 |
| Bulgaria | 4.6 | 7.7 | 4 |
| Croatia | .. | 8.3 | 4 |
| Czech Republic | 2.8 | 6.4 | 4 |
| Estonia | 2.2 | 5.3 | 2 |
| Georgia | 4.2 | 8.7 | 4 |
| Hungary | 2.6 | 5.8 | 6 |
| Kazakhstan | 4.3 | 8.7 | 4 |
| Latvia | 3.9 | 7.6 | 4 |
| Lithuania | 3.3 | 7.2 | – |
| Poland | 3.1 | 6.8 | 4 |
| Republic of Moldova | 4.2 | 8.4 | .. |
| Romania | .. | 7.7 | 6 |
| Russian Federation | 3.8 | 8.6 | 8 |
| Slovakia | 4.1 | 7.3 | 4 |
| Slovenia | .. | 5.0 | 2 |
| The former Yugoslav Republic of Macedonia | 3.1 | .. | 8 |
| Turkmenistan | .. | .. | 4 |
| Ukraine | 3.4 | 8.4 | 4 |
| Uzbekistan | 4.4 | 9.2 | 4 |
| **Mean** | 3.7 | 7.6 | 4.3 |

*Source:* World Development Report (WDR), Transparency International (TI) and Neumann indexes.

of corruption. To avoid awkwardness in interpretation, I have re-scaled all the indices in table 4.2.1 so that low values imply low levels of corruption.)

It should be kept in mind that the WDR and TI indices are based on people's perceptions, as opposed to objective measures of corruption, and perception can differ from reality. However, two points are worth noting. First, for many questions such as how corruption affects foreign investment, perception – and thus perhaps our measure – is what actually matters. Second, despite the very different sources of the surveys, the pair-wise correlations among the subjective indexes as well as with the more "objective" Neumann index are very high.

## 4.3 Corruption and economic growth

Corruption can affect the growth rate of an economy through its effect on domestic and foreign investment, as well as through other channels. In addition, it can also affect the quality of economic growth. The evidence for these statements is reviewed below.

### (i) Rate of economic growth

In a cross-country regression of the total investment/GDP ratio, averaged over 1980-1985, on a constant and an index for lack of corruption, the point estimate of the slope in Mauro's study[250] is 0.012. This

---

[248] D. Kaufmann, A. Kray and P. Zoido-Lobaton, *Aggregating Governance Indicators*, World Bank Working Paper, No. 2195 (Washington, D.C.), 1999.

[249] It was used first by A. Ades and R. Di Tella, "National champions and corruption: some unpleasant interventionist arithmetic", *The Economic Journal*, Vol. 107, No. 443, pp. 1023-1042, July 1997.

[250] P. Mauro, "Corruption and growth", *Quarterly Journal of Economics*, Vol. 110, No. 3, August 1995, p. 696, table IV.

shows that investment and corruption are negatively correlated. The quantitative effect of corruption can be illustrated by taking literally the point estimate and the corruption ratings. Unfortunately, because most transition economies are not included in Mauro's sample, the illustration has to be drawn from countries in other regions. If the Philippines, say, could reduce its corruption level to that of Singapore, other things being equal, it would have been able to raise its investment/GDP ratio by 6.6 percentage points (=(6.5-1)x0.012). This is quite a substantial increase in the investment ratio.

Using a data set of bilateral flows of foreign direct investment in the early 1990s, from 14 major source countries to 41 host countries, Wei[251] studied the effect of corruption on host countries' ability to attract foreign investment. He employed a modified Tobit framework[252] that takes into account the fact that some host countries attract hardly any FDI from certain source countries. Controlling for the size and level of development of the host country, for historical/linguistic links and for geographic proximity between the source and host countries, he found that corruption in the host countries was negatively associated with foreign investment (the coefficients on corruption and the host country tax rate being -0.09 and -1.92, respectively). Taking these point estimates at face value, and using the corruption ratings in table 4.2.1, it could be said that a rise in corruption from the level in Singapore to that in India is equivalent to raising the marginal tax rate by over 20 percentage points.

Many transition and developing countries offer substantial tax incentives to lure multinational firms to locate in their countries. For example, China offers all firms with foreign investment an initial two-year tax holiday plus three subsequent years at half the normal tax rate. This research suggests that such countries would have attracted just as much or even more foreign investment without any need for tax incentives if they had been able to get domestic corruption under control.

In fact, Wei argues that, contrary to a cursory reading of the news, China is an underachiever as a host of direct investment from five major source countries (France, Germany, Japan, the United Kingdom and the United States), once account is taken of its size, proximity to some major source countries and other factors. Wei suggests that the high level of corruption in China may very well have contributed to this.[253]

If corruption is negatively associated with domestic investment and reduces foreign investment, one would expect that it would also be negatively associated with the rate of economic growth. Mauro examined how the conditional growth rate (that is, the growth rate given the country's starting point and population size in a Solow-Barro style cross-country growth regression framework) is affected by corruption: he found that the data did in fact reveal such a relationship.

To illustrate the quantitative effect, let me take the point estimate in column 6 of table VII in Mauro's paper. If Bangladesh reduced its corruption level to that of Singapore, its average annual growth rate of per capita GDP over 1960-1985 would have been higher by 1.8 percentage points. Assuming its actual average growth rate was 4 per cent a year, its per capita income by 1985 could have been more than 50 per cent higher.[254] Using an instrumental variable approach (where ethno-linguistic fractionalization is the instrument for corruption), as in column 8 of table VII in Mauro's paper, there would be an even larger effect of corruption on growth, although the result is barely significant at the 15 per cent level.

### Why is corruption so taxing?

Why is corruption so damaging to economic activities relative to an equivalent tax system? The answer lies in the nature of corruption. Unlike a tax, it is inherently secretive and arbitrary. The implicit contract between the briber and bribee cannot be enforced by a reliable court system. Shleifer and Vishny[255] theorized that countries with a more disorganized pattern of corruption would be particularly inhospitable to economic growth. Wei[256] shows that, after holding the level of corruption constant, countries with a more disorganized corruption structure – as measured by the dispersion of the corruption ratings by the respondents – receive significantly less foreign direct investment.

Discretion by officials and, consequently, the uncertainty faced by firms and private citizens are crucial characteristics of corruption. That is why bribery in a corrupt society and the fees paid to lawyers in a relatively clean society are not equivalent.

### (ii) Firm-level evidence

Most of the studies reviewed in this section are based on cross-country regressions. It is important to stress that significant coefficients in these regression are evidence of a correlation between the level of corruption and other variables (such as the rate of economic growth, investment or the composition of public expenditure), but they do not necessarily imply that corruption is a cause of variation in the latter. On an *ex-ante* basis, it is plausible to argue that changes in corruption (particularly in the

---

[251] S.-J. Wei, *Why is Corruption So Much More Taxing Than Tax? Arbitrariness Kills*, NBER Working Paper, No. 6255 (Cambridge, MA), November 1997.

[252] Ibid., see appendix for details.

[253] S.-J. Wei, "Why does China attract so little foreign direct investment?", in T. Ito and A. Krueger (eds.), *The Role of Foreign Direct Investment in East Asian Economic Development* (Chicago, University of Chicago Press, 2000).

[254] If the actual growth rate was assumed to be lower (at, say, 3 per cent a year) the result would have been an even greater improvement in 1985 per capita income as a result of reducing its corruption level.

[255] A. Shleifer and R. Vishny, "Corruption", *Quarterly Journal of Economics*, Vol. 108, No. 3, August 1993, pp. 599-617.

[256] S.-J. Wei, *Why is Corruption...*, op. cit.

subjective perception of corruption) can be *caused* by changes in income level, in investment and so on. Besides, good things tend to go together. It is possible that something else causes investment and income to rise and that this could be correlated with corruption even if corruption does not cause either investment or income to change. In illustrating the results from other studies, I have invoked a thought experiment of the type such as "if we could reduce corruption from the level in country X to that in Singapore, variable Y would rise by Z per cent". In fact, one might argue that the statement "if variable Y (say income level) rises by Z per cent, then the level of corruption in country X could be reduced to the Singapore level" was equally plausible.

Some of the studies reviewed do employ instrumental variable regressions. For example, Mauro[257] used ethno-linguistic fractionalization as an instrument for his corruption measure (this particular instrument has been followed in many subsequent studies) and shows that corruption instrumentalized by this variable has a negative effect on economic growth. This is one step closer to establishing causality. But the validity of the causal influence depends on the validity of the instruments. For example, ethno-linguistic fractionalization could slow down growth for reasons unrelated to corruption, e.g. by increasing the risks of ethnic conflict and civil wars. In that case, the correlation between ethno-linguistic fractionalization and growth would not be evidence that corruption lowers the growth rate.

To establish causality in relationships involving corruption, it would be very useful to supplement cross-country regressions with some event studies in which some of the determinants of corruption are subject to discrete change. For example, from time to time, some countries may experience "exogenous" regime changes – such as a military coup overthrowing a democratic government, or the reverse, a new democracy emerging from a previous dictatorship. If we believe that these changes should exogenously increase and decrease the extent of corruption, then, studying the growth rate, investment, or other variables of interest before and after the regime change may provide useful information about the effects of changes in corruption.

Aside from the issue of the direction of causality, one should also note that, across countries, the broad attributes of public governance and public institutions (for example, the rule of law, the strength of civil groups, press freedom, the educational level of civil servants and corruption) tend to be correlated. This makes the problem of isolating the effect of corruption more challenging if not infeasible.

The studies reviewed so far are mostly based on country-level observations and cross-country regressions.[258]

As just noted, isolating the effects of corruption from the other attributes of public institutions and determining the direction of causality are difficult in cross-country regressions.

While the evidence clearly shows that domestic investment, foreign investment and economic growth are lower in more corrupt countries, a version of the "virtuous bribery" story can still be heard. In particular, some say that bribes often work like "grease", easing the wheels of commerce. In a country that is rife with bad and heavy regulations, the offering of bribes to circumvent bad government control is like deregulation, and hence can be seen as good.

Kaufmann and Wei[259] argue that this view is true only in a very narrow sense when bad regulation and official harassment are taken as exogenous. Officials often have lots of leeway to customize the type and amount of harassment of individuals firms. Tax inspectors may have room to overreport taxable income.[260] Fire inspectors can decide how frequently they need to return to check fire safety in a given year. Taking account of these, Kaufman and Wei built a simple model in which bureaucrats set up red tape and bureaucratic obstacles in order to extract bribes and stop only when firms start to exit (by not investing or by fleeing to foreign countries). Furthermore, the outside options of the firms differ either because of the characteristics of their industry or because of the type of investor (foreign versus domestic). In this case, they show that bribery across firms is not only positively correlated with the nominal red tape in the rule book, but can be positively correlated with the *effective* red tape (e.g. the amount of time wasted in securing a permit after having paid a bribe). It is not that paying bribes causes an increase in red tape, but rather, the scale of bribery and red tape is simultaneously determined by the same set of firm characteristics.

Using data from a survey of nearly 2,400 firms in 58 countries, Kaufmann and Wei show that, even within a country, managers of the firms that pay more bribes on average waste more, rather than less, time in negotiating with government officials. This evidence supports the idea of "tailored harassment" and "endogenous obstacles," and thus rejects the hypothesis of beneficial "grease." It is important to stress that the evidence does not suggest that individual firms can do better by not offering bribes. They cannot, given the environment. However, collectively, all firms can do better if there is an exogenous constraint on all firms' ability to bribe.

---

[257] P. Mauro, "Corruption and growth", loc. cit., pp. 681-712.

[258] B. Smarzynska and S.-J. Wei, *Corruption and Composition of Foreign Direct Investment: Firm-level Evidence*, NBER Working Paper, No. 7969 (Cambridge, MA), October 2000.

[259] D. Kaufmann and S.-J. Wei, *Does "Grease Money" Speed Up the Wheels of Commerce?*, NBER Working Paper, No. 7093 (Cambridge, MA), April 1999 (also released as a World Bank Policy Research Working Paper, No. 2254).

[260] J. Hindriks, M. Keen and A. Muthoo, "Corruption, extortion and evasion", Facultes Notre-Dames de la Paix (Belgium), University of Essex (Colchester) and Institute for Fiscal Studies (London), mimeo, 1998 (first version, July 1996).

Thus, for example, the OECD convention on combating bribery in international transactions, that went into effect in February 1999, may not only reduce bribery but also the level of bureaucratic red tape.

One problem with the Kaufmann and Wei study is that the observations on bribery are "inferred" from the survey respondents' answers as to their perception of the level of corruption. Svensson[261] extends this research in a significant way by utilizing a direct firm-level measure of bribery in Uganda. He showed that bribes are positively related to the firms' profitability (which can be instrumented by industry and location dummies) and negatively related to a measure of investment irreversibility. Both findings are consistent with the hypothesis that harassment and the demand for bribes are related to firms' underlying characteristics, and thus reject the "efficient grease" hypothesis.

Using the same Uganda firm-level observations, Fisman and Svensson[262] revisited the question posed by Wei.[263] They found that an increase in the bribery rate is associated with a reduction in the firm's growth rate about three times as large as an equivalent increase in tax.

### (iii) Quality of growth

In addition to reducing the rate of economic growth, corruption can also have a negative effect on the quality of growth as corruption can lead to a misallocation of resources. If the allocation of bank loans and government investment funds are based on personal connections rather than the competence of firm managers, they may be channelled into unnecessarily riskier projects (without necessarily higher returns). Indeed, Wei and Zhang[264] found that countries with more serious corruption problems tend to have more volatile investment-to-GDP ratios and more volatile growth rates.

By the same logic, corruption may also divert government spending away from socially desirable sectors, such as education and health, and towards those that permit more opportunities for taking bribes, such as weapons acquisition. Tanzi and Davoodi[265] carried out a systematic study of the effect of corruption on government finance. There are several important findings:

- corruption tends to increase the size of public investment (at the expense of private investment) because many items of public expenditure lend themselves to manipulations by high-level officials to obtain bribes. (One should note that the causality may also go the other way – i.e. more government expenditure may provide more opportunities for corruption);
- corruption skews the composition of public expenditure away from running costs and maintenance towards expenditure on new equipment;[266]
- corruption skews the composition of public expenditure away from needed health and education funds because, relative to other public projects, these areas are more difficult for officials to extract bribes;[267]
- corruption reduces the productivity of public investment and of a country's infrastructure;
- corruption may reduce tax revenue because it compromises the government's ability to collect taxes and tariffs, although the net effect depends on how the nominal tax and other regulatory burdens were chosen by corruption-prone officials.[268]

Let us illustrate the implications of some of the Tanzi-Davoodi findings by looking at the effect of a change in corruption on a variety of indicators, averaged over 1980-1995. An increase in corruption from the Singapore level to that in Pakistan would increase the public expenditure/GDP ratio by 1.6 percentage points (column 2 of Tanzi-Davoodi's table 1) and reduce the government revenue/GDP ratio by 10 percentage points (column 2 of Tanzi-Davoodi's table 2).

An increase in corruption reduces the quality of roads, and increases the incidence of power outages, telecommunication faults and water losses. Specifically, an increase in corruption from the Singapore level to the Pakistan level would be associated with a 15 per cent increase of roads in bad condition, after controlling for a country's level of development and its public investment to GDP ratio (column 2 in Tanzi-Davoodi's table 5).

---

[261] J. Svensson, *Who Must Pay Bribes and How Much? Evidence from a Cross-secton of Firms,* World Bank Working Paper, No. 2486 (Washington, D.C.), 1999.

[262] R. Fisman and J. Svensson, "The effects of corruption and taxation on growth: firm-level evidence", Columbia Business School and the World Bank, mimeo, 1999.

[263] S.-J. Wei, *Why is Corruption ...,* op. cit., and idem, "How taxing is corruption on international investors?", *Review of Economics and Statistics,* Vol. 82, No. 1, February 2000, pp. 1-11.

[264] S.-J. Wei and L. Zhang "Crony capitalism and volatility of business cycles", Brookings Institution, mimeo, 2001. See also S.-J. Wei and R. Zeckhauser, "Dark deals and dampened destines: corruption and economic performance", *Japan and the World Economy,* 1999.

[265] V. Tanzi and H. Davoodi, *Corruption, Public Investment and Growth,* IMF Working Paper WP/97/139 (Washington, D.C.), October 1997. See also S. Gupta, H. Davoodi and R. Alonso-Terme, *Does Corruption Affect Income Inequality and Poverty?,* IMF Working Paper WP/98/76 (Washington, D.C.), May 1998.

[266] On this point see also R. Klitgaard, *Tropical Gangsters* (New York, Basic Books, 1990).

[267] P. Mauro, "The effects of corruption on growth, investment and government expenditure: a cross-country analysis", in K. Elliot (ed.), *Corruption and the Global Economy* (Washington, D.C., Institute for International Economics, 1997), also found that corruption tends to skew public expenditure away from health and education, because they are more difficult to manipulate for bribes than are other projects.

[268] D. Kaufmann and S.-J. Wei, op. cit. See also S. Johnson, D. Kaufmann and A. Shleifer, "The unofficial economy in transition", *Brookings Papers on Economic Activity,* 2 (Washington, D.C.), Fall 1997, pp. 159-239; S. Johnson, D. Kaufmann and P. Zoido-Lobaton, "Regulatory discretion and the unofficial economy", *American Economic Review,* Vol. 88, No. 2, May 1998, pp. 387-392.

## 4.4 Corruption in a globalizing world

The world economy is increasingly globalized. This is not just a cliché but a fact. Several pieces of evidence suggest that the gap between countries with good governance and those with severe problems of corruption is being widened by globalization. The benefits of globalization, such as foreign direct investment and the associated transfer of technology and managerial know-how, tend to go mostly to the less corrupt countries. On the other hand, the risks of globalization, such as volatile international capital flows and associated currency and financial crises, tend to pose a greater threat to countries with serious corruption problems. Hence, anti-corruption becomes more important as globalization deepens.[269]

### (i) Corruption reduces the benefits of globalization: a specific example on transition economies

A major channel through which the benefits of globalization spread to developing and transition economies is inward foreign direct investment. The previous section has already reviewed a study on how corruption has reduced inward FDI based on bilateral FDI data.[270] More recently, there has been a micro study of corruption and FDI that focuses specifically on the transition economies in Europe.[271] Some features of that study are worth highlighting here. First, it covered a relatively large number of the transition economies (22 in all). Second, it utilizes a unique firm-level FDI data set based on a survey conducted in 1995 by the European Bank for Reconstruction and Development (EBRD). Firm-level studies are generally rare, for the obvious reason that firm-level data are more difficult to assemble. Third, the paper examines two questions simultaneously: first, does corruption reduce the total volume of FDI in the transition economies? This may be labelled the "volume effect"; and, second, does corruption in eastern Europe and the former Soviet republics alter the entry mode of the multinational firms? This may be labelled the "composition effect".

The statistical analysis reveals a strong "volume effect": the more corrupt transition economies (such as Azerbaijan or Uzbekistan) are significantly less able to attract foreign direct investment than the less corrupt economies (such as the Czech Republic or Estonia). This holds after allowance for the influence of the host country's size, labour costs, tax rates, trade openness and other characteristics.

In addition, there is also a strong support for an intriguing "composition effect": once a foreign firm decides to enter a transition economy, whether it forms a joint venture with a local partner or maintains full ownership control depends on the level of local corruption as well as the level of technological sophistication of the investing firm. Corruption makes a country's business transactions and the interaction between the government and the business sectors less transparent. A local partner may help the foreign firm to acquire the local licences and permits, and otherwise negotiate with the bureaucratic maze at a lower cost. Hence, other things being equal, a foreign firm may find it more advantageous to use a local partner in a host country the more serious the corruption problem. This is confirmed by the data.

On the other hand, a foreign firm with sophisticated technology may worry about the leakage of the technology by the local partner. This could reduce the profits of the foreign firm even in third markets. In a corrupt host country, there is less chance that the foreign firm will get adequate protection from the local court system in the event of an unlawful leakage of its technology by the local partner. Therefore, other things being equal, a foreign firm with more sophisticated technology is less likely to invest in the more corrupt host country, and if it does it is more likely to prefer the wholly-owned mode of operation.

### (ii) Corruption increases the risks of globalization

Crony capitalism is sometimes mentioned as a possible contributor to the 1997-1998 emerging market currency crisis, but systematic evidence for this is generally lacking. However, it has been shown[272] that corruption tends to influence the composition of capital flows into a country and to make it more dependent on international bank loans as opposed to international direct investment. Such a composition leaves the country more vulnerable to currency crises triggered by a sudden shift in international investors' sentiment. Thus, this is one possible channel through which corruption may increase a country's propensity to run into a currency crisis.

Corruption may also raise the likelihood of a financial crisis through other channels. The financial sector is weak in many countries in the recent crisis. Might corruption be implicated? Corruption could obscure the meaning and reliability of publicly disclosed accounting numbers and, as noted already, it may also skew financial resources away from the most efficient uses towards the efficient but politically well-connected firms.

---

[269] A summary of this argument can be found in S.-J. Wei, "Corruption and globalization", Brookings Institution, *Brookings Brief*, No. 79, April 2001 (www.brook.edu).

[270] S.-J. Wei, "How taxing is corruption on international investors?", *Review of Economics and Statistics*, Vol. 82, No. 1, February 2000, pp. 1-11.

[271] B. Smarzynska and S.-J. Wei, op. cit.

[272] S.-J. Wei, "Local corruption and global capital flows," *Brookings Papers on Economic Activity*, 2 (Washington, D.C.), 2000, pp. 303-354; S.-J. Wei and Y. Wu, *Negative Alchemy? Corruption, Composition of Capital Flows and Currency Crises*, NBER Working Paper, No. 8187 (Cambridge, MA), March 2001.

Using a clever data set that measures the strength of Indonesian firms' connections to Suharno and his family, it has been shown[273] that the stock market valuation of the politically well-connected firms tended to lose value sharply each time there was a rumour about the health problems of Suharno. This suggests that the market does not believe that the resources allocated to these firms are justified except for the abnormal returns associated with their political connection. On the basis of data from the 1997 GCR survey,[274] there is a clear correlation: corrupt countries are more likely to have inadequate government supervision of the financial system and are also more likely to have vulnerable banks. There is also evidence that more corrupt countries tend to have more volatile stock returns, more inside trading and smaller capital markets.[275]

### 4.5 Corruption control: what can be done?

Because corruption is a crime in most countries' penal codes, it is common to emphasize the role of law enforcement in fighting corruption. While there is no question that law and law enforcement are important, it should be emphasized that it is at least as important to look into the root causes of corruption, the institutional environment and the incentive structure under which corruption thrives.

To understand the effectiveness of anti-corruption strategies, it is important to recognize that the prevalence of corruption in an economy is the consequence of a combination of opportunities and incentives. Robert Klitgaard's succinct formula[276] that "corruption = monopoly + discretion – accountability" can be understood as a special case of this general principle. To reduce corruption effectively, both the opportunities that allow corruption to flourish as well as the incentives that encourage it must therefore be attacked. The opportunities include the scope of the governmental role in the economy and the ability of civil society to monitor the actions of government officials. The incentives include both "positive" ones such as civil servant salaries and recruitment procedures and "negative" ones such as the degree and likelihood of punishment for corrupt acts.

### (i) Opportunities derived from government's role in the economy

Without compromising the need to recruit moral people as government officials, economists are never tired of pointing out the importance of minimizing the institutionalized opportunity for officials to take bribes. The more discretion government officials have over the operation of business or the lives of citizens, the more likely it is that corruption will occur and flourish, other things being equal. Labyrinthine government regulations create fertile grounds for government officials to extract rents, whereas an economy where government's role is minimal is less likely to breed corruption.

This point is almost elementary. If a firm must acquire a licence and pay a tariff before it can import certain goods from abroad, then the officials who decide who gets a licence and tariff exemptions have the opportunity to extract bribes. If no licence or tariff is needed, no firm would pay a bribe before importing.

Tanzi's excellent survey[277] offers a number of concrete descriptions of where the opportunity for corruption may arise as a result of government (over)regulation. For example, in the taxation area, he pointed out that the more difficult it is to understand the laws, the more likely there is to be corruption; and the more discretion given to tax administrators over the granting of tax incentives, the determination of tax liabilities and the selection of audits and litigations, the more likely there is corruption.

Similarly, the size of government spending and the procedures used in allocating it also significantly affect the opportunities for corruption. Also, if a government is involved in providing certain goods and services at subsidized prices – for example, foreign exchange, credit, public housing, educational opportunities or water and electricity – then officials with the duty to decide also have the opportunity to pocket a fraction of the implicit subsidy (i.e. the difference between the market value of the goods or services and the price the government is asking), in the form of bribes extracted from the recipients of the subsidized goods or services. Both Mauro[278] and Kaufmann and Wei[279] show that the indexes of corruption and of government regulation are positively correlated.

Among the transition economies in Europe, some have been pursuing more active industrial policies than others. Such policies, by their very nature, involve discretion on the part of government officials, in terms of which industry to support, which firms within a given industry to support, how to allocate subsidized loans, grants, tariff rebates, and so on. It has been argued that, logically, industrial policies can promote corruption as well as investment.[280] Using indices of corruption and industrial policy across a number of countries, it has been shown that corruption is indeed higher in countries with more active industrial policy.[281] The negative effect of corruption

---

[273] R. Fisman, "It's not what you know ... estimating the value of political connections", Columbia Business School, mimeo, 1998.

[274] S.-J. Wei and S. Sievers, "The cost of crony capitalism", *The Asian Competitiveness Report 1999*, World Economic Forum (Geneva), 1999, pp. 50-55.

[275] J. Du and S.-J. Wei, "Crony capitalism and financial market volatility", Harvard University, mimeo, 2000.

[276] R. Klitgaard, op. cit.

[277] V. Tanzi, *Corruption Around the World*: ..., op. cit.

[278] P. Mauro, "Corruption and growth", loc, cit., pp. 681-712.

[279] D. Kaufmann and S.-J. Wei, op. cit.

[280] A. Ades and R. Di Tella, "National champions and corruption: ...", loc. cit.

[281] Ibid.

induced by the industrial policy appears to be large (probably of the order of 56 per cent to 84 per cent of the direct beneficial effect), and therefore should not be neglected in any cost-benefit analysis of industrial policies.

Gatti has confirmed that more open economies tend to have lower levels of corruption.[282] Furthermore, she shows that while the share of imports in GDP is not a significant explanatory variable for corruption (after controlling for other variables including population), the average tariff is. She interprets this as evidence that the direct policy distortion rather than the absence of foreign competition is more important in inducing corruption. One question that needs further research is whether high tariffs are erected for the purpose of extracting bribes (rather than being an exogenous cause of corruption). Some models suggest that this is possible.[283]

Svensson has reported evidence that some countries that receive generous foreign aid (which is determined for geopolitical reasons) tend to see their level of corruption rising.[284] As a consequence, the economic lot of the people in these countries may not be made better off (and can be made worse off).

Based on information concerning the transition economies in Europe, a recent paper[285] suggests that the variation in the level of corruption among the transition economies is best explained by an index of the depth of structural reform – a composite of privatization, price and financial sector liberalization, and competition policies.

Before leaving this subsection, it should be pointed out that while less discretion by government officials reduces the scope for corruption, we are not advocating the abolition of all regulations. Many of them, and even a degree of bureaucratic discretion, serve useful functions in society. The point is that we should be mindful of the implications for corruption when designing government regulations.

## (ii) Civil servant recruitment and promotion system

The moral character and quality of government officials are another very important determinant of the extent of corruption in a country. The quality of the bureaucrats, in turn, is closely related to how they are recruited and promoted. In countries where nepotism and patronage are rampant, or government posts are sold explicitly or implicitly, bureaucrats will be less competent and less well-motivated because success depends on advantages gained by connections or on bribing superiors rather than merit; as a result, they will be very vulnerable to corruption. The German sociologist Max Weber made this point amply clear.[286]

Indices of the degree to which recruitment and promotion for civil servants in 35 countries is meritocratic (as well as their average wages relative to private sector alternatives) show that the cross-country ratings provided by the International Country Risk Guide are significantly related to the way civil servants are recruited and promoted.[287] Meritocratic recruitment is most important for reducing corruption, followed by meritocratic promotion and security of employment.

## (iii) Compensation for civil servants

It has been long recognized that it is naive to give people power, pay them a pitiful wage and then expect them not to use their power for personal gain. Recognizing this, Singapore, starting in the 1960s under the leadership of then Prime Minister Lee Kuan Yew, and Hong Kong, starting in the late 1970s, began to pay their civil servants well, sometimes above their best alternative in the private sector. It is often noted, with or without approval, that Singapore's cabinet ministers' salaries are pegged to those of the CEOs in the largest multinational firms in the world. The pay of Singapore's Prime Minister is several times that of the United States President. Many scholars (and the governments of Singapore and Hong Kong) contend that this wage policy is in an important factor behind the very low corruption levels in these two economies. (Singapore, as noted above, is rated as one of the least corrupt countries in many surveys.)

The view that high salaries for civil servants help to deter corruption is certainly not restricted to Asia. For example, the low level of corruption in Sweden during the 1870-1970 period is partly attributed to the fact that high-level government administrators earned 12-15 times the salary of an average industrial worker.[288]

Systematic examination of the statistical evidence on the connection between corruption and public sector wages is a relatively recent undertaking. Although the cross-country regression study cited above did not find

---

[282] R. Gatti, *Explaining Corruption: Are Open Countries Less Corrupt?*, unpublished World Bank Working Paper (Washington, D.C.), 1999. See also A. Ades and R. Di Tella, "Rents, competition and corruption", *American Economic Review*, Vol. 89, No. 4, 1999, pp. 982-993; C. Bliss and R. Di Tella, "Does competition kill corruption?", *Journal of Political Economy*, Vol. 5, No. 105, December 1997, pp. 1001-1023; S.-J. Wei, *Natural Openness and Good Government*, NBER Working Paper, No. 7765 (Cambridge, MA), June 2000.

[283] See for example, D. Kaufmann and S.-J. Wei, op. cit.; J. Svensson, *Who Must Pay Bribes ...* op. cit.

[284] J. Svensson, "Foreign aid and rent-seeking", *Journal of International Economics*, Vol. 51, No. 2, August 1998, pp. 437-461.

[285] G. Abed and H. Davoodi, *Corruption, Structural Reforms and Economic Performance in the Transition Economies*, IMF Working Paper WP/00/132 (Washington, D.C.), July 2000.

[286] M. Weber, *The Theory of Social and Economic Organization* (London, The Free Press of Glencoe, 1947).

[287] J. Rauch and P. Evans, "Bureaucratic structure and bureaucratic performance in less developed countries", University of California-San Diego and University of California-Berkeley, mimeo, 1997.

[288] V. Tanzi, *Corruption Around the World: ...*, op. cit., quoting an unpublished paper by A. Lindbeck, "Swedish lessons for post-socialist countries", 1998.

robust support for the role of high salaries,[289] the World Bank and an IMF Working Paper by Van Rijckeghem and Weder do report evidence that countries with poorly paid public officials tend to have higher levels of corruption.[290]

What is important here is not the absolute level of civil servants' wages, but their values relative to the best private sector alternatives. In the IMF Working Paper, given the constraint of data availability, the average civil servant's pay was measured relative to the average wage in the manufacturing sector. However, the true private sector alternatives for senior government officials with comparable skills and responsibilities are likely to be paid a lot more than the average wage in the manufacturing sector. But as the manufacturing sector's average wage is the only one available on a consistent cross-country basis there is a potential measurement error in the denominator. As for the numerator, one should note that the civil servants' wage data were found by the authors, but as fringe benefits for civil servants (e.g. free housing, maids, and expense accounts) can be large relative to official salaries in many countries, there may also be measurement errors in the numerator as well. In fact the study assumes that, across countries, the manufacturing wage and the salaries of the private sector alternative of government officials are highly positively correlated and that fringe benefits plus official wages are highly correlated with the civil servants' official wages.

The IMF study found a negative and statistically significant correlation between the public sector's relative wage and the extent of corruption involving government officials. Based on their point estimates, one can calculate, for each country in their sample, the ratio of public to private sector wages that is needed in order to reduce the corruption to the Singapore level, which has the lowest corruption grade (this is called the "warranted relative wage" below). Since their sample does not include the transition economies in Europe, we cannot report the results with these economies as examples. Nonetheless, it may be instructive to reproduce that part of their table 6, which reports the actual versus the warranted relative wage for a few countries in their sample. Like all other similar projections in this paper, the numbers in table 4.5.1 are meant to be illustrative and not to be taken literally.

A few points are particularly worth noting from table 4.5.1. First, to eradicate corruption (or to reduce it to the Singapore level), public sector pay needs to be raised by a substantial margin (sometimes by 500 per cent or even 900 per cent). Second, we do not know for sure if the warranted increase would raise the salaries of

TABLE 4.5.1

**How much should civil servants' legal pay be increased to lower corruption?**

| Country | Public sector relative to manufacturing sector wage Actual (1) | Calibrated ratio to reduce corruption to Singapore level (2) | Needed increase in public sector pay (per cent) (3) |
|---|---|---|---|
| Singapore | 3.49 | 3.49 | – |
| Hong Kong | 1.79 | 2.85 | 59 |
| Korea | 1.91 | 7.08 | 271 |
| India | 1.09 | 5.40 | 395 |
| Kenya | 0.90 | 5.36 | 496 |
| Sri Lanka | 0.85 | 5.07 | 496 |
| Turkey | 0.92 | 5.38 | 498 |
| Colombia | 0.64 | 4.87 | 660 |
| Mexico | 0.50 | 5.04 | 908 |
| Ghana | 0.63 | 6.77 | 975 |

*Source:* The first two columns are from table 6 in C. Van Rijckeghem and B. Weder, *Corruption and Rate of Temptation: Do Low Wages in the Civil Service Cause Corruption?*, IMF Working Paper WP/97/73 (Washington, D.C.), June 1997. Column (3) is the author's calculation based on the first two columns and simply applies the Van Rijckeghem-Weder calculation literally.

government officials above their private sector alternatives. If they do, there is a serious equity issue even if these governments have the money (or have the ability to transform most of the currently illegal bribes into the incremental taxes needed to finance the increase in civil servants' legal pay). Third, if civil servants are paid a higher salary than their private sector alternatives, many people may pay a bribe to be chosen for these public jobs. So the high pay policy itself may create new types of corruption. Fourth, extortion and bribe-taking may have become part of the bureaucrats' work culture and habit, so that increased legal pay may not do much to reduce corruption, at least initially.

Fortunately, one need not draw such a pessimistic conclusion from this exercise if one realizes that the public sector wage is but one of the elements in a successful anti-corruption campaign. We now turn to another important component.

### (iv) Other elements in an anti-corruption strategy

The ability of a country to detect acts of corruption and to prosecute those guilty of committing them is essential in any fight against corruption.

There are several channels through which the capacity for detection and punishment is realized. Let me mention seven of them here:

(i) an independent and impartial judicial system;

(ii) an official anti-corruption agency such as Hong Kong's Independent Commission Against Corruption;[291]

---

[289] J. Rauch and P. Evans, op. cit.

[290] World Bank, *World Development Report 1997* (Washington, D.C.), and C. Van Rijckeghem and B. Weder, *Corruption and Rate of Temptation: Do Low Wages in the Civil Service Cause Corruption?*, IMF Working Paper WP/97/73 (Washington, D.C.), June 1997.

[291] For a discussion of Hong Kong and Singapore's anti-corruption measures along this and other lines see J. Quah, "Singapore's experience in

(iii) the existence of grassroots "watchdog" organizations;

(iv) a telephone "hot line", such as those in Mexico and the United Kingdom that allow citizens to complain directly to the government;

(v) public opinion surveys such as those carried out by the Public Affairs Centre in Bangalore, India or by the World Bank's Economic Development Institute in other countries that register the public's attitude, particularly that of the poor, towards corruption;

(vi) freedom of the press to bring to light any official corruption;

(vii) democratic institutions that serve the dual purpose of enabling corrupt officials to be thrown out of office by the populace and protecting those individuals and organizations that dare to expose corrupt officials.

All of these channels are potentially important. There are some case studies and much anecdotal evidence that demonstrate both their effectiveness in specific countries and time periods, and suggestions on how to implement them.[292] It is possible that the extra revenue collected by the government as a result of the actions of the various anti-corruption bodies can exceed their cost.

While the importance of these channels is intuitively straightforward, so far there has been very little systematic statistical analysis of their relative importance for a broad sample of countries. This will be a very fruitful area for future research.

One question that has received some statistical attention is the relationship between decentralization (the devolution of power from central to local governments) and corruption. In terms of logic, decentralization could reduce corruption if it can help to increase the accountability of government action. However, it could also increase corruption if the propensity and scope for rent-seeking are greater at the local than at the central level. While the theoretical prediction is ambiguous, Fisman and Gatti found that countries with a higher degree of fiscal decentralization (a larger share of total government expenditure by local governments) tend to be those with lower levels of perceived corruption.[293] This is the first regression study of this question, so it is very valuable. Of course, the central-local government relationship may not be well captured conceptually by the expenditure shares, so this paper is unlikely to be the last word on the question.

### (v) International pressure

There are two kinds of international pressure that can be brought to bear on the corruption problem. First, international organizations such as the United Nations Development Program, the World Bank, the International Monetary Fund, the EBRD and the like, can provide persistent moral persuasion as well as technical assistance[294] to induce or help countries in their fight against corruption. Various conferences on good governance and corruption organized by the UNDP, the World Bank and so on are also useful. Cutting off loans, or threatening to do so, by the IMF or World Bank on the grounds of corruption in recipient countries may be even more effective at the margin in some cases.

The second channel is concerted international effort to criminalize the offering of bribes by multinational firms to host countries' officials. So far, the United States is the only major source country of international direct investment that has such a law – The Foreign Corrupt Practices Act (FCPA) of 1977 – that prohibits its companies from bribing foreign officials. For most other major source countries in the OECD, not only is it not illegal to bribe foreign officials but until very recently, it was also tax-deductible.[295] The United States law has not been very effective in reducing corruption in foreign countries, mainly because companies from other countries are too eager to pick up the business that the United States firms lose because of the law.[296] Corruption-prone foreign officials do not feel enough pressure to change their behaviour even if they are genuinely interested in attracting foreign investment into their countries. An international treaty that bans foreign corruption could strengthen the collective ability of all major multinational firms not to pay bribes. They are more likely to resist the demand for bribes if they can be confident that they will not lose business to their competitors as a result.

We should, however, not entertain any romantic hopes as to the effectiveness of international pressure. First, the mandates of almost all international governmental organizations place some limits on how far the anti-corruption objective can be pursued in their activities. If the World Bank were to suspend lending to countries with severe corruption ratings according to

---

curbing corruption", in A. Heidenheimer, M. Johnson and V. LeVine (eds.), *Political Corruption: A Handbook* (New Brunswick, NJ, Transaction Publishers, 1989), and idem, "Controlling corruption in city-states: a comparative study of Hong Kong and Singapore", paper presented at the conference, The East Asian Miracle: Economic Growth and Public Policy, Stanford University (Palo Alto, CA), 25-26 October 1993.

[292] For example, see the cases presented at the Ninth International Anti-Corruption Conference in Lima, Peru, September 1997.

[293] R. Fisman and R. Gatti, "Decentralization and corruption across countries", Columbia Business School and the World Bank, mimeo, 1999.

---

[294] Proper procurement guidelines are an example of this.

[295] Britain has a 1906 law that can be interpreted as prohibiting its firms from bribing foreign officials, but essentially it is not enforced.

[296] Hines found that United States firms do invest less in more corrupt countries. J. Hines Jr., *Forbidden Payment: Foreign Bribery and American Business After 1977*, NBER Working Paper, No. 5266 (Cambridge, MA), September 1995. Wei found that United States firms are not very different from those from other OECD source countries in this regard, and hence United States firms' behavior may not be attributable to the FCPA. S.-J. Wei, *Why is Corruption...*, op. cit. An article in the *Wall Street Journal*, "Greasing wheels: how U.S. concerns compete in countries where bribes flourish?", 29 September 1995, suggests that some firms may indeed be evading the requirements of the law.

Transparency International, it would have to stop half or more of its loans. That is not realistic as it would contradict its other very important objectives and possibly question the survival of the organizations.

Second, and more importantly, domestic efforts and domestic institutions ultimately determine the success of any anti-corruption program. If government officials do not intend seriously to reduce corruption, they will simply not request a loan from an international organization that requires corruption reduction as a prerequisite. So while international pressure is useful and should be applied whenever and wherever possible, it should be regarded as a complement to other domestically-based reforms.

### (vi) Political economy considerations and "special governance zones"

It is observed that following a price liberalization or exchange rate stabilization in a developing country, the finance minister or the prime minister often has to leave office involuntarily. This suggests that economically efficient reforms can be politically risky for individual political leaders. Similarly, comprehensive reforms that are necessary to reduce corruption can also be politically risky. In addition, anti-corruption reform can be expensive, as discussed in section 4.5(iii) above. Finally, even if we are sure that we know why corruption is low in Singapore and Sweden (which is a big "if"), it is quite another story to convince a highly corrupt country to do what Singapore or Sweden is doing in their country. Local culture, history and institutions matter, and a combination of these considerations often results in political inaction.

Are there reform proposals that can deal with these kinds of political economy considerations better than the usual comprehensive national reform programme? A "special governance zone" (SGZ) is one possibility.[297] An SGZ is an enclave within a country within which a comprehensive set of reforms can be undertaken ahead of the rest of the country. It is small enough for the perceived political risk to be smaller than a national reform and for a given amount of financial resources to make a bigger difference (for example, it would be possible to raise the civil servants' salaries all the way to their appropriate level). And it is explicitly an experiment: a "blueprint" based on international experience can be fine-tuned to fit local conditions. The initial success in an SGZ would not only provide a model for the rest of the country, but it would also put pressure on political leaders in other regions to adopt similar measures to reduce corruption. Because comprehensive reforms can be made within an SGZ, it has distinct advantages over an alternative partial reform proposal that focuses on a particular function of government (e.g. tariff collection).

## 4.6 Concluding remarks

While one may quote examples of firms and people benefiting from paying a bribe or the opportunity to pay a bribe, the evidence surveyed here suggests that the overall effect of corruption on economic development is strongly negative. Recent, systematic research finds that corruption is negatively related to a number of "desirable items" (such as income levels). There are several channels through which corruption hinders economic development. They include lower levels of both domestic and foreign direct investment, overblown government expenditure, a distorted composition of government expenditure away from education, health and the maintenance of infrastructure, towards less efficient but more manipulatable public projects. As globalization deepens, countries with severe corruption problems are likely to be left increasingly behind. Not only do they tend to benefit less from globalization but they are also more vulnerable to the risks of globalization. Therefore, anti-corruption measures are increasingly important.

The fight against corruption has to be multi-fronted. While laws and law enforcement are indispensable, countries serious about fighting corruption should also pay attention to reforming the role of the government in the economy, particularly in those areas that give officials discretionary power, as they tend to be hotbeds of corruption. Recruiting and promoting civil servants on the basis of merit, and paying them a salary competitive with private sector alternatives, will help to attract high-quality, moral civil servants. International pressure on corrupt countries, including criminalizing the bribing of foreign officials by multinational firms, is useful. But the success of any anti-corruption campaign ultimately depends on the reform of domestic institutions in corrupt countries.

Political economy considerations are important for a successful strategy to get underway. A special governance zone within a country may help to reduce the perceived political risk, make it financially more affordable, and allow more scope for local adaptation. In other words, it allows the political leaders to get away from the narrow choice between embarking on a risky nationwide reform and doing nothing. It can therefore help to enhance the probability of an initial success that can then generate momentum for further reforms.

---

[297] Suggested in S.-J. Wei, "Special governance zones: a practical entry-point for a winnable anti-corruption strategy", paper presented at the 11th International Anti-Corruption Conference (Durban, South Africa), 1999.

# DISCUSSANTS' COMMENTS ON PROFESSOR WEI'S PAPER

## 4.A  Krassen Stanchev

It is a pleasure and an honour to comment on Professor Wei's challenging paper. It provides a survey of different approaches to corruption and attempts to outline a policy mix to combat it. It is difficult to isolate transition economies from the overall research field but in these comments I intend to suggest some transition peculiarities, which might be taken into account and thus support Professor Wei's suggestions as to what can be done.

### General beliefs

Honesty is the best policy, because it is better than any other. In the economy, such a policy eliminates or reduces government control over prices, profits and contracts. The aggregate effect is believed to be lower transaction and contract enforcement costs and increased economic freedom. At the same time, a provisional Kantian *replique* to the original Franklin's proverb would be: yes, honesty is the best policy because it is better than any politics.

Honesty is to be sought in relations between the state, its representatives, i.e. bureaucrats, and entrepreneurs. Entrepreneurs, within the existing freedom of choice, aim to reduce the above-mentioned costs (and thus increase their profits). The bureaucrats also aim at maximizing their specific utility. Unlike entrepreneurs they are non-profit organizations, and they finance themselves, at least partially, through the distribution of funds collected from others (mostly taxes) and/or grants.[298]

### Transition and corruption surveys

If we look at the relationship as an exchange, on the bureaucrats' side there is a supply of goods and services, which are reported as a share of total government expenditure. On the entrepreneurs' side lies an objective to supply the same or similar goods and services, not as redistribution but as a search for paying customers.

An important principle in analysing corruption is the conviction that on the market both buyer and seller give less than they get. In other words, if anti- and non-market mechanisms of the exchange were "greasing", they would tend to decrease transaction costs. Mr. Wei's question, whether corruption is grease or sand, is in a sense pre-answered. As market outsiders, "bureaucrats" are theoretically doomed to block economic exchange. Corruption surveys tend to raise empirical evidence for common theoretical hypotheses.

Transition (in central and eastern Europe (CEE)) is about establishing, or re-establishing: a) bureaucrats, b) entrepreneurs, and c) the relations between the two, including the supporting political background for the bureaucrats (i.e. the political parties). Transitions seem to launch multifaceted processes of mutual impacts. Thus, a regulation imposing barriers to market entry that supposedly influences the number of competing entrants, and the distribution of opportunities, may serve as a systemic source of corruption, but the development of competitive companies, of their clusters, trust structures and critical mass of influence may lead, over time, to an amendment of the regulation and the lifting of the barriers. Mr. Wei's paper provides some guidance as to the expected effects of such developments.

### Honest policies and economic freedom

It seems that the expert community believes also that the basics of honest policy and freedom coincide: countries tend to have a similar rank for indices of both economic freedom and the perception of corruption. Wei's analysis does not deal with this issue as such, but the correlation is fairly obvious, whatever index of economic freedom is used for comparisons with corruption levels. Although economic freedom is measured relatively objectively, while the corruption indexes measure perceptions, the country correlations between the two are significant, recurrent and frequent. The leading 20 countries in terms of economic freedom (measured by the indexes of the Freiser Institute, the Economic Freedom Network and the Heritage Foundation) are also at the lower end of the scale of perceived corruption.

Comparisons among CEE and the EU accession countries are likely to provide more insights into the interdependencies between proper reforms, FDI, and greater integrity of the front-running countries vis-à-vis the reform and accession laggards.

I can summarize the economic sense of the above reasoning in the following way. In a single exchange between a bureaucracy and an entrepreneur, the rent (price) for getting past the bureaucrats might result in either a positive or negative outcome, i.e. the transaction intermediated by the bureaucrats is either successful or not. For a sequence of transactions, however, the quality of the environment deteriorates due to the mechanism of supply and demand of the most favourable (and least cost) conditions getting blocked. The cumulative effect on net investment impedes growth and decreases wealth. Wei's paper summarizes the evidence for the above, adds new facts and details, and sketches a set of policy proposals to prevent corruption and support economic growth.

---

[298] For a detailed description see W. Niskanen, Jr., *Bureaucracy and Public Economics* (Aldershot, Edward Elgar, 1994), pp. 15-23.

Transition and accession reforms that fail to establish economic freedom and/or freedom-improvement mechanisms tend to make investment more expensive and reforms less transparent, thus increasing the risk of transition failures.

### Structural policies

In transition economies, there may be policy mixes where governments attempt to simultaneously pursue incompatible policies: liberalization and price adjustments to levels of developed market economies, but at the same time preserving the "license requiring state" (LRS), and exercising control over entrepreneurship and investment within the country via non-financial constraints. The LRS can be defined as one in which the *bureaucracy* functions as an agent for collecting rents and quasi-taxes. The performance of the bureaucracy may differ from one policy area to another. The transparency of a country's economic arrangements should presumably be reflected in the relations between the different areas of structural reform. Here are some examples.

In the field of privatization Bulgaria has the most complicated legislation in CEE, as well as a controversial method of implementation and a lack of consistency. In Poland, privatization was essentially delayed while other policies and "Greenfield" investments (domestic and FDI) compensated for the lack of structural change.

In taxation, Bulgaria (since 1997) has had a system which is consistent and simpler, flatter and more impartial than, for example, those of Poland and Slovakia.[299]

Bank liberalization (understood as privatization, opening up to competition, sound central banking and the enforcement of creditors' rights) was somewhat delayed, for instance, in the Czech Republic and Slovenia, but fiercely and one-sidedly (free entry without foreign institutions and lack of supervision and protection of creditors' rights) enforced in Bulgaria.

In all these cases the overall impacts on growth and prosperity surfaced with a considerable time lag. Something similar holds for the development of corruption, with presumably very different costs in the three countries.

However, in all these cases, it makes little sense to try to determine what is a cause of what: whether corruption undermines reforms and wealth creation, or the transition policy constellation causes the former. The important thing is to get a critical number of policies right.

### Bulgaria's key intervention principle before 1997

In order not to offend anybody I will take my own country as an example. Between 1990 and 1997, the key principle of institutional reform in Bulgaria[300] was flexible or no regulation when using other people's and taxpayers' money but overregulation of the use of private properties and savings. The former is to be found in the rules of the banking sector and in the manner in which state owned enterprises are run. The latter are protected by entry barriers in the form of licencing, permit procedures and other requirements, applied with discretion and at high cost to private sector enterprises. The cost to the latter of complying with these regulations in Bulgaria (1999-2000) was 168 man-days and $5,556. The cost of *not* complying was 50 man-days and $2,105. In other words, compliance was considerably more expensive than breaking the law. The in-kind costs alone of all enterprise sectors in 1999 were equal to 12 per cent of GDP, of which 4.2 per cent of GDP were the costs to small and medium enterprises.

The number of permits for separate business activities established by the new laws (i.e. excluding minor acts, instructions, executive orders, etc.) in Bulgaria (1989-2000) was constantly rising before 2000:

|      | Newly enforced permits | Number of permits in place |
|------|------------------------|----------------------------|
| 1989 | 2                      | 2                          |
| 1990 | 4                      | 6                          |
| 1991 | 3                      | 9                          |
| 1992 | 1                      | 10                         |
| 1993 | 5                      | 15                         |
| 1994 | 6                      | 21                         |
| 1995 | 21                     | 42                         |
| 1996 | 13                     | 55                         |
| 1997 | 10                     | 65                         |
| 1998 | 21                     | 86                         |
| 1999 | 20                     | 106                        |
| 2000 | 20 (-26)               | 100                        |

*Source:* Institute of Market Economies (Sofia).

### Some conclusions

It is a common truism that what is good for the economy and prosperity is good for curbing corruption. The question is where this honest policy mix will come from?

Imagine a government has a programme for the creation of a "competitive economy driven by a viable private sector"; this programme is backed by the IMF (is equipped with the necessary set of sticks and carrots); but no honest policy is possible without measures to reduce the government stake and interference in the economy. The key question is to what extent such policies have won "the hearts and the minds" of the "bureaucracy", whether the habits of integrity and the commitment to downsize bureaucratic involvement in economy will persist after the international financial institutions have left.

If public-private sector relations and legitimate means of financing of political parties are not yet established, an imaginative quick fix of corruption in a transition economy might put the entire system in jeopardy, eventually imposing considerable risks to business and prosperity.

Also, there may be situations in which a government, despite notorious deficiencies in transparency, is supported internationally for non-economic reasons, e.g. political risks, threat of war or as a

---

[299] M. Bak, J. Rencko and K. Stanchev (eds.), *Need for Deregulation of Tax Systems in CEE: A Comparative Study of Bulgaria, Poland and Slovakia* (Sofia, Institute for Market Economics, 1998) (http://www.ime-bg.org).

[300] For a comprehensive study of the systemic sources of corruption in Bulgaria see Y. Ganchev, A. Yonkova and K. Stanchev (eds.), *Administrative Barriers to Business Activities* (Sofia, Institute for Market Economics, 2000).

counterbalance to other external risks. I think the last 10 years in the Balkans can provide plenty of such examples. In principle, these perceptions of non-economic risks (albeit with economic impacts) suggest that corruption might sometimes be a cost-effective alternative to other unfavourable developments. These situations must be isolated and separately researched and dealt with. Then, I suppose, the link between transition and corruption would become more visible while the level of perceived corruption could be expected to be somewhat lower.

## 4.B Ivo Bićanić

Economic corruption has always existed but until recently mainstream economists did not consider it a topic worth discussing in its own right. Economic theory had no place for it and even those fields of research which dealt with similar topics simply left it out (e.g. studies of the unofficial economy). But the fact that economists ignored corruption did not make it less important. Indeed quite the opposite: economists have realized that "real world" outcomes in many economies depend on corrupt practices and that for many environments it simply cannot be left out.

So under the influence of events over the last decade economists have turned their attention to corruption. Theory has designed models and analytical tricks which could start handling it (assymetric information, incomplete contracting, rent-seeking, etc.) while empirical research increasingly grappled with how to measure this "soft", and by definition unreported, variable into quantitative analysis. Thus the economics of corruption entered "respectable economics" and became part of the shift in economics which included a host of similar real world topics (democracy and democratic deficits, social capital and business culture, inequality and rent-seeking, etc.).

With time the sophistication of the analysis of corruption increased. How far it has advanced in the field of quantitative analysis is admirably shown in this paper by Shang-Jin Wei. His paper goes as far as one presently can in applying rigorous economic analysis and deriving robust results to the study of corruption. The regressions, surveys and quite a few causalities in the paper show how economists can now discuss in "hard" terms what was hitherto left to the "soft" approach. In this sense one extremely important result in the paper, worth stressing because of its policy implications, is that under no circumstances is there justification for corruption. Allowing for any level of corruption is a second best solution. This is an important finding given the often argued defence of corruption as a means of bypassing red tape and unreasonable bureaucratic interference (the former SFR of Yugoslavia even had a term: "useful malpractices") or as a cultural tradition of doing things.

In spite of these welcome results much of the study of corruption still remains in the "soft" area and it is useful to remember its importance. One such area of corruption could be called "small-scale corruption" (as opposed to "grand corruption"). Small-scale corruption concerns both the business and consumer side of the economy as well as civil servants and public administration. On the supply side it consists of the small- or medium-sized entrepreneur paying the tax inspector who checks his books, the coffee house owner paying the sanitary inspector to turn a blind eye, the street vendor paying the policeman to let him sell, and many similar cases. On the demand side it's the shopper paying the customs official to look the other way, the patient paying medical staff for better treatment or for some queue-related service (either to jump, or to avoid being jumped, in the surgery queue) and many similar cases. A recent study conducted in many countries by Gallup showed that in Croatia the highest level of corruption is in the customs service, the police and the medical profession. Other studies of Croatia show that for public services it is an expected way of going about obtaining one's entitlements.

"Small-scale" corruption is thus something going on as part of everyday, ordinary lives and one is involved in it just to manage and muddle through in "untidy economies" (another "soft" but useful term with a strong intuitive meaning). It does not create big pay-offs and certainly is not at the top of the list of anti-corruption campaigns; it does not make anyone fabulously rich and generally is out of the limelight. However, very importantly, the difference between "small corruption" and "grand corruption" is not only a matter of scale: there is a qualitative difference. It has a completely different role in "untidy" economies, where it is structurally incorporated, from tidy ones where it is negligible. Thus, in this respect one cannot talk a continuum of economies but of a point of discontinuity among them. Small-scale corruption does have three important consequences which make it stand out and justify special attention. First, it has become part of the social capital of untidy economies. Second, it is incompatible with tidy economies. Thirdly, it is extremely difficult to control and stamp out.

The first feature is that "small-scale" corruption is "structurally incorporated", i.e. it is an integral part of social capital. As a result it is persistent, difficult to eliminate and a major hurdle. Croatia can again provide an example of this (this small south-east European economy is not being cited because of any special importance: the same is applicable in most economies, in transition, in Africa, South America or South-East Asia). During the 1980s, as socialism was disintegrating, a whole generation of small-scale entrepreneurs emerged, both in the socialized sector and outside it creating a fuzzy edge to the socialized sector. These entrepreneurs were closely linked to the socialized sector (indeed often formally employed there) but were often involved part-time in unofficial economic activities and frequently in corruption. Earning quasi-rents in a badly administered economy with incomplete and unregulated markets was very lucrative. They developed a remarkable talent for survival and admirable business imagination. Before laws were passed they already knew how to bend them their way. With the onset of transiton they were best poised to benefit from the disorganization: the public administration did not improve, the scope of unregulated markets increased and they had the know-how. And they did benefit, not according to the new rules of a contractual market economy, transparency and competition, but by doing what they knew best with a high dependence on corrupt and unofficial practices, small as well as medium scale. Now, 10 years later, they are the captains of industry and have made respectable fortunes by any standards. However their behaviour has also led to path dependency and their mixture of a corrupt and contractual economy remains dominant and not stigmatized. Similarly, with understaffed and underfunded public services and an erosion of entitlements, consumers were increasingly making the kind of side payments to which they were already accustomed.

This structural incorporation of small-scale corruption has an important policy consequence. Recalling Wei's conclusion, that corruption is always a second best, raises the central issue which is not purely academic for those living in untidy economies. That is, the question of how to dismantle an economy with widespread "large and small corruption". Again Croatia can provide an example. Croatia has now had two experiences of dismantling economies with high levels of overall corruption. In the early 1990s it was degenerate socialism and now it is crony capitalism. The first dismantling was an unneccesary failure and regarding the second the jury is still out and seems likely to be so for quite some time: and the likely verdict is far from clear.

Drawing from this experience it seems important to distinguish between large- and small-scale corruption in three important senses. The first is the problem of the "biblical first stone", the second the issue of "trickle down" and the third the dependence on "homegrown efforts".

The significance of the first point becomes clear if one is reminded of the "biblical first stone". Who can throw it now that "we are all corrupt" on a small scale and when we all depend on it to muddle through? Not less important for those who would take the initiative (the politicians) and those who would implement any crackdown (the state administration), there are some major participants in small-scale corruption while organizing the donors runs into the problem of organizing any numerous and diverse group whose perceived benefit is not that large (the case is similar to that of tariffs in international trade: a few large gainers versus many small losers).

Regarding the second two points, again the Croatian experience can be used as an example. In both cases of dismantling corruption the effect on small corruption has been limited. This is not true of "grand corruption" which in many respects seems easier to deal with. Putting the most visibly corrupt politicians and businessmen who have not fled the country into jail is not that difficult (small numbers, a lot of public attention and support, sufficient funds, fulfilling election promises). In many respects it is similar to winning elections. In fact those are the "easy shots". Dealing with the corruption which permeates every pore of the economy (and society and the minds of people), which involves the vast majoritiy of the population, is a hard, slow upward effort with few or no spectacular stories to capture public attention, and the whole effort has no clear ending, numerous opponents and many setbacks. Furthermore, experience shows that due to the discontinuity of the two there is not much "trickle down" either. Showcases of "grand corruption" have not had major effects on "small corruption"; dealing with one does not automatically solve the other. But regarding Croatia's second transition (a task facing many transition economies not in the "first wave"): with only a limited impact on small corruption crony capitalism cannot be dismantled, social capital cannot start growing and the moral economy cannot be revived. (I thank my colleague Vojmir Franičević for pointing to the importance of this: like social capital it is yet another vague term that economists shy away from but are increasingly realizing that it is unavoidable in "real world economics".) The advantage noted by Tanzi that market efficiency would improve with heightened ethical standards does not materialize.

The third point regarding "homegrown efforts" is not less important. It is linked to the interrelationship between size and corruption. While there may not be a direct link between the share of corruption and the size of the economy, we do have large very corrupt economies and small tidy ones, and the relationship may be relevant for efforts to dismantle it. It is simply a matter of the amount and variability of the available human capital. Small economies simply have fewer resources which are less varied. Fewer judges, policemen and civil servants, fewer managers and businessmen, etc. (Again an example from Croatia from a different field but it can illustrate the constraint: Croatia is a small country of just over four million – it has five monetary economists and there are simply not enough for each party to have one.) Not only is it difficult to replace one lot of civil servants with another with the required expertise (small corruption does not require ignorance, just low ethical standards) but it is impossible to import the human capital required for dismantling small corruption. So small economies have to make do with the people they have (skeletons in the cupboard included). This places extra burdens on the dismantling of small corruption. While showcases of "large-scale" corruption can be monitored, administrative foot-dragging discovered and perhaps even human and social capital imported to deal with it, dealing with "small-scale corruption" is much more difficult.

Success in dismantling small-scale corruption is central to transforming an untidy economy into a tidy one and, with it, to the success of economic transition and of sustaining economic growth with consequent economic convergence. There is another reason for concern regarding the likely success of the dismantling process. Of the four economic pathologies so aptly distinguished more than 30 years ago,[301] two, the economics of the underground and the economics of overstrain, are linked to social capital, but the other two, the economics of overlord and overchange, are linked to the process of dismantling. The process of dismantling crony capitalism and extensive small corruption can itself lead to new pathologies of corruption and thus be counterproductive.

---

[301] R. Bićanić, "How not to develop a country: an essay in economic pathology", *Turning Points in Economic Development* (The Hague, Mouton, 1972).

# CHAPTER 5

# GOVERNANCE IN TRANSITION: THE CHALLENGE OF SUBVERTING CORRUPTION

*Antoni Z. Kamiński and Bartlomiej Kamiński*[302]

## 5.1 Introduction

Corrupt behaviour, commonly defined as abuse of public office for private gain, has no regard for borders or cultures. Although corrupt practices are a universal phenomenon, their pervasiveness varies widely across countries. In some countries, the incidence of corruption is low and limited to petty corruption, in others it is large scale and systemic, distorting incentives, eroding institutions, and redistributing power and wealth to the undeserving.

Corruption appears to be inversely correlated with the level of economic development – affluent societies are less corrupt than poor societies. There are reasons to believe that corruption paralyses political development and preserves non-democratic regimes.[303] Recent research has produced empirical estimates of its negative impact on economic growth. Hence, no matter whether one deals with petty, large-scale or systemic corruption, governments committed to increasing economic welfare have to address this issue.

But corruption is not the cause of poor governance. The causative links appear to be the reverse: these are rather dysfunctional institutions, a lack of transparency in political decision-making, and the expectation of high payoffs combined with the absence of behavioural ethics that breed corruption and corrode the political system. Measures restricting the role of the state in the economy, by removing excessive government controls and simplifying administrative procedures, can both improve governance and reduce corruption. On the other hand, increasing the criminal penalties for bribe-taking or the pay of public officials without addressing the problems that plague a country's institutional landscape are unlikely to reduce the incidence of corruption.

In the absence of institutional reforms, new laws and regulations aimed at curbing corruption often have perverse effects. In regimes not fully based on the rule of law, they may increase the potential for corruption. Under the conditions of non-democratic regimes, they have often been used as instruments of selective repression and as an alternative to reform.[304] Politics, rather than observance of the spirit and the letter of law, drives their use. Examples abound, ranging from the current experience of China in fighting corruption to the earlier experience of Soviet-style communist countries. Both cases provide support to the observation that corruption cripples political development and perpetuates non-democratic regimes.

The last two decades have witnessed worldwide convergence in what constitutes the right institutional design and policies for survival and development in the contemporary global economy, often referred to as the second global economy, in contrast to the late nineteenth century world economy. The institutional design consists of democracy as the way of political organization of society, and markets as the way to organize economic activity. Democracy, once regarded as an impediment to accumulation and growth, is now seen as improving the quality of decision-making and, more generally, governance. Opening to the world and maintaining macroeconomic stability is among the policies regarded as necessary to take advantage of the opportunities offered by globalization. Last but not least, governments are willing to accept external disciplines in some areas of their economic policy-making in return for the benefits derived from participation in various international organizations (e.g. the World Trade Organization).

Paradoxically, the shift towards democracy and markets creates more demanding tasks for governance.

---

[302] A revised version of a paper presented at the UNECE Spring Seminar, *Creating A Supportive Environment for Business Enterprise and Economic Growth: Institutional Reform and Governance* (Geneva), 7 May 2001. The authors would like to thank Martin Loken, Jorge Braga de Macedo, Mark Pieth, Beata Smarzynska and Peter Voitsekhovsky for their very helpful comments. The usual caveats apply.

[303] L. Whitehead, "High-level political corruption in Latin America: a "transitional" phenomenon?", in J. Tulchin and R. Espach (eds.), *Combating Corruption in Latin America* (Washington, D.C., Woodrow Wilson Center Press, 2000), p. 108.

[304] S. Huntington, *Political Order in Changing Societies* (Yale and London, Yale University Press, 1968), p. 66.

As the cases of the United States, on the one hand, and the Korean financial meltdown in 1997, on the other hand, amply demonstrate, the freeing of markets calls for more sophisticated regulatory structures. Markets cannot deliver improved economic performance if they are not supported by complex institutional arrangements assuring competition, predictability and the enforcement of property rights and, in consequence, low transaction costs. Similarly, democracy will fail to deliver improvements in the quality of governance if it is not supported by an expanding civil society, powerful mass media and accountability of those in power. The two outcomes are interconnected: improvements in governance will usually lead to more competitive and open markets. The constitutional design of the state should be such as to make sure that the political environment is conducive to citizens' participation and the development of civil society, and that the economic environment encourages private business activity.

While mature democracies have most of the ingredients required to assure a high quality of governance, their existence cannot be taken for granted in developing countries, including societies in transition from communism. They still lack a developed civil society and independent mass media (TV in particular). In consequence, the accountability of those in power leaves in most cases much to be desired. The choice of electoral rules and, more broadly, of a grand political institutional design has often created conditions not only conducive to the widespread abuse of power and rent extraction by policy makers and bureaucrats but also to the effective blocking of the institutional changes that would improve governance.

Governance is used here in the broad sense of assuring the functioning of a market-based democracy – better governance means more competitive markets, decreasing prospects of higher taxes in the future, stronger civil society, greater accountability of those in public service and less obscure rules separating the public from the private sphere. Corruption is a symptom of poor governance. Fighting symptoms is always less effective than directly addressing their roots. Establishing a national anti-corruption agency may be a good way to fight corruption,[305] but only to the extent that its mandate is limited to a well-defined area of public policy (e.g. public procurement). Success in containing corruption does not necessarily bring improvement in governance if its underlying institutional design curbs efficiency and innovation.

Thus, the emphasis in designing anti-corruption strategies, especially in transition societies, should be first and foremost on the improvement in governance. Its main objectives should be defined as the modernization of governance structure conceived as (1) the development of an efficient, accountable system of governance; and (2) the building of institutions to support the development of competitive markets. These two objectives are interrelated. In fact, when complemented by outward-oriented policies, they constitute what, according to the emerging consensus on economic development, is the best response to the twin challenges of globalization and governance that reflect respectively "… the constraints resulting from international exchange and property rights."[306]

While this paper addresses the issues of governance and corruption in general terms, the focus is on post-communist countries. But despite the shared legacy of communism, they are neither a distinct group vis-à-vis countries that had not experienced communism nor are they a homogenous group. The same institutional design of communism has produced very distinctive experiences in the central and east European countries (CEECs)[307] with implications for subsequent transitions. Communism does not seem to be a defining feature of the incidence of corruption. CEECs[308] have made much greater progress towards establishing market-based democracies than the former republics of the Soviet Union (hereafter CIS) excluding the Baltic states. The level of corruption is significantly lower in the CEECs than in the CIS economies, which as a region has the highest incidence of corruption in the world.[309] But within both groups, especially among the CEECs, there is considerable variation in the levels of perceived corruption.

The high variation of corruption among post-communist countries testifies to their different historical experiences. It seems that the historical experience has contributed to distinct strategic choices being made at various stages of transition from communism and, consequently, different outcomes in terms of structures of governance, i.e. economic regimes with their respective political arrangements. The CEECs mostly opted for a fundamental break with their communist past, with their populations willing to accept high short-term, as it turned out, costs.[310] They adopted a radical approach to transform their economic regimes, whereas the CIS chose the path of gradualism. The former opened their economies to competition and encouraged the entry of

---

[305] J. Pope and F. Vogl, "Making anti-corruption agencies more effective", *Finance & Development*, Vol. 37, No. 2, June 2000, pp. 6-9.

[306] J. Braga de Macedo, "Converging European transitions", *The World Economy*, Vol. 23, No. 10, November 2000, pp. 1335-1365.

[307] G. Ekiert, *The State Against Society. Political Crises and Their Aftermath in East Central Europe* (Princeton, Princeton University Press, 1996).

[308] CEECs include 10 central European applicants to the EU: the Czech Republic, Estonia, Hungary, Poland and Slovenia (the so-called Luxembourg Group); and Bulgaria, Latvia, Lithuania, Romania and Slovakia (the Helsinki Group).

[309] World Bank, *Anti-corruption in Transition. A Contribution to the Policy Debate* (Washington, D.C., The World Bank, 2000), p. XI.

[310] People in the Baltic states were particularly exposed to hardship because of the drastic contraction in the supply of energy by Russia during the first two years after independence. Once, however, they introduced their own convertible currencies, they were able to base commercial relations with Russia on a sound economic basis. This experience contrasts dramatically with the experience of CIS net energy importers.

new firms, whereas in the latter gradualism in tackling first-generation reforms created enormous opportunities for enrichment to a few with good access to power.

The rest of this paper is structured as follows. Section 5.2 examines the links between governance and corruption in the context of strategic political and economic decisions made during the transition from central planning. Section 5.3 discusses the political underpinnings of the capture of public policies and administrative corruption. It presents the ingredients of an anti-corruption strategy. Section 5.4 links the emergence of the issue of corruption to prominence in various international forums to the convergence of economic thinking on what policy measures lead to sustained economic growth, combined with the rediscovery of the role of institutions in economic growth. It argues that the alternative to globalization and improved governance is a "vicious cycle" of economic devolution and corruption. Section 5.5 concludes.

## 5.2 Corruption, governance and transition

Corruption is part of a larger phenomenon typical of any organized social activity. All purposeful social action depends for its success on the will of participants to conform to the rules adopted by the organization or the norms and modes of behaviour prevailing in society. Conformity to rules is a necessary condition for the survival of a social order – no matter whether this is an economic order governed by the market or a political order governed by the state. Organizations seek to assure individual and group conformity by using a combination of positive incentives, basic motivation and negative incentives. The balance may be difficult to strike as the use of one set of incentives may undermine other organizational goals by deterring, for instance, the will to produce. But to do it properly effective instruments of control must be in place. And – as Amartya Sen reminds – so must norms of conduct, whose "… importance may be particularly attached to the conduct of people in positions of power and authority."[311]

When individuals and groups in an organization follow objectives other than those formally established, the organization finds itself in crisis. A private company must assure that participants cooperate to achieve its overall objectives; when they lose sight of its basic mission and start following their own interests, the negative impact on profits will soon become visible. The same is true of any other organization, be it the army, church or an academic institution. Corruption belongs to the broad category of phenomena when formal institutional rules become inactive and people's activities become guided by concerns external to the organization's mission.

We discern corruption within this category by reserving the concept for the public domain. To speak meaningfully of the public domain, there must exist a distinction between the public and the private. The public domain has to emerge as an autonomous institutional system effectively claiming loyalty and obedience on the part of citizens. This aspect of the problem has been well grasped by Samuel Huntington. According to him, "Corruption is one measure … of the absence of effective institutionalization. Public officials lack autonomy and coherence, and subordinate their institutional roles to exogenous demands".[312] Thus, whatever the root of the problem, the occurrence of corruption always indicates weakness in the institutionalization of the public sphere. The public sphere is invaded, or "captured" as World Bank publications call it, by interests and motives alien to it – personal, family or private cliques.

A mature political system with a well-developed system of rules and rule enforcement is indispensable for economic development, simply because property rights must be protected, contracts guaranteed, administrative powers over the private sector limited, their use predictable, and last but not least the overall cost of public administration must be tolerable. In the economic sphere, the state then acts – to borrow apt terms from Shleifer and Vishny[313] – as the *helping hand* rather than the *grabbing hand*. Corruption undermines the ability of the state to deliver such services. Corruption means that public officials prey on the private sector, while some businessmen conspire with public officials to prey on public resources. This is the product of a bad policy environment and a major impediment to its improvement.

The task of creating a rule-governed polity and economy is, therefore, threefold. First, there is the question of the constitutional design of the state; second – of the political culture of the society; and third – the maturity of the civil society. All three areas are interrelated. In all of them post-communist societies suffer from serious weaknesses, although – as we shall see below – their intensity varies across countries reflecting their different historical experience. This section begins with an assessment of the pervasiveness of corruption in post-communist societies followed by an examination of its roots and its links with the rate of progress in implementing a new constitutional design of the state.

### (i) Corruption in post-communist societies and progress in transition

There are strong public perceptions of pervasive corruption in the CEECs and the CIS alike. As the collapse of communism also involved the contraction, if not outright collapse, of central controls that had already been strongly eroded during the final stages of the

---

[311] A. Sen, *Development as Freedom* (New York, Vintage Anchor, 2000), p. 277.

[312] S. Huntington, op. cit., p. 59.

[313] A. Shleifer and R. Vishny, *The Grabbing Hand: Government Pathologies and their Cures* (Cambridge, MA, Harvard University Press, 1998).

disintegration of central planning in the 1980s, corruption was bound to increase following its demise. The institutional vacuum created by the collapse of the communist state and the slow emergence of a "new" state compatible with the market economy has established a fertile environment for public sector corruption and, in many cases, private sector crime.

The double transition, in the political and economic realms, opened unprecedented opportunities for those in power to enrich themselves in an environment mostly devoid of any rules defining the boundaries between private and public interest. The two were fused under the communist regime, and the notion of "conflict of interest" had not been part of the political code of conduct. On the contrary, in line with the Soviet tradition of uniting political and economic power, public officials saw nothing wrong in occupying political office and owning newly privatized companies.[314] They also regarded the appropriation of public resources to themselves and their cronies as a standard practice in democracy. Private greed has dangerously undermined the credibility of public institutions in most post-communist societies.

### (a) Perceptions of corruption

Various measures of corruption seem to confirm these perceptions, albeit post-communist countries are by no means a homogenous group. First, there is a clear-cut difference between the CEECs and the CIS. The index, assuming values between 0 (no corruption) and 1 (maximum corruption) and summarizing 12 different indices of international corruption, suggests much higher levels of corruption in the CIS than in the CEECs.[315] In fact, the CIS as a group has the highest value of this composite index among eight groups of countries. OECD (excluding its most recent entrants – the Czech Republic, Hungary, Mexico, Poland and the Republic of Korea) has the lowest perception of corruption. CEECs rank third after OECD and South-East Asia, and above the Middle East and North Africa, but the difference between the CEECs and the latter two regions is minuscule. Other regions included in the analysis were, in rank order, Latin America, sub-Saharan Africa and South Asia. Hence, the CEECs as a group seem to have levels of corruption in line with those in most other developing countries, whereas the CIS countries stand out.

Second, there seems to be a significant variation in terms of perception within both groups of countries. In terms of Transparency International's corruption perception indices (CPI), the levels of corruption are significantly lower among countries of the Luxembourg group, i.e. the first five CEECs invited to negotiate accession to the EU in July 1997, than of the Helsinki group invited to begin accession negotiations in December 1999 (table 5.2.1). Within the latter, Romania stands out with the perceived level of corruption putting it in the same league as the "less corrupt" CIS countries. While no CPI values are available for three central Asian republics – Tajikistan, Turkmenistan and Uzbekistan – one suspects that together with Kazakhstan and Kyrgyzstan they would rank in the middle of the list.

Third, post-communist transition societies do not emerge as clusters in the Transparency International rankings of corruption perception. If anything, their respective ratings seem to be mostly determined by their level of development in terms of GDP per capita, albeit with some qualifications. Some of them rank higher than their GDP per capita would predict.[316] For instance, Estonia, ranked 27th in terms of the CPI, Slovenia (ranked 28th) and Hungary (ranked 32nd) are regarded as "cleaner" than two more prosperous EU members – Greece (ranked 35th) and Italy (ranked 39th). On the other hand, Russia (82nd) and Ukraine (89th) are in the same league as Angola, Kenya or Nigeria, i.e. countries with significantly lower GDP per capita.

The level of GDP appears to be a relatively good predictor of perceived corruption among post-communist countries. The coefficient of correlation between GDP per capita (column 7 in table 5.2.1) and the values of the CPI (column 2 in table 5.2.1) is quite high at 0.76. Hence, post-communist countries with higher levels of GDP per capita tend to have lower levels of perceived corruption. Again, this does not differentiate them from other societies that did not experience communism.[317]

In other words, the communist past of transition societies cannot be easily discerned from the values of the CPIs. This finding is surprising considering that the communist institutional design fused economy and polity, and rejected organizational principles that have traditionally provided powerful checks on corruption, i.e. the rule of law, democracy and civil society in democratic societies.[318] Communist regimes had sought actively to eradicate skills and motives determining society's

---

[314] V. Coulloudon, "The criminalization of Russia's political elite", *East European Constitutional Review*, Vol. 6, No. 4, 1997, pp. 73-78.

[315] World Bank, *Anti-corruption in Transition ...*, op. cit., p. XIV.

[316] There is clearly a correlation between a country's level of corruption and its development. Basic social and economic data (level of GDP, education, openness, religion) may explain between 85 and 90 per cent of the variation in the CPI. T. Persson, G. Tabellini and F. Trebbi, *Electoral Rules and Corruption*, CEPR Discussion Paper, No. 2741 (London), March 2001. But there are still large variations in the incidence of corruption among countries at roughly similar levels of economic development. D. Kaufman, "Corruption: the facts", *Foreign Policy*, Summer 1997.

[317] Some authors see corruption as an idiosyncratic feature of post-communist societies. R. Rose, W. Mishler and C. Haerpfer, *Democracy and Its Alternatives: Understanding Post-Communist Societies* (Cambridge, Polity Press, 1998). These data and subsequent analysis suggest that this is not the case. If there is something that sets them apart from other countries at similar levels of development, it is the lost opportunity to create out of the ashes of communism governance structures minimizing the potential for corruption, effectively mediating conflicts and promoting economic growth. It is this which stands out – not the incidence of corruption.

[318] For an extensive discussion of the institutional structure of communism (or state socialism) along these lines see B. Kamiński, *The Collapse of State Socialism* (Princeton, Princeton University Press, 1991).

capacity to self-organize and make institutions based on democratic accountability work.[319] The existence of these institutions is crucial to subvert corruption, while their absence breeds corruption.

This legacy of communism seems to explain the ranking of Russia or Ukraine on a par with countries at much lower levels of economic development, but it does not explain the rankings of the CEECs, especially those from the Luxembourg group. Russia and Ukraine, with two decades more under communist regimes than the CEECs, have considerably higher levels of corruption than countries at a similar level of economic development. This is clearly not the case of the CEECs.

The legacy of communism can only be traced through the extent to which central planning suppressed economic development and thereby institutional development. This seems also to undermine Mancur Olson's explanation of the collapse of central planning. According to his argument, centrally planned economies imploded because the scale of corruption became so great that the "centre" ran out of resources, as interest groups (e.g. state enterprise managers colluding with their superiors and with other managers) organized and diverted resources away from the productive economy.[320] Since central planning disintegrated most in countries such as Hungary and Poland, they should have faced the greatest challenge of corruption after the collapse of central planning. The available evidence suggests that they did not.

### (b) Beyond perceptions: World Bank's diagnosis of corruption

Perceptions, however, are merely perceptions, and as such they may not necessarily reflect reality. Moreover, they are usually based on opinions of external observers and investors rather than those of local people. But as we shall see below, they converge to a large extent with corruption diagnosed on the basis of surveys of locals. Last but not least, perception indices do not unbundle various patterns and forms of corruption.

Several World Bank studies make an attempt to go beyond mere perceptions. They identify forms of corruption, and seek to capture qualitatively the extent of corruption.[321] The studies use data from the 1999 Business Environment and Enterprise Performance Survey based on interviews of firm managers and owners in the CEECs and CIS countries.[322] The data are used to capture two kinds of corrupt behaviour: *administrative corruption*, as revealed in the percentage of firms' revenue allotted to illicit, illegitimate payments to public officials, and *state capture*, measured by the percentage of firms directly affected by the "purchase" of laws and regulations. The index of state or economy capture is a simple average of the proportion of firms directly affected by the "sale" to private interests of parliamentary votes, Presidential decrees, central bank mishandling of funds, court decisions in commercial cases, court decisions in criminal cases, and illicit contributions to political parties and election campaigns.[323]

Three interesting observations can be derived from examining both indices of corruption presented in table 5.2.1 and portrayed in chart 5.2.1. First, contrary to expectations that countries would be clustered in two extreme quadrants – low/high administrative corruption and low/high state capture – they split almost evenly into four quadrants divided in terms of high/low levels of administrative and high/low levels of state/economy capture – seven, five, four and five (chart 5.2.1). Not surprisingly, five CEECs from the Luxembourg group are in the south-west quadrant (low administrative corruption, low state capture). Belarus, which is yet to begin the journey towards competitive markets, and Lithuania from the Helsinki group, are also in this quadrant. At the other extreme, there are five countries with very high scores for both indices of corruption – two Transcaucasian economies (Azerbaijan and Georgia), the Republic of Moldova, Romania and Ukraine (north-east quadrant). *Captive* economies include Bulgaria and Slovakia from the Helsinki group, Croatia, Latvia and Russia, although the latter is a borderline case with administrative corruption slightly below 50 per cent (south-east quadrant). Economy *capture* in two central Asian countries (Kazakhstan and Uzbekistan) and Albania and Armenia is relatively low whereas *administrative corruption* is very high (north-west quadrant).

Second, it seems that the incidence of corruption in countries located in the north-east and south-east quadrants is much more serious and more difficult to eradicate than in countries located in the north-west quadrant. High scores on both indices (north-east quadrant) indicate "systemic" corruption cutting across all levels of government. Countries falling in this group include Azerbaijan, Georgia, the Republic of Moldova, Romania and Ukraine. High scores on state capture suggest that the system acts massively and deliberately to set the rules of the game in ways that maximize rent-seeking behaviour by those enjoying political power. Under these circumstances, corruption feeds on itself by fuelling counterproductive, corruption-generating, regulations.

---

[319] A. Kamiński, "Res publica, res privata", *International Political Science Review*, Vol. 12, No. 4, 1991, pp. 337-351.

[320] M. Olson, *Power and Prosperity. Outgrowing Communist and Capitalist Dictatorships* (New York, Basic Books, 2000).

[321] World Bank, *Anti-corruption in Transition ...*, op. cit; and J. Hellman, G. Jones and D. Kaufman, *Seize the State, Seize the Day: State Capture, Corruption and Influence in Transition*, World Bank Policy Research Working Paper, No. 2444 (Washington, D.C.), September 2000.

[322] Depending on the size of a country, the sample varied between 125 and 550 firms. For details of the survey, see J. Hellman et al., op. cit.

[323] J. Hellman et al., op. cit., p. 9.

TABLE 5.2.1

**Measures of corruption and governance in the CEECs and the CIS**
(Per cent)

|  | EBRD (2000) indices of progress in transition [a] (1) | Corruption perception index, 2000 [b] (2) | State capture index [c] (3) | Administrative corruption index [d] (4) | Aggregate corruption index [e] (5) | PPP GNP per capita in 1999 (dollars) (6) | FDI over 1997-1999 as per cent of GDP in 1999 (7) |
|---|---|---|---|---|---|---|---|
| Albania | 69 | 65[e] | 39 | 70 | 55 | 3 240 | 4 |
| Bosnia and Herzegovina | 51 | .. | .. | .. | .. | .. | .. |
| Croatia | 85 | 41 | 66 | 19 | 43 | 7 260 | 12 |
| The former Yugoslav Republic of Macedonia | 79 | .. | .. | .. | .. | 4 590 | 2 |
| **Luxembourg Group** | 94 | 30 | 23 | 31 | 27 | 9 795 | 12 |
| Czech Republic | 95 | 35 | 27 | 44 | 35 | 12 840 | 17 |
| Estonia | 95 | 26 | 24 | 28 | 26 | 8 190 | 21 |
| Hungary | 100 | 29 | 17 | 30 | 23 | 11 050 | 11 |
| Poland | 95 | 37 | 29 | 28 | 29 | 8 390 | 10 |
| Slovenia | 87 | 27 | 17 | 25 | 21 | 16 050 | 3 |
| **Helsinki Group** | 82 | 45 | 56 | 42 | 49 | 6 417 | 11 |
| Bulgaria | 80 | 43 | 68 | 37 | 53 | 5 070 | 17 |
| Latvia | 84 | 44 | 73 | 25 | 49 | 6 220 | 20 |
| Lithuania | 84 | 44 | 27 | 49 | 38 | 6 490 | 18 |
| Romania | 74 | 52 | 74 | 56 | 65 | 5 970 | 13 |
| Slovakia | 87 | 43 | 59 | 44 | 51 | 10 430 | 4 |
| **CIS Group** | 61 | .. | .. | .. | .. | .. | 4 |
| **Central Asia** | 62 | .. | .. | .. | .. | 2 803 | 9 |
| Kazakhstan | 74 | 50 | 29 | 54 | 42 | 4 790 | 16 |
| Kyrgyzstan | 74 | 65 | 71 | 93 | 82 | 2 420 | 13 |
| Tajikistan | 57 | .. | .. | .. | .. | .. | 4 |
| Turkmenistan | 34 | .. | .. | .. | .. | 3 340 | 8 |
| Uzbekistan | 51 | .. | 15 | 77 | 46 | 2 230 | 2 |
| **Transcaucasian** | 66 | 75 | 59 | 85 | 72 | 2 455 | 36 |
| Armenia | 69 | 60 | 17 | 81 | 49 | 2 360 | 20 |
| Azerbaijan | 52 | 100 | 100 | 100 | 100 | 2 450 | 64 |
| Georgia | 77 | 65 | 59 | 75 | 67 | 2 540 | 16 |
| **European** | 60 | 58 | 66 | 55 | 61 | 5 503 | 3 |
| Belarus | 39 | 37 | 20 | 23 | 21 | 6 880 | 1 |
| Republic of Moldova | 74 | 58 | 90 | 70 | 80 | 2 100 | 14 |
| Russian Federation | 67 | 71 | 78 | 49 | 64 | 6 990 | 3 |
| Ukraine | 61 | 100 | 78 | 77 | 78 | 1 160 | 4 |

*Source:* Derived from data in EBRD, *Transition Report 2000* (London), November 2000; World Bank, *Anti-Corruption in Transition. A Contribution to the Policy Debate* (Washington, D.C., The World Bank, 2000) and *2001 World Development Indicators* (Washington, D.C., The World Bank, 2001); Transparency International website.

[a] Indices of progress in transition in the CEECs and the CIS derived from EBRD. For each country, eight "grades" (ranging between 1 and 4+) of the progress made in creating the foundations for competitive markets were summed with "pluses" and "minuses" treated as plus or minus 0.5. For each assessed area the maximum grade (4+) reflects the standards typical of advanced industrial economies. The index was normalized in terms of the most advanced among the transition economies. The higher the value of the index the more advanced a country is in terms of its transition to a mature market economy.

[b] The corruption perception index (CPI) as reported by Transparency International multiplied by 10 and inverted. The value of the index equal to 100 denotes the country (among the CEECs and the CIS) with the highest perception of corruption.

[c] The measure of state capture (measured by the percentage of firms directly affected by the "purchase" of laws and regulations) as reported in J. Hellman, G. Jones and D. Kaufman, *Seize the State, Seize the Day: State Capture, Corruption and Influence in Transition*, World Bank Policy Research Working Paper, No. 2444 (Washington, D.C.), September 2000, inverted and multiplied by 100 and standardized in terms of the most "corrupt" country. The closer the value of index is to 100, the more corrupt a country is in relation to other CEECs and CIS economies.

[d] The measure of administrative corruption (bribes as a share of annual revenues of firms) as reported in J. Hellman, et al., op. cit., inverted and multiplied by 100 and standardized in terms of the most "corrupt" country. The closer the value of index is to 100, the more corrupt a country is in relation to other CEECs and CIS economies.

[e] The aggregate corruption index is the simple average of state capture and the administrative corruption indices.

Third, after the collapse of communism almost all transition economies faced the task of building entirely new governance structures and mechanisms. With their policy capacity unbalanced and accountability structures underdeveloped, one would expect them to score high on both counts of corruption – decentralized administrative corruption and centralized state capture. Weak state structures seem to favour both kinds of corruption,

**CHART 5.2.1**

**Administrative corruption and state capture in transition economies**

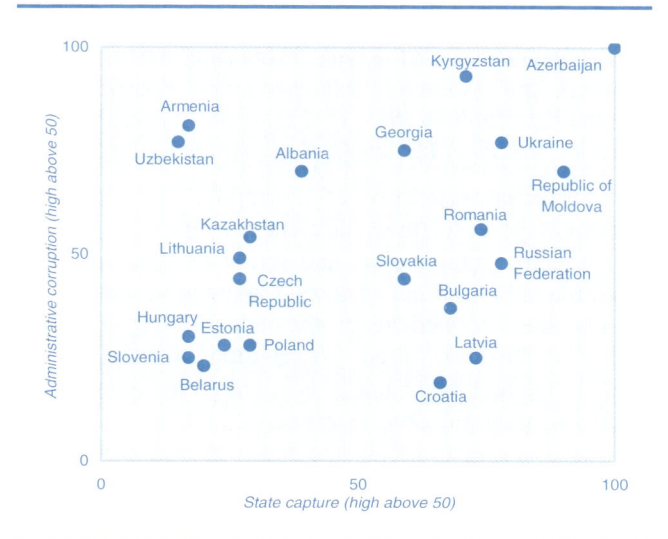

[Scatter plot with x-axis "State capture (high above 50)" from 0 to 100 and y-axis "Administrative corruption (high above 50)" from 0 to 100. Countries plotted include: Kyrgyzstan, Azerbaijan, Armenia, Uzbekistan, Albania, Georgia, Ukraine, Kazakhstan, Republic of Moldova, Lithuania, Romania, Slovakia, Russian Federation, Czech Republic, Bulgaria, Hungary, Estonia, Poland, Latvia, Slovenia, Belarus, Croatia.]

Source: Derived from data in table 5.2.1.

whereas strong centralized states might be expected to create a more fertile ground for state capture than for administrative corruption. "Decentralized" bribe-taking seems to fit countries with weak states – Albania and Armenia. The reasons for low *state capture* in Kazakhstan and Uzbekistan may relate to the high concentration of political power combined with the relative autonomy of central government. However, one would expect that low levels of *administrative corruption* should accompany effective central controls. But this is so if society and the authorities do not regard bribe-taking as a serious offence.

It is interesting to note that "bundled" corruption indices, i.e. the average of *administrative corruption* and *state capture*, give assessments of the incidence of corruption across transition economies which are very similar to those generated by Transparency International's CPIs.[324] This suggests that foreign investors – major respondents to Transparency International surveys – have largely adequate assessments of the "friendliness" or "hostility" of the business environment and foreign direct investment (FDI) may provide a relatively good indication of the business climate including governance.

Indeed, as empirical research shows, foreign investors avoid countries characterized by high corruption.[325] A cursory examination of the data in table 5.2.1 (column 7) seems to corroborate these findings.

Although FDI, cumulated over 1997-1999, as a per cent of GNP in 1999 is highest in the Transcaucasian countries, with Azerbaijan the most corruption prone with the highest share of all, one should bear in mind two facts. First, the data on FDI do not distinguish FDI in natural resources, which tends to be much less sensitive to the business climate. The bulk of investments in Azerbaijan have been in the energy sector. Second, for less developed countries GNP data tend to underprice the value of non-tradeables in relation to estimates based on purchasing power parity (PPP). In consequence, GNP is "artificially" raised in more developed countries and diminished in less developed countries. Nonetheless, the correlation between the CPI and the FDI/GNP ratio is positive, albeit not very strong (+0.41).

These comments notwithstanding, the data clearly show that the Luxembourg Group attracted relatively more foreign investment than countries with much higher levels of corruption.[326] Accounting for 16 per cent of the population and 33 per cent of the region's GNP in 1999, it attracted 58 per cent of all FDI inflows to post-communist countries over 1990-1999. Within the Luxembourg Group three countries that are clearly at the top – the Czech Republic, Estonia and Hungary[327] – are also ranked with the lowest levels of corruption. Thus it would appear that foreign investors are well informed as to which countries should be avoided because of high levels of corruption.

### (c) The incidence of corruption and progress in transition

Corruption is a symptom of bad governance and a simple juxtaposition of EBRD transition indicators and World Bank corruption indices demonstrates this point. Table 5.2.1 gives aggregate indices for each country based on EBRD "grades" (with pluses and minuses treated respectively as +/-0.5) in the following areas of governance: enterprises (large-scale privatization, small-scale privatization, governance and enterprise restructuring); markets and trade (price liberalization, trade and foreign exchange system, competition policy); and financial institutions (banking reform, securities markets, non-bank financial institutions). We treat the sum of "grades" as an indicator of the overall progress achieved by a country. The values shown in table 5.2.1 (column 1) are normalized in terms of the best performer (i.e. the country with the highest total of scores in 2000), which is set at 100.

---

[324] The value of the correlation coefficient is 0.88, and the regression line (y = 0.85x + 7.9) explains 77 per cent of the variation.

[325] B. Smarzynska and S.-J. Wei, *Corruption and the Composition of Foreign Direct Investment: Firm-Level Evidence*, World Bank Policy Research Working Paper, No. 2360 (Washington, D.C.), June 2000.

[326] One country stands out. Despite its superb geographical location and low CPI, Slovenia attracted relatively little FDI in the 1990s. This was because of a deliberate policy to keep foreign investors at bay. B. Kamiński, "Foreign trade and FDI in Hungary and Slovenia: different paths – different outcomes", *Transition*, Vol. 9, No. 6, December 1998.

[327] In terms of cumulative FDI per capita over 1990-1999, Hungary with $1,817 tops the list, the Czech Republic was second ($1,529), and Estonia third ($1,213).

The EBRD indices measure only the institutional aspect of the transition, i.e. the progress made in building institutions and implementing policies supporting competitive markets. Since they do not include assessments of the sustainability of macroeconomic policies and taxation, they do not catch all aspects of governance. Yet, since the transition economies have mostly restored macroeconomic stability, the "total" index provides an overall assessment of progress achieved in establishing market-friendly institutions and policies.

Similarly, the extent of *state capture* should be expected to be positively correlated with slower progress in institutional development. But the values of the *state capture* index do not seem to offer a good indication of actual or past resistance to market reforms. Even a very perfunctory examination of the values of the EBRD index and of *state capture* suggests great discrepancies. Kyrgyzstan scores extremely high on both measures of corruption, but – according to the EBRD – has one of the most advanced economic regimes among the CIS countries. Armenia has a much lower incidence of *state capture* than Georgia, but Georgia seems to have moved faster in implementing market reforms. On the other hand, Belarus and Uzbekistan clearly belong to a different realm of authoritarian regimes with administratively run economies. Both countries rank very low in terms of progress in transition relative to the other transition economies, but they also rank low in terms of corruption, especially in terms of *state capture*.[328]

The absence of concurrence between the *state capture* index and the EBRD index of progress in transition does not erode the former's usefulness as a diagnostic tool. Consider that an "uncaptured" state, provided there is strong commitment within the political class to reform the administration and to eliminate corruption, will achieve these objectives without much difficulty. While the unbundling of corruption offers interesting insights into the factors blocking reforms and development, the two measures of corruption might be used together when assessing the weight of corruption in impeding liberalization. Administrative corruption provides an indirect indication of the state's capacity to implement reform. If it is weak, then no matter whether the state is "captured" or not, the state administration will ignore the reform measures. Hence, in order to address the question of whether corruption blocks reforms, one should use both indices simultaneously.

Have countries with lower levels of corruption, as measured by the combined or "bundled" corruption indices, i.e. the average of *administrative corruption* and *state capture,* also made greater progress in transition toward competitive markets as measured by the EBRD "total" index? Even a quick examination of the data in table 5.2.1 suggests that the two sets are correlated with the exceptions of Belarus and Uzbekistan. If these two countries are excluded, the correlation coefficient is negative and high (0.88), i.e. the greater the progress in transition the lower the levels of corruption (chart 5.2.2).

Different measures of corruption in transition economies point to three different clusters of countries: least affected, intermediate and seriously affected. The least-affected group includes the Luxembourg group and Croatia; the intermediate group includes four countries from the Helsinki group (excluding Romania) and Kazakhstan; and the seriously-affected group includes the other CIS economies and Romania.

The differences between the "corruption" and "transition" rankings are interesting as they may reveal weaknesses in both classifications. Using the EBRD aggregate index, Slovakia is in the most advanced group in transition (above or equal to 84 per cent), Georgia advances to the intermediate group (above or equal to 76 per cent), and Kazakhstan drops to the least advanced group (below 75 per cent). It may be that firms participating in the World Bank survey overstated the incidence of corruption in Slovakia, while the EBRD overstated the progress in transition achieved by Georgia.[329] Clearly, something is missing in the World Bank and EBRD assessments of corruption and progress in transition respectively.

### (d) Ignored dimensions: "takeover" of state and criminalization of the economy

Neither the World Bank's corruption measures nor the EBRD's transition indices seem to perform well when dealing with cases of extreme corruption. The EBRD indices fail to catch the real meaning of privatization when the state itself has been privatized. This seems to have been the case in Russia when following the 1996 presidential election the oligarchs – the product of ill-designed privatization – had "… captured Yeltsin, his successive governments and the political process".[330] But other CIS countries also witnessed the fusion of criminal, political and business worlds.[331] Under

---

[328] In the case of Belarus the accuracy of the World Bank assessments is questionable. First, President Lukashenka and his circle run the state as a private domain, which may represent an extreme case of *state capture*. Second, according to one of the present author's interviews with businessmen with sound experience of Belarus, the Transparency International index does not adequately capture the incidence of corruption there. Last but not least, there may be nothing for sale as politicians in their dual role as businessmen guard their monopolistic positions in order to maintain high levels of rent extraction.

[329] Surveys often tend to skew the results "… merely by asking questions". J. Gole, "Public opinion polls as an anti-corruption technique", *LGI Newsletter 1*, No. 1 (Budapest), 1999, as quoted in W. Miller, A. Grodeland and T. Koschechkina, *A Culture of Corruption. Coping with Government in Post-communist Europe* (Budapest and New York, Central European University Press, 2001), p. 22. In consequence, they exaggerate the importance of corruption.

[330] L. Wolosky, "Putin's plutocrat problem", *Foreign Affairs*, Vol. 79, No. 2, March/April 2000, p. 25.

[331] M. Ellman, "The social costs and consequences of the transformation process", UNECE, *Economic Survey of Europe, 2000 No. 2/3* (United Nations publication, Sales No. E.00.II.E.28) p. 134.

**CHART 5.2.2**

**Progress in transition (governance) and corruption**

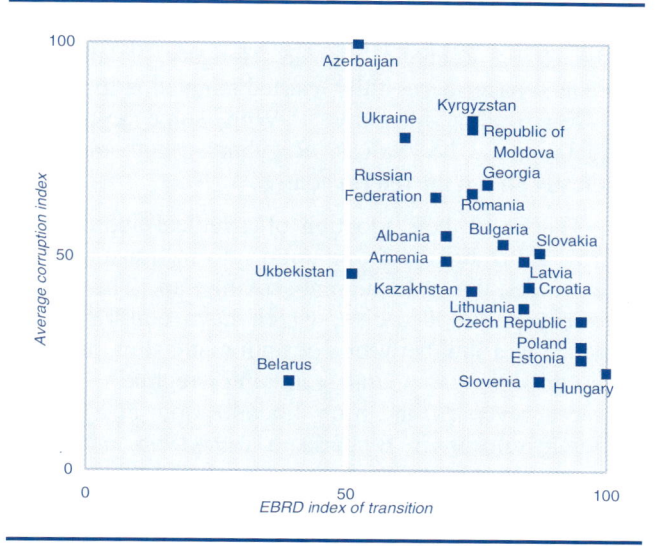

[Scatter plot with "Average corruption index" on y-axis (0-100) and "EBRD index of transition" on x-axis (0-100). Countries plotted: Azerbaijan (~100), Ukraine, Kyrgyzstan, Republic of Moldova, Russian Federation, Georgia, Romania, Albania, Bulgaria, Slovakia, Armenia, Uzbekistan, Latvia, Kazakhstan, Croatia, Lithuania, Czech Republic, Poland, Estonia, Belarus, Slovenia, Hungary.]

Source: Derived from data in table 5.2.1.

these circumstances, indices of *state capture* also lose their relevance as the issue ceases to be one of its extent – there is nothing left for *sale* when everything has been acquired. Despite a complete capture of a state and its economy by a narrow interest group, the Bank's *state capture* index will yield zero values as respondents will state that regulations and politicians cannot be bought.

Similarly, the progress achieved, for instance, in interest rate or price liberalization – areas graded in the EBRD indices – have little meaning when the economy is criminalized, i.e. when there is the rule of lawlessness, when businessmen hire criminals for contract enforcement and have to buy "protection". Securities markets may be established and banks may be reformed and thus obtain high scores in the EBRD transition measures. Yet, if banks do not enjoy the trust of their clients, this index conveys little information.[332] If a firm cannot survive without bribing local officials, price liberalization will do little to boost competition and economic growth.

The broader point is that state bureaucracies, albeit often barely functioning, do not provide "public goods" but create "public bads". The examples of the latter include conditions allowing officials to prey on businesses and, on the other hand, to collude with businessmen to make money by eliminating competition – domestic and foreign alike. In consequence, the most corrupt are also the most successful under these conditions.[333]

Without taking into account the political economy circumstances peculiar to each country, both the World Bank and EBRD indices may often be misleading. For instance, Azerbaijan ranks highest in the incidence of corruption and – excluding Belarus and Uzbekistan – lowest in terms of progress towards establishing competitive markets. Yet, to the best of our knowledge, not a single banker or businessman has been killed in business disputes, and its high-ranking officials have not been arrested in the west with millions of dollars in their pockets. In other words, the economy of Azerbaijan seems to be much less criminalized than many other CIS countries, including Russia and Ukraine, and the state has not been taken over by a group of oligarchs. Contrary to warnings implicit in the EBRD and World Bank indices, foreign investors have not been discouraged – total FDI per capita over 1990-1999 of $383 in energy-abundant Azerbaijan was well above levels in resource-rich Russia ($111) or Kazakhstan ($272), not to mention Ukraine with just $64.

Thus, while EBRD and World Bank indices as well as CPIs perform the important and very useful function of generating peer pressure on governments to reduce corruption and improve governance, the implied rankings and values of the respective indices should be cast against the political economy circumstances peculiar to each country.

### (ii) Strategic political choices during transition

The shift from central planning toward markets and democracy – provided it is accompanied by the establishment of rules separating *res publica* from *res privata* – tends as a rule to reduce the potential for corruption by undercutting the opportunities to abuse public office for private gain. The shift to the "market" establishes the conditions for the emergence of civil society – ultimately the most effective device to control corruption. This shift, if properly designed, replaces a myriad of counterproductive regulations, designed to create opportunities for rent-seeking and bribe-extraction, with a core of laws and regulations establishing the framework for a rule-based, public-private interface. This involves, among other things, removing assets from state ownership and replacing direct state intervention in the economy with a panoply of institutions, to regulate conduct in the goods, services, labour, asset and financial markets.

None of these, however, reduces corruption by default. In fact, ill-designed economic reforms responding to the vested interests of corrupt elites may significantly increase corruption. Examples include "gradualism" in the removal of central controls or hasty privatization in many transition economies. Similarly, political transitions may fail to trigger the emergence of public activities independent from state controls, i.e. the vibrant civil society. The common denominator of the failure to downsize corruption is continued penetration of the economic realm by politics. Its most visible symptom

---

[332] The Russian banking system provides a good illustration. For an excellent study of its fall and demise in the 1990s see J. Johnson, *A Fistful of Rubles. The Rise and Fall of the Russian Banking System* (Ithaca and London, Cornell University Press, 2000).

[333] A. Åslund, "The trouble with economic reforms in Ukraine", paper presented at the conference *Towards a New Ukraine: Meeting the Next Century*, University of Ottawa (Ottawa), 2-3 October 1998.

is the pursuit of material gains through political rather than business career. The inherent danger in such a situation is that it may lead to *pirate capitalism*.[334]

The choice of the approach to political and economic transitions sheds light on the variations in corruption and the quality of policy environment. The strategic choices that countries faced in the aftermath of the collapse of communism can be divided into two groups. First, the decision, which approach – radical or gradual – to take towards first-generation economic reforms, falls into the first group. Two choices concerning the constitutional structure of the state belong to the second group: (i) the choice between presidential and parliamentary government; and (ii) the choice of the system of representation. The strategic choices made in any given country depended on the strength of the opposition, i.e. the support it had in society, as well as on the prevailing pattern of relationships between the communist regime and society.

### (a) Radical versus gradual approach

"Gradualism versus shock therapy" was a hotly disputed issue in the early 1990s. Advocates of gradualism criticized "shock therapy" for the allegedly excessive costs that it would impose on the economy and the population. According to this view, the slow liberalization of prices, of the conditions of entry to the private sector, and of the conditions of access to the domestic market would prevent the collapse of output and assure a smooth transition to a market-based economy.

Subsequent developments have demonstrated that this was a false dilemma, at least in the case of the approach towards first-generation reforms aimed at assuring macro-stability and the shift from a "supply-constrained" to "demand-constrained" economy, to borrow Janos Kornai's terms. Hyperinflation, falling output and external crisis eventually compelled countries that had initially rejected this approach to adopt measures similar to those implemented by radical reformers.

While reforms in the CEECs' stabilization-cum-transformation programmes differed, they all – excluding Romania – opted in favour of a radical approach toward first-stage reforms rather than the gradualism pervasive among the former Soviet republics during the initial stages of transition.[335] With the stroke of a pen they liberalized prices, privatized distribution, introduced small privatization, removed quantitative restrictions on imports, liberalized tariffs and introduced convertibility of the domestic currency initially limited to current account transactions. More importantly, the best performers – that is, the Czech Republic, Estonia, Hungary, Poland, and Slovenia – have successfully resisted temptation to reverse the liberalization measures. With some exceptions (Bulgaria and the Slovak Republic), they have all steadfastly stayed the reform course.

Thanks to their adoption of a radical approach to macro-stabilization and economic reform, most CEECs have avoided the systemic rent-seeking trap. There were some exceptions, however, which have confirmed the existence of a link between corruption and slow progress in establishing a liberal economic regime.[336] For instance, reversals in price liberalization and stalled economic reforms in Bulgaria in 1994-1996, under the guise of populist promises, corrupted the privatization process and resulted in significant transfers from the state budget to rent seekers. Romania was also subject to similar developments. A deep economic crisis has helped to overcome the opposition of vested interests in Bulgaria and put the country back on the reform track.[337]

The former Soviet republics (excluding the Baltic states) initially adopted the doctrine of gradualism. However, spiralling inflation and prolonged recession eventually compelled them (except Belarus and Uzbekistan) in the second half of the 1990s to follow the earlier path of the CEECs, thus demonstrating that gradualism had never been a viable policy approach to first-generation reforms.

Neither has it been a viable option for second-generation, structural reforms aimed at establishing a rule-based public-private interface supporting a market economy. In fact, gradualism has created fertile ground for systemic rent-seeking and has been responsible for the emergence of powerful vested interests opposing liberalizing reforms not only in the economy but also in the political realm. It has created a vicious cycle of partial reforms contributing to economic deterioration and growing opposition to reforms. In the transition economies that followed a gradualist approach, the winners effectively suspended the economies in a *"bad" equilibrium* offering them spoils at the expense of high costs imposed on their respective societies.[338]

Incomplete liberalization of prices – usually justified in populist terms to protect the poor – has created opportunities to obtain trade licences to export

---

[334] "The kind of skills and knowledge that will pay off will be a function of the incentive structure inherent in the institutional matrix. If the highest rates of return in a society are piracy, than organizations will invest in knowledge and skills that will make them better pirates ...". D. North, "Towards a theory of institutional change", in W. Barnett, M. Hinich and N. Scofield (eds.), *Political Economy* (Cambridge, Cambridge University Press, 1993), p. 63.

[335] Bulgaria initially adopted a radical approach to stabilization and then reversed it. The financial crisis in 1996 led to the change of government and, consequently, of the approach to economic reforms in 1997.

[336] D. Kaufman and A. Kaliberda, "Integrating the unofficial economy into the dynamics of post-socialist economies: a framework of analysis and evidence", in B. Kamiński (ed.), *Economic Transition in Russia and the New States of Eurasia* (Armonk, NY, M.E. Sharpe, 1996).

[337] *Bulgaria. The Dual Challenge of Transition and Accession*, World Bank Country Study (Washington, D.C.), 2001.

[338] J. Hellman, "Winners take all: the politics of partial reform in post-communist transition, *World Politics*, Vol. 50, No. 2, 1998, pp. 203-234.

subsidized goods by those well-connected to political elites. Since the difference between domestic and world prices was often huge, this created opportunities for those with good government connections to obtain large rents. Similarly, central controls over credits and interest rates allowed enterprises to borrow at attractive rates – such funds could then be siphoned off to related private enterprises or used to finance purchases of subsidized goods for subsequent sale in international markets.

The results of the emergence of this rent-seeking regime were twofold: rent-seeking became more profitable than productive activities with all the negative consequences for economic growth so aptly described by Krueger;[339] and the emergence of powerful interest groups with a stake in preserving the existing economic regime. Although corruption and side-payments clearly exist also in the CEECs, in those that adhered to a radical approach to first-generation reforms, rent-seeking did not crowd out activities aimed at restructuring inefficient industries or starting up new companies able to compete in domestic and international markets.

Yet a good initial choice of a radical approach to first-generation reforms does not necessarily make second-generation reforms (regulatory and government reforms) immune to bad governance and rent-seeking. On the contrary, second-generation reforms are much more difficult and complex. Their crux is the establishment of a rule-based public-private interface. This task involves at least four critical steps. First, a framework for sustainable macroeconomic stability has to be established. Among other things this involves the introduction of transparent, financially sound revenue and expenditure mechanisms in public finance. Second, it requires the privatization of state assets and the elimination of state micro-management. Third, it requires creating the framework for guiding private sector behaviour and enforcing property rights. Fourth, it must make the framework effective by improving the capacity, integrity and oversight of the civil service as well as that of the judiciary.

Given its institutional complexity, the establishment of a rule-based public-private interface is time consuming. Considering the political resistance and the different sets of winners and losers that each of these steps is bound to produce, it cannot take place with the stroke of a pen. Each step changes the distribution of power and most of them (especially privatization) offer enormous opportunities for both enrichment and impoverishment. Failure or policy mistakes in one step usually has a negative impact on a country's economic performance. Even one of the boldest reformers among the CEECs – the Czech Republic – has not avoided a reversal in economic growth due to incomplete regulatory reforms (box 5.2.1).

While the choice of the approach to the first-generation economic reforms has turned out to be a major determinant of the subsequent path of transition, other policy decisions have also clearly played a role. In addition to the various factors accounting for the 1997 Czech crisis, another example is the approach to decentralization of the state and the corresponding reform of public finance. The establishment of a viable tax base for local governments combined with transparent rules of local finance have created a pro-business climate in local communities in Poland.[340] Assured of a predictable share in corporate and personal income tax revenues, local governments have actively sought investment and avoided erecting barriers to private business activity. This contrasts rather sharply with the predatory behaviour of local governments wholly dependent on centralized financing in several CIS countries as well as in some CEECs.

To summarize, the policy of massive institutional change imposed upon the society within a short span of time permits the problem of systemic friction that a gradual approach must face to be overcome. Provided that the group that initiates it has a clearly defined set of goals and actively seeks broad popular support, "shock therapy" will generate a virtuous cycle of reforms.[341] Gradual approaches to first-generation reforms, on the other hand, have produced a vicious cycle of reforms. They have given an opportunity to the privileged of the old regime to retrench and successfully readapt to the new conditions. They were also able to enrich themselves through exploiting the opportunities created by partial liberalization and privatization. The paradox is that, with the economic situation deteriorating and a growing gap between rich and poor, public opinion has often turned against reforms; opposition parties (frequently dominated by former communist parties) would often form coalitions with the *nouveaux riches* or oligarchs, largely responsible for the economic malaise, to stall the reform process.

### (b) Political choices: electocracies and democracies

The declared objective of simultaneous political and economic transitions in most post-communist countries was a market-based democracy. Communism and central planning, or whatever was left of it, was to be replaced by the greatest institutional innovation of the eighteenth century that provided a solution to the opposition between public and private interests. The innovation was Adam Smith's "invisible" hand combined with Montesquieu's system of checks and balances among the legislative, executive and judiciary branches of government.[342]

---

[339] A. Krueger, "The political economy of a rent-seeking society", *American Economic Review*, Vol. 64, No. 3, 1974, pp. 291-303.

[340] J. Hicks and B. Kamiński, "Local government reform and transition from communism. The case of Poland", *Journal of Developing Societies*, Vol. XI, No. 1, 1995, pp. 1-20.

[341] For instance, the Balcerowicz team, during their first term in office in 1989-1991, possessed a good sense of what it wanted to achieve. It also initially had strong public support: that, however, was not cultivated and quickly vanished.

[342] A. Kamiński, "Res publica ...", op. cit.

> **Box 5.2.1**
>
> **The 1997 Czech foreign exchange crisis and governance**
>
> Despite success in creating stability and predictability in the macro-policy environment and transferring assets to the private sector, continued direct state intervention through the banking sector, combined with the absence of well-designed laws protecting private property and regulating capital markets, contributed to the exchange crisis of 1997 and the subsequent economic recession.
>
> The exchange rate of the koruna, against a basket of currencies (the deutsche mark – 65 per cent, and the dollar – 35 per cent), displayed minimal variation between 1991 and March 1996. Following the adoption of a wider band, volatility in the exchange rate dramatically increased. At its peak in February 1997, the koruna rose to 6 per cent above its central parity, but fell to 15 per cent below its previous parity in November 1997.
>
> What caused this huge swing? It is tempting to blame the foreign exchange crisis on a speculative attack by international short-term investors on the koruna. But this would fall well short of addressing the underlying causes of the crisis. Leaving aside domestic demand pressures, which under more favourable circumstances might have been financed by foreign capital inflows, the run against the koruna in Spring 1997 was due to an emerging consensus that structural microeconomic reforms had advanced much less than was generally assumed. The sharp decrease in portfolio equity flows was triggered not only by the apparent overvaluation of the koruna, but also, if not predominantly, by the slow progress in microeconomic restructuring and in establishing well-functioning financial markets.
>
> Portfolio investors lost confidence in spite of the koruna's depreciation, which should have made the purchase of Czech assets more attractive. Two factors seem to account for the declining interest of portfolio investors: the narrowing range of attractive investment opportunities and institutional weaknesses in the financial markets. The two are connected, simply because an active and well functioning capital market helps to ensure the efficient behaviour of firms. But the way in which the Czech capital markets were initially created made it impossible for them to fulfil this standard function. Stock trading was occurring in three separate markets. With shares dispersed and flaws in the institutional design of the Czech equivalent of mutual funds, the Investment Privatization Funds, an effective supervision of management was unlikely. Other weaknesses included a lack of transparency, massive insider trading and poor protection of minority shareholder rights.[1] These clearly pointed in the direction of microeconomic restructuring and rent-seeking opportunities as the major causes of the malaise.
>
> Furthermore, for many large, previously state owned firms, the banking sector had replaced the state budget as a source of financing.[2] With banks still controlled by the state, political criteria rather than strict economic considerations of a firm's creditworthiness and its anticipated profits drove this financing.
>
> One might thus conclude that the koruna crisis was the result of an earlier misreading by international investors of the reform process in the Czech Republic. Despite rhetoric to the contrary, sizable parts of the economy had not been restructured. Despite privatization, weak direct monitoring by shareholders and non-transparent capital markets failed to change managerial behaviour in large firms. Moreover, a large amount of key assets remained in the hands of the state, including banks and public utilities, as well as stakes remaining in privatized and strategic enterprises. The subsequent contraction of GDP suggests that institutional shortcomings of Czech reforms had slowed down industrial restructuring and weakened productivity gains.
>
> ---
>
> [1] P. Mertlik, "A case study: the Czech privatization and subsequent structural changes in capital ownership and property rights", in UNECE, *Economic Survey of Europe, 1998 No. 2* (United Nations publication, Sales No. E.98.II.E.18) chap. 5; F. Turnovec, "Privatization and transparency. Evidence from the Czech Republic", in J. Holscher (ed.), *Financial Turbulence and Capital Markets in Transition Countries* (New York and London, St. Martin's Press, 2000).
>
> [2] C. Buch, "Privatization, ownership structure and transparency: comment on Frantisek Turnovec", in J. Holscher (ed.), op. cit.

But within these broad institutional guidelines, there is considerable room for diverse solutions. The most general choice concerns the distribution of power between the legislature and the judiciary followed closely by the choice of an electoral system. The solutions chosen have an impact on the quality of decision-making and the policy environment. Strategic thinking manifests itself in the readiness of the designers of the new political system to take into account the criteria of accountability and legitimacy together with governability. It is interesting to note that these crucial decisions for the future of post-communist societies were usually made in a haphazard way without any deeper reflection about their long-term consequences under the specific historic and social conditions. Not a single post-communist country had a serious debate about the choice of a political architecture. Thus, the most important societal decisions received in most cases the least attention. While the conspicuous absence of any serious constitutional debate remains one of the puzzles of the transition from communism, this nonchalant attitude towards its political ramifications has negatively affected both political and economic transitions.[343]

---

[343] Bruce Ackerman aptly remarked that: "Work on the text of the constitution provides the winning movement with the opportunity to engage in collective effort aimed at defining basic principles and mobilizing wide social support for key initiatives". B. Ackerman, *The Future of Liberal Revolution* (New Haven, Yale University Press, 1992), p. 51.

The balance of influence of the various political groups that came to prominence following the collapse of communism shaped strategic choices during the initial stages of transition. In several countries the political transition unfolded under particular duress. For instance, ethnically fuelled territorial conflicts and violent domestic crises affected political developments in three Transcaucasian states – Armenia, Azerbaijan and Georgia. A Russian-supported separatist movement in Transdniestria has left the Republic of Moldova largely non-viable as a state. Although the threat to central Asia from Islamic rebels has not led (except in Tajikistan over 1992-1998) to the ethnic strife and political volatility characteristic of the Transcaucasian states, it probably contributed to the progressive institutionalization of authoritarian, one-man rule in the region. Last but not least, except in Slovenia, the bloody disintegration of what was Yugoslavia did not create an environment friendly to democracy. These adversities notwithstanding, all but two post-communist countries (Tajikistan and Turkmenistan) have held at least one largely free and fair election.

But not all countries have made the progress that would be deemed sufficient to sustain the democratic transition. In particular, not a single CIS country seems to have reached this phase. They are "electocracies" – to borrow an apt phrase from Karen Dawisha[344] – rather than democracies. And as Linz and Stepan[345] remind us in their "electoralist fallacy", free elections are a necessary but not a sufficient condition for democracy. Except for Slovenia and, for different reasons, Bosnia and Herzegovina, the former Yugoslav republics have had their share of "electoralist" experience. Croatia, in the aftermath of President Tudjman's death, moved swiftly from electocracy to democracy. The Federal Republic of Yugoslavia and the Former Yugoslav Republic of Macedonia seem to be in transition from electocracy to democracy. These two countries have recently made the first strides in this direction. So has Slovakia since the 1998 parliamentary elections (followed by the 1999 presidential elections) removed the autocratic President Meciar from power.

In contrast, other CEECs have met the EC's Copenhagen Summit political conditions (viable democracy and protection of minorities) to be included in two pan-European projects – the expansion of NATO and the enlargement of the EU. By the late 1990s, all the CEECs had completed the democratic transition with politics – to borrow the criterion for consolidated democracy from Linz and Stepan: "… no longer dominated by the problem how to avoid democratic breakdown".[346] The prospect as well as the process of accession to the EU has not only provided guidance to institutional transformation, enhancing economic efficiency and growth, but also set limits on the political activities of ruling elites. As the experience of Bulgaria and Slovakia seems to suggest, the lack of progress in European integration was one of the factors contributing to the defeat of anti-reform coalitions in the 1997 elections. Yet, except for a broad requirement of a viable democracy protecting minority rights, the European Council, as the guiding body of the EU, has left open the choice of political architecture to acceding countries.

Two aspects of political evolution have had a particular bearing on the path of transition and consequently on the quality of governance and the potential for corruption: the electoral system and the type of political regime as captured by the balance between executive, judiciary and legislature.

### (c) Electoral system

The choice of the system of representation together with the mode of electing representatives to the parliament is one of the decisions that are considered as constitutional, whether or not they are part of such a document. This choice predetermines the relationship between polity and state and has a crucial impact on the development of the party system. The accountability of the rulers to the ruled and the legitimacy of the political regime depend to a large extent on it. The choice of electoral system seems to be crucial for the process of transition and the accountability of the "rulers" to the "ruled". It determines the ability of society to choose and control its leaders and, *pari passu*, its government. "Not only are electoral systems the most manipulative instrument of politics; they also shape the party system and affect the spectrum of representation".[347]

There is no universal prescription for a good electoral regime.[348] At the general level, we tend to prefer a plurality or majoritarian (first-past-the-post) system awarding a seat to candidates receiving the highest share of votes in a constituency to proportional representation (PR) systems with voters choosing between different

---

[344] K. Dawisha, "Electocracies and the Hobbesian fishbowl of post-communist politics", in S. Antohi and V. Tismaneanu (eds.), *Between Past and Future. The Revolutions and their Aftermath* (Budapest, CEU Press, 2000).

[345] J. Linz and A. Stepan, *Problems of Democratic Transition and Consolidation* (Baltimore, Johns Hopkins University Press, 1996), p. 4.

[346] Ibid, p. 5.

[347] G. Sartori, *Comparative Constitutional Engineering. An Inquiry Into Structures, Incentives and Outcomes* (New York, New York University Press, 1994), p. IX.

[348] For an important exchange of views on the effects of different electoral systems see A. Lijphart, "Constitutional choices for new democracies", *Journal of Democracy*, Winter 1991; idem, "Democratization and constitutional choices in Czechoslovakia, Hungary and Poland, 1989-1991", *Sisyphus*, Vol. VIII, No. 1, 1992; idem, "Double-checking the evidence", in L. Diamond and M. Plattner (eds.), *The Global Resurgence of Democracy* (Baltimore and London, Johns Hopkins University Press, 1995). See also Q. Quade, "PR and democratic statecraft", *Journal of Democracy*, Summer 1991; and G. Laydeyret, "The problem with PR", *Journal of Democracy*, Summer 1991.

party lists.[349] The fundamental problem here involves the choice between an environment creating greater opportunity for minority parties and proportionality of representation (whatever this may mean) and one ensuring more intense competition and favouring a quick shift to a two-party political system. The plurality system is more likely to produce stability and accountability of government but frequently at the cost of alienating significant fragments of society.[350] It also seems to reduce the potential for the earlier discussed state *capture*.

Plurality accompanied by single-member districts is more likely than PR to establish a direct link between performance in office and prospects for re-election. This in turn generates direct accountability to voters and greater reluctance by politicians to become involved in corrupt practices. Constraints on politicians' behaviour in office are different under the PR system. They have to take into account the preferences of their party "elders" who decide, rather than their respective constituencies, whom to include in the party list. This significantly reduces accountability to voters and makes corruption more likely. While the links between electoral systems and corruption have not been extensively analysed empirically, a recent cross-country study provides strong empirical support to the view that plurality leads to a lower incidence of corruption as captured by CPIs.[351]

Under certain conditions in transition societies, however, adoption of the majoritarian system may turn out to be detrimental to democracy. This is so when society is disintegrated and new, non-communist parties are weak. The majoritarian formula, in these conditions, then enables a well-organized post-communist party (usually calling itself "social democrat") to stay in power for a long time at a significant cost to the development of democracy. The majoritarian system becomes viable only if civil society is sufficiently well developed. Then, its qualities may come to the fore.

A PR system suffers from several weaknesses, which are particularly troublesome during the initial stages of transition. First, it offers disproportional influence to small parties. This in turn contributes to instability in government and reduces the effectiveness of government when it must implement difficult structural changes.[352] Second, as was argued earlier, it increases the potential for corruption, as officials owe their careers to the party hierarchy rather than voters. Their pay back is finding ways to fund their political parties. The inherent danger of this arrangement is the perpetuation of conditions friendly to rent-seeking though "corrupting" laws and regulations, i.e. enacting laws which expand the opportunities for corruption. These factors may injure the overall credibility of the system, damaging its legitimacy in the eyes of society. These considerations are behind the rising movement in Poland against the PR system.

Political circumstances unique to a country rather than well-considered arguments were decisive in the choice of the electoral formula. Where there was a strong institutional presence of the communist party, single member districts and the majoritarian formula were selected as the electoral solution.[353] On the other hand, high social mobilization combined with a lack of trust in elites favoured PR electoral systems. Most governments (18 of the 27 post-communist states) chose total or partial PR rather than the majoritarian formulae.[354]

From the perspective of the particular interests of leaders of political parties, PR is a very attractive proposition. While the plurality system provides stability of government during its term in office, it also allows for the change in election outcomes to occur much more easily than PR.[355] Hence, during periods of uncertainty and high social mobilization, PR may ensure a greater predictability of outcomes for the political elites. But even during periods of relative stability, PR seems to ensure the stability of a political class. Governments may change but the same politicians tend to return to the cabinet. Furthermore, when memories of the communist system bring about unfavourable associations, communist successor parties prefer PR for it saves them the embarrassing confrontation with their own history.

Within the PR system, the problem that has far-reaching implications for the development of political parties concerns the choice of the threshold level, i.e. the minimum percentage of votes required to have any representation in parliament – the size of electoral districts, and the method of "transforming" votes into seats. Low thresholds increase the number of parties in parliament, reduce the incentives for pragmatic compromise, and make it difficult to curb the overblown ambitions of individual politicians. Thus, to use Albert Hirschman's terminology, PR may favour exit over voice and loyalty. Instead of debating and compromising, politicians create new mini-parties calculating, as in the case of Poland, that they may pass the 5 per cent threshold in parliamentary elections, or at least a threshold of 3 per cent which will assure them public financing for the next term. Poland began with no

---

[349] Among the authors who have taken a similar position on this issue are M. Weber, "Politik als Beruf", *Gesammelte zur Sociologie und Sozialpolitik* (Tubingen, J.C.B.Mohr (Paul Siebeck), 1921); J. Schumpeter, Capitalism, Socialism and Democracy (London, Allen & Unwin, 1966); K. Popper, "Who should govern?", *The Economist*, 23 April 1988; G. Sartori, op. cit., to name just a few.

[350] A. Lijphart, *Electoral Systems and Party Systems: A Study of Twenty-Seven Democracies, 1945-1990* (Oxford, Oxford University Press, 1994), p. 144; P. Norris, "Choosing electoral systems: proportional, majoritarian and mixed systems", *International Political Science Review*, Vol. 18, No. 3, July 1997, p. 301; M. Pinto-Duschinsky, "How to get rid of a bad government?", *Times Literary Supplement*, 25 September 1998.

[351] T. Persson et al., op. cit.

[352] L. Balcerowicz, *800 dni. Szok kontrolowan* (800 days. Controlled shock), Polska Oficyna Wydawnicza BGW (Warsaw), 1992, p. 108.

[353] Such countries included the electocracies of central Asia as well as Belarus, The former Yugoslav Republic of Macedonia and Ukraine.

[354] K. Dawisha, op. cit., p. 296.

[355] M. Pinto-Duschinsky, op. cit.

threshold in 1991, with the result that there were nearly 30 parties in parliament and a government coalition consisting of five parties. The problem of governability led to new elections in 1993, this time with a 5 per cent threshold (7 per cent for a coalition of small parties). Poland was no exception. Between the first and second parliamentary elections, in all the post-communist countries, the average threshold increased from 3.9 per cent to 4.6 per cent.[356]

The percentage of voters' casting votes for parties that do not receive a single seat in the legislature has also increased. The average of almost 17 per cent of wasted votes, as compared with 6 per cent for all proportional representation elections held in western Europe in the postwar era, is striking for one major reason:[357] a significant proportion of these voters may lose faith in the democratic process thereby eroding trust in the system. Furthermore, by giving post-communist parties a much larger influence than implied by the proportion of the total vote they received (e.g. in elections in Russia), they have indirectly contributed to the emergence of strong presidencies in several CIS countries. As Karen Dawisha notes, "... the institutionalization of post-communist parties, while not foreordaining the failure of democracy, nevertheless does little to strengthen the prospect of its success."[358]

The usual choice in the method of transforming votes into seats is that between the d'Hondt and Sainte-Lague formulas. Large districts, like low thresholds, favour small parties at the cost of big ones. The d'Hondt formula favours the big parties, whereas the Sainte-Lague formula tends to overrepresent the small ones. In consequence, small parties favour low thresholds, large electoral districts and the Sainte-Lague formula, whereas big parties, on the contrary, opt for high thresholds, small districts and the d'Hondt formula. Let us note that the more effectively the interest of the big parties is manifest in the adopted solutions, the more the composition of the parliament will approach that produced by the plurality system without, however, increasing the accountability of government.

Thus, on the one hand, to enable the government to function, parliaments had to adopt thresholds, and on the other hand, administrative constraints imposed on electoral competition by political elites in most post-communist countries produced a massive waste of votes. This seems to have choked the development of democracy and institutionalized post-communist parties. The latter have blocked the space that otherwise might have allowed genuine social democratic parties to emerge.

While the communist successor parties had a fairly large apparatus in the regions and an initially narrow but stable electorate, the non-communist parties that won the first democratic elections in most CEECs did not possess a stable constituency. Their popularity grew only as a result of mistakes committed by their competitors. For the parties established outside of the communist party system, the strategic choice was either to seek support from local and regional civic groups and associations or to communicate their message to the electorate at large through the mass media. The PR system together with large electoral districts favours the second option. Parties with unstable electorates prefer the media message to direct negotiations with grass roots organizations. Media campaigns are costly, and money is a problem.

Party financing has become the soft belly of political parties in the post-communist world, and elsewhere too. Pressed with this problem, parties use privatization policies, public procurement and political control over the public administration to create new illicit sources of financing. Violation of legal norms is an unavoidable practice in the struggle for power, while political power offers ways to cover up these violations. In this respect, all political parties are willing to cooperate in not disclosing their practices.[359]

What are the implications of this discussion for combating corruption and improving governance? While specific answers vary depending on the political economy circumstances of each country, sacrificing the stability of governance and accountability associated with plurality systems for fuller representation achieved under PR (with low thresholds) may be worthwhile under a weak civil society subjugated by the ex-communist *nomenklatura*. However, when a certain level of participation in civic life is attained, plurality systems provide a better chance for the efficiency and accountability of governments.

At the local level, a better solution is the majoritarian formula with relatively small districts provided that two conditions are met. First, that decentralization of government accompanies transition, and second, that public finance is redesigned to reflect the transfer of power from central to local governments. This was the solution in place in Poland over 1991-1998, which combined majority representation with decentralization of public finance.[360] The recently observed deterioration in the quality of management of local finances ("excessive" salaries, embezzlement of public funds, etc.) has coincided with the change in electoral procedures. While prior to 1998, constituencies of less than 40,000 elected their representatives under the majority electoral formula, the majority rule now applies only to districts with fewer than 20,000 people. In larger

---

[356] K. Dawisha, op. cit., p. 297.

[357] In the 1995 elections in Russia only four of 43 parties received more than the 5 per cent threshold and 49 per cent of the total vote was wasted, K. Dawisha, op. cit., p. 302.

[358] K. Dawisha, op. cit., p. 300.

[359] A. Kamiński, "Corruption under the post-communist transformation", *Polish Sociological Review*, Vol. 118 (2), 1997, pp. 106-108.

[360] J. Hicks and B. Kamiński, loc. cit.

districts, PR has replaced majority procedures. It thus appears that the extension of proportional electoral procedures has led to a decrease in voter turnaround, an increase in the incidence of corruption, and a fall in the quality of self-governance.

### (d) Presidential vs. parliamentary government

The choice between presidential and parliamentary systems of government was the most natural that faced political elites in the post-communist states. But there was no single case of serious constitutional debate. The outcome came by default indicating the weight of history and the special features of the political landscape in each of the post-communist countries. Its most striking feature is that the CIS countries, together with the countries of the former Yugoslavia, with the notable exception of Slovenia, have adopted a presidential system of government, whereas the CEECs (excluding Romania) moved towards parliamentary government. Under conditions of strong political pluralism articulated through the institutions of parliament, it was very difficult to establish a strong presidential regime.

The duality of parliament and presidency characterizes many presidential regimes in CIS countries. The president, as a chief executive, is largely independent of the parliament, and has the right to issue decrees that need no legislative approval. Communists and nationalists – in large part because of the adopted PR formula – control the legislature, which, usually expressing the popular sentiment, opposes liberal-democratic reforms. More recently, the alliance of the biggest winners from the transition with the communists, both groups opposed to changes that would increase domestic competition, has erected barriers to reform. This may be a highly unstable *equilibrium*, as Russia's experience shows. However, as long as the parliament serves merely as a valve for venting social frustration while the president is able to perform his role effectively, this potentially unstable solution may survive and eventually result in the emergence of genuine democracy due to the gradual evolution of social norms and values. But in many cases the parliament does not limit itself to a passive role. This, if combined with an ineffective executive power, leads to an irresponsible parliament and corrupt presidential power.

### (e) Implications for corruption

Three points are worth emphasizing. First, the presidential systems in the CIS countries operate in an institutional environment with weak legislatures and civil society institutions. The parliaments do not have sufficient support to resist the aspirations of strong presidents in charge of powerful administrations involving the army and security services – both untouched by serious reforms. The president's office has become the last line of defence for the administration against the unpredictable intrusion of political parties. The constitution was imposed upon the parliament by the president. It is impossible to find there any mechanism of checks and balances: the powers of the president exceed by far those of the legislature and the judiciary combined. But despite the concentration of power in the executive branch, "presidential" states remain relatively weak in their capacity to implement reforms and revive the economy. This, combined with an unsound civil service, criminalization of the economy and strong anti-rule-of-law constituencies[361] breed corruption and do not augur well for good governance. Under these circumstances, the fight against corruption easily becomes a mechanism of selective repression and a tool of political infighting.

Second, transition has created embryonic ties between the emerging private sector and the state administration. Examples abound especially in the CIS countries where leading "businessmen" made their fortunes "... in collusion with [bureaucracy], which shielded them from competition".[362] Both groups would be opposed to any attempt to eliminate the sources of their wealth.

Third, the political evolution in the post-communist world seems to give credence to Juan Linz's view that parliamentary governments are more effective than presidential ones in establishing stable democracy.[363] Parliamentary democracies also seem to coincide with lower levels of corruption, although they require complex and mature institutional arrangements assuring the accountability of political elites to society at large.

Yet, the universality of this proposition may be debatable. The success of parliamentary government should not mislead anybody into thinking that this has produced a responsible and efficient system of the rule of law. In many cases, it has rather resulted in a situation of weak executive power, weak legislative power and weak political parties with very strong informal cliques in control of the system. This institutional weakness is due to the fact that these fundamental parts of the state organization, overwhelmed by informal party connections, have been unable to gain sufficient autonomy to develop their own missions and procedures. Under conditions of such excessive informality, corruption becomes a natural way of solving conflicts between ideals and material interests. In both cases, these political systems are far from an ideal situation of a strong civil society in a state with strong executive and legislative powers.

### (iii) Conclusion

The communist political, economic and cultural project was anti-western, both in the logic of its

---

[361] S. Holmes, "Crime and corruption after communism: introduction", *East European Constitutional Review*, Vol. 6, No. 4, 1997, pp. 69-70.

[362] A. Åslund, "The trouble with economic ...", op. cit.

[363] J. Linz and A. Stepan, op. cit.

intellectual content and in its ideology. Due to its revolutionary nature, communist regimes had either physically or socially eliminated the old intelligentsia, that is, the social class most attracted by and most familiar with western civilization. It also eliminated, or drastically curtailed, the most active elements in other social classes, albeit the level of devastation varied across countries. In consequence, post-communist societies were poorly endowed with the institutions and human capital capable of handling either the economic or political dimension of transition.[364] On the other hand, however, they had some important assets in terms of overall educational levels of their respective populations vis-à-vis countries at similar levels of economic development.

Nevertheless, the transition was difficult throughout the post-communist world especially for countries devoid of civil society in the period preceding the communist takeover and for those with longer periods of communist rule.[365] The liberal-democratic transition in the CEECs was, in cultural and psychological terms, easier than in the CIS. An additional boost to the former's effort was given by the close interest of west European countries in promoting political and economic stability in the region close to its borders. The goal of joining NATO and the EU – which is enjoying high levels of popular support in the CEECs – has become the "strategic" goal of governments compelling them to comply with "western" institutional standards.

Yet, one should not conclude from the above that the need for improvement in governance and combating corruption should be removed from the agenda of the CEECs. First, with the disintegration of central controls during the last stage of communism, these societies became highly vulnerable to corruption. In fact, the extent of corruption in post-communist countries may be higher than indicated by various measures discussed above. Consider the following: they all went through privatization programmes of unprecedented scope – their legacy in many countries is mistrust of government and, often, corruption of politics. Most of them are still coping with the legacy of legal acts designed to use the "rule of law" as a tool for the exercise (rather than the restraint) of power by the ruling elites. In practically all of them, institutions are weak, the competence of public administration wanting and anti-rule-of-law constituencies relatively strong. In a few of them, the "enforcement of property rights", abdicated by the state, has been takeover by criminal groups. These institutional weaknesses have provided ample opportunities to small, well-knit informal groups capable of exploiting their strategic positions in the government to enrich themselves and to strengthen their hold over power.

Second, the CEECs have to cope with the legacy of issues overlooked during the initial stages of transition. They mostly ignored the fact that liberal democracy is a concept that embraces the organization of both the economy and the polity. As Douglas North observed, "… it is the structure of political and economic organization which determines the performance of an economy at the incremental rate of growth in knowledge and technology".[366] Post-communist reformers should have paid at least as much attention to the design of the political system as they did to the problems of privatization and economic transformation. They did not, however, as the organization of the state, the condition of public administration and the judiciary was of no interest to the reformers.[367] During the initial stages of transition, the preoccupation with macro-stability and privatization seems to have pushed institutional issues from the agenda of those involved in the external assistance effort. Only recently, with prodding from the European Commission,[368] have they begun to realize that in a liberal democracy economy and polity are closely related.

A sound administration is a crucial asset for such a complex task as the transformation of a post-communist economy. Of course, this was absent in the post-communist administrative landscape, but a lot more could have been done had the reformers been aware of how important is an efficient and reliable administration.

## 5.3 Governance and combating corruption

The development of an efficient, accountable system of governance goes beyond reforms aimed at uprooting corruption in public administration or public finance. It calls for change in the governance structure to

---

[364] B. Kamiński, "The legacy of communism", *East-Central European Economies in Transition*, study papers submitted to the Joint Economic Committee, Congress of the United States, United States Government Printing Office (Washington, D.C.), 1994.

[365] It is often emphasized that civil society in east-central and eastern Europe had never been strong, with the possible exception of Czechoslovakia. This opinion seems to be wrong-headed at least with respect to the CEECs (with the possible exception of Romania) and the former European republics of the USSR. If we only take into account the history of the region during the nineteenth century, there is no doubt that Germany as a whole, and Prussia as the dominant part of it, were able to develop an effective, albeit authoritarian, version of the rule of law. The Austrian Empire made impressive progress towards a parliamentary central government combined with decentralization of the state and a full-blown local self-government. The Russian Empire succeeded in reforming the public administration. It also created a fairly sound judiciary as well as an effective educational system. If one can easily accept the thesis about the economic and social backwardness of Russia, the importance of this country to the development of the arts and sciences in the nineteenth and twentieth centuries, as well as the intellectual quality of its elite, cannot be rejected. Russia before the First World War was part of Europe and, by and large, followed the European path. The southern part of the post-communist space, the Balkans, is another problem: it had a long history of subjection to Ottoman rule. But here also the model to follow was the west of Europe, and the European example influenced Turkey as well. Thus, the trends in Europe at the turn of the nineteenth century were more uniform than perhaps ever after.

[366] D. North, *Structure and Change in Economic History* (New York, Norton, 1981), p. 19.

[367] B. Ackerman, op. cit., p. 46.

[368] European Commission, *Enlargement Strategy Paper. Report on Progress Towards Accession By Each of the Candidate Countries* (Brussels), 1999.

ensure that the public interest is not captured by narrow private interest. While the capture of policy by private interests occurs in mature democracies, societies with weak, recently established structures are particularly prone to it. This is so because the constraints placed on the behaviour of public officials and politicians by civil society as well as the mass media, judiciary, etc. remains low. And so does the penalty levied on politicians caught making decisions benefiting their private interest at the expense of the broader public good.

Another reason is that the weakness in major branches of the government is often compounded by limited administrative capacity. Weak parliaments are unable to impose upon the participants its own rules protecting the public interest and to monitor the executive branch. The simultaneous weakness of the legislature and the executive favours a party system that consists of networks of informal cliques preying upon the public sector rather than institutions designed to represent the interests of the electorate. Under such conditions, the accountability of public office holders becomes doubtful, at best, and wanting, at worst.

The absence of accountability of politicians and public officials may also erect a strong barrier to the emergence of institutions effectively supporting the development of competitive markets. Competitive markets reward the best performers, whereas distorted markets reward those with political influence. Without the sanction of the ballot, they will block measures that threaten to remove market distortions on which they thrive. In transition economies that implemented only partial reforms (most members of the Commonwealth of Independent States and Bulgaria until 1996), the biggest winners, rather than the losers, sought to "... stall the economy in a *partial reform equilibrium* that generates concentrated rents for themselves, while imposing high costs on the rest of society".[369] They would find support among the losers, as they objected to liberalization of the economy that would undermine their monopolistic positions.

With the political system effectively immune to popular pressure (*voice*), only a rapid deterioration in the economy and the involvement of external actors can tip the economy from its state of *partial reform equilibrium*. Economic crisis usually forces the government to seek external assistance, but this does not automatically guarantee the shift from a *bad equilibrium*. It will depend on the conditions attached to assistance and government's commitment to their implementation. Both are often wanting, and the room for manoeuvre for external actors is limited. Furthermore, they cannot act as a substitute for government, but without their involvement no meaningful progress can take place.

Successful transition from central planning to competitive markets entails a shift from the fusion of a monopartisan state in control of both society and economy, a characteristic of communist regimes, to a rule-based interaction between the public and private sectors enforced by law.[370] It requires several tasks all aimed at changing the *public-private interface*. They all share one common denominator – separation of the public and private spheres, the establishment of the rule of law and a leaner, more transparent government.[371]

The process of post-communist transition requires the simultaneous reorganization of the state and of the economy. Within the former, conditions must be created to allow for a plurality of political parties to compete for control over the government, for the establishment of legislative controls over the executive, and for the emergence of an independent judiciary. In the economic domain, there are at least four interwoven tasks that are necessary to institute a viable public-private interface. The first and the most immediate task relates to the establishment of stability and predictability in the macro-policy environment. The second involves the transfer of assets from state to private ownership combined with the elimination of direct state intervention based on ad hoc rules. The third task is to build the political framework with laws, regulations and administrative procedures immune to "capture" by narrow interest groups and supportive of competitive markets. The minimization, if not removal, of *state capture* that was discussed earlier falls into this task, although we shall give it a broader interpretation. The fourth task is to make sure that the framework actually works, i.e. that policies are implemented and rules observed. This can be accomplished through increasing the capacity, integrity and oversight of the civil service, and its credibility vis-à-vis society at large. Subverting *administrative corruption* belongs to this task. In the post-communist transition, economic and political transformations go hand in hand.

Crucial to the establishment of viable governance structures is to establish the framework that would ensure competition in both the political and economic spheres and separate the two through well-defined rules. This is probably the most critical step in building a good governance structure, but it is also an important step in curtailing the opportunities for corruption. The common denominator of the failure to reduce corruption is the continued penetration of the economic realm by politics, and of the political realm by business interests. Its most visible symptom is the pursuit of material gain through a political rather than a business career, and in parallel, making profits through political connections. The solution to this problem requires an adequate institutional

---

[369] J. Hellman, loc. cit., pp. 204-205.

[370] For a discussion of fusion as a major characteristic of the socialist institutional design of the state see B. Kamiński, *The Collapse of ...*, op. cit.

[371] S. Rose-Ackerman, "Is leaner government necessarily cleaner government?", in J. Tulchin and R. Espach (eds.), *Combating Corruption in Latin America* (Washington, D.C., Woodrow Wilson Center Press, 2000), pp. 87-106.

design with an underlying culture that permits the distinction and endows it with a normative meaning. The competition system must be designed as a function of the objectives it should serve.

Once the institutional design for effectively mediating between various competing vested interests is in place, two other issues emerge. The first relates to policy implementation – unaccountable officials may be tempted to abuse their position for private gain and distort the process of policy implementation. The second concerns the existence of viable external checks on public officials. Some of these checks may work at the office level, others may be provided by specialized control agencies. Civil society, including the mass media, can help to subvert corruption and improve the quality of decision-making at all levels of government.

### (i) Capture of policies by narrow interest groups: democracies versus electocracies

Money in politics is always corrupting. Generally one cannot prove quid pro quos. It is obvious that legislators are more attentive to the views of those who give money, and those with money have plenty of reasons to support or obstruct legislative initiatives. Otherwise, they would not pay. The challenge is how to minimize the chances that policies will be captured to the detriment of national welfare.

The classic Madisonian tradition of American democracy perceives interest groups, which Madison called "factions", not only as natural but also, when specific institutional conditions are met, as agents potentially beneficial to democracy because they tend to check and balance each other. The division of powers within the state, a plurality of political parties competing for power, and a multitude of interest groups seeking to influence public decision makers prevents any single faction from dominating the state. This is the essence of the mechanism of checks and balances – a dynamic political equilibrium. To work, it requires a well-structured system of competition, i.e. a mature constitutional democracy.

What will happen, however, when the political space is filled with groups that approach the state in an instrumental fashion, as an external entity that may be preyed upon? These groups may have all appearances of political parties but, deprived of a republican tradition, they treat the idea of *raison d'état* or of *public good* as an empty phrase. In order to function in politics they need money that they raise by preying upon the state. They compete in elections, but once the electoral campaign has passed they cooperate closely with their adversaries in making illegal deals. Each party knows enough about the others to feel safe, and each knows that collecting "dirt" on the others is the most effective insurance policy. The "dirt collection" mechanism operates not only in relations between political parties, but also in the selection of party cadres: to trust a colleague one must have "dirt" on him.

This is a checks and balance system of sorts, but it works in ways different from the one devised by Montesquieu and Madison. Under such a system, although politicians are elected, their accountability to the electorate is very limited. The social environment in which they function makes them act against the public interest.

The picture described above characterizes, albeit to varying degrees, a number of post-communist states, CEECs and CIS alike. There is a constitutional division of powers, there are competitive elections, an independent judiciary and a relatively free press. In constitutional terms, most of the post-communist states, particularly among the CEECs, could be considered mature democracies, but the least we can say is that they are not mature. Karen Dawisha has mockingly called them "electocracies".[372] The idea that the term conveys is formality of procedure.

At the root of the problem is the starting point of the post-communist transition. As we have mentioned time and again, communism was a system that never tolerated autonomous centres of power. Under communism politics was a zero-sum-game: acquisition of autonomy by any outside centre implied a loss of power to the party. Not only were formal organizations and associations supposed to be under the full control of the party, but also individuals were supposed to be at the party's full disposal: whether he or she was a judge, a doctor, a priest or a taxi driver he or she owed loyalty first of all to the party. This ideal was nowhere fully achieved, the closest to it being the USSR between 1930 and 1953. Nonetheless, communism never tolerated self-organization. Thus, the demise of communism came to societies hardly prepared for life in a democracy. Hungary and Poland may have been different in this respect, but not enough to make the transition an easy task.

Thus, at the start of the transformation the post-communist countries had weak civil societies, an old institutional system in demise, and the new one *in statu nascendi*. There was a state administration, a system of justice and parliamentary elections. The state administration was in many respects shaped by its communist past, but it still maintained some bureaucratic values. The judiciary was surely mutilated by the old regime, although it still claimed respect for the traditional values of the institution.

The coalitions that came to power in the aftermath of free elections were composed – depending on the country – of different mixes of new faces and old-timers, but usually with little organizational experience. Many administration officials with former communist connections were replaced by people without any administrative experience or skills. In most post-communist countries no effort was made to develop

---

[372] K. Dawisha, op. cit.

programmes aimed at establishing a competent civil service. Instead, each party and coalition in power tried to infuse the administration with its protégés. This resulted in the politicization of administration and in growing corruption. Bad administration makes implementation of any policy difficult. It also seeks protection in a legal mess to which it willingly contributes, thus strengthening the hand of anti-rule-of law constituencies.

The new regimes also neglected the state of the judiciary. In most post-communist countries, justices were formally guaranteed immunity, while their salaries were kept at a very low level. Professional autonomy serves well the exercise of justice but only when legal corporations are strong enough to ensure the maintenance of professional standards. In the absence of effective professional controls, the autonomy of judicial procedures and the immunity of the judges will not protect the legal community against corruption. An opportunity for impressive gains facilitated by weak controls made many judges responsive to the temptation of fast enrichment. In consequence, the system that was supposed, among other things, to uphold contracts and property rights has remained ineffective even in most advanced CEECs.[373]

Furthermore, the public prosecution service is often part of the state administration and manifests all the drawbacks of the system as a whole. In most cases, chief prosecutors with wide powers of control over the activities of subordinate prosecutors are politicians. Such interference leads to the decline of professional standards and the demoralization of the personnel.

Thus, the question arises: what can be done to minimize *state capture?* In other words, how can established rules of behaviour be changed? In the absence of a commitment by the political class to reform itself, it would seem that change could come from two sources – civil society and external actors. External actors should support the "voice" of grass roots movements. Once corruption has become the norm and permeated decision-making structures, civil society can play an important role in subverting state capture. But outside-state activities are likely to have a more immediate impact in "democracies" rather than in "electocracies". In the absence of civil society, external actors seem to be the most important source of influence for the latter.

As for external actors, the distinction has to be made between CEECs seeking accession to the EU and CIS countries. EU candidate countries are subject to annual assessments of their progress towards accession, while the CIS are not subject to similarly strong "peer pressure". The Commission is often highly critical. For instance, it has stated that "... corruption, fraud and economic crime are widespread in most candidate countries, leading to a lack of confidence by the citizens and discrediting the reforms".[374] Individual country reports usually voice similar complaints about the low quality of public administration, an overloaded and corrupt judiciary, etc. Convergence to the *acquis communautaire* is another lever for improvement in the quality of governance. For instance, procurement and state aid (extra-budgetary revenues and expenditures) are among the domains particularly vulnerable to corruption: once they are brought under EU rules, the opportunities for rent-seeking will drastically fall.

Yet, in spite of the Copenhagen political criteria for accession (functioning democracy and protection of minorities), subsequently enshrined in the Treaty of Amsterdam, no serious analysis by the European Commission of the flaws in the political structures of the candidate countries has so far accompanied these assessments. A detailed appraisal of political arrangements, however, including the degree of freedom of the mass media from the point of view of good governance and the opportunities for corruption (i.e. mainly *state capture*), could set in motion a virtuous circle of change in the candidate countries.

Other external actors with a potentially huge impact on the quality of governance in both the CEECs and the CIS are the multilateral financial institutions. Since around 1997 both the IMF and the World Bank have recognized the importance of subverting corruption and improving governance.[375] The programmes of both institutions have included measures liberalizing the economy, strengthening the budgetary process and the treasury system, establishing independent central banks, privatizing state owned assets and improving legal systems. By slashing various state-run preferential schemes, excessive regulation of private business activity, non-transparency and non-accountability, these measures have improved management and limited the opportunities for bribes or kickbacks.

But in the absence of a strong government commitment to establishing a good policy environment and a viable judicial system, their immediate impact is rather negligible, as the persistently high levels of corruption in many post-communist countries seems to indicate. The system of justice presents a particularly pressing problem. First of all, the commercial courts – which are key to the functioning of a market economy since they are responsible for the enforcement of

---

[373] European Commission, op. cit.

[374] Ibid.

[375] For instance, the World Bank has produced, at the request of governments, a number of country studies assessing corruption, provided loans for public administration reforms (including tax and customs) and kept the issue alive through its research and publications and policy dialogue with member countries. Both institutions provide technical assistance to improve governance in their respective areas of expertise. T. Wolf and E. Gurgen, *Improving Governance and Fighting Corruption in the Baltic and CIS Countries: The Role of the IMF*, IMF Economic Issues, No. 21 (Washington, D.C.), July 2000.

contracts and private property rights – are generally perceived as either highly corrupt or inefficient, or both. Second, the system of justice is a critical factor in fighting corruption in all other areas of the state and the economy. Without it, effective sanctions for corruption are unlikely to materialize.

While the only effective way to deal with *state capture* is to establish an effective system of checks and balances with three strong branches of government (executive, judiciary, legislature) as well as clear rules for the public-private interface, there are several "small" measures that may reduce corruption. These include, for instance, full disclosure of personal finances by public officials; the disclosure of sources of contributions to political campaigns; a ban on public officials and civil servants combining political/administrative positions with positions in business including firms in which the state has equity; and a ban on the simultaneous participation of an elected official in other branches of government.[376]

Addressing corruption at the centre seems to be of paramount importance in addressing *state capture*. As Amartya Sen succinctly noted, "Corrupt behaviour in "high places" can have effects far beyond the direct consequences of that behaviour, and the insistence on starting at the top does have reasoning behind it".[377] Values and norms propagated from the top count, and the disappearance of corrupt behaviour there can weaken the hold of corruption elsewhere, in part also, due to the ensuing collapse in the supply of laws and regulations generating opportunities for corruption. "In trying to alter a climate of conduct, it is encouraging to bear in mind the fact that each vicious circle entails a virtuous circle if the direction is changed."[378] But the question remains whether the direction can change without addressing the institutional design itself.

### (ii) Combating administrative corruption

Most post-communist countries face the problem of a large, inefficient and underpaid civil service with weak professional incentives. A weak state administration cannot defend its integrity and impartiality in the face of political pressures. Personnel policies, based on political rather then meritocratic criteria, promote partisanship and incompetence. The difficulty is even more acute at the level of the local administration since decentralization of a corrupt system of government brings about decentralization of corruption. The situation becomes, therefore, even less manageable. In what follows, our focus is on the identification of measures that address the issue of reducing the potential or opportunities for corruption.

While it takes a lot of effort and time to uproot corruption sustained from above by *state capture,* a political elite already committed to combating corruption may push for implementing strategies focused on selected areas of the public domain. Courses of action may differ depending on the issue area. But to be effective, they all should start by addressing the opportunities or the potential for corruption due to the regulatory environment and the motives for corruption. The potential for corruption increases with the expansion in the supply of regulations, and is also a function of their wording and complexity.

Because of the legacy of overregulation, taking stock of the existing laws and regulations is the first step of the "deregulation" effort.[379] They should be subject to several tests, which might include: What purpose do they serve? Do they unnecessarily hinder business activity? Do they provide protection to firms as well as citizens against harassment by the state? Do they offer easy mechanisms of appeal against administrative decisions? Are they endowed with procedures limiting excessive discretion by the administration? Are they simple and transparent?

Because of the legacy of the misrule of law and weak administrations in post-communist countries, reviewing the existing laws from the point of view of their wording in terms of offering excessive discretion to the administration is a good point of departure. Excessive official discretion derives from the loose wording of laws and regulations, the absence of clear criteria for administrative practices, and complex regulations.

The same concerns should apply to new laws and regulations. They should be subject to similar scrutiny taking into account the opportunities for corruption, the administrative costs involved to both firms and individuals while simultaneously seeking to minimize the opportunities for rent-seeking. This involves observing a couple of rules. First, regulations should be introduced only when they are needed for important reasons to ensure competition, the protection of health or the environment or, in general, to prevent the production of "public bads". Other reasons may derive from commitments under international treaties or from participation in various international organizations.[380]

---

[376] For instance, a member of Parliament serving in the executive branch blurs the boundary between legislative and executive functions. In consequence, the democratic mechanism of control by the legislature over the executive may become weaker and – with the absence of a clear dividing line between bureaucratic and political elites – the opportunities for corruption become more pervasive.

[377] A. Sen, op. cit., p. 278.

[378] Ibid.

[379] A. Åslund and M. Dmitriev, "Economic reform versus rent-seeking", in A. Åslund and M. Brill Olcott (eds.), *Russia After Communism* (Washington, D.C., Carnegie Endowment for International Peace, 2000), note that in Russia more than 60 agencies inspect businesses and virtually all economic activities are subject to not one but multiple licences. These regulations serve as a tool to extract bribes and create a basis for the tyranny of bureaucrats.

[380] Some of them may be quite demanding especially those related to membership in the WTO. For the advanced countries with systems compatible with international conventions the WTO merely obliges them to apply their domestic rules fairly at the border. This is clearly not the case for most developing countries including CEECs and CIS members of the WTO. For a discussion of the barriers they face see J. Finger and P. Schuler, *Implementation of Uruguay Round Commitments. The*

Aside from these, "deregulation" should bring considerable economic benefits through, at least, transforming a shadow economy into a formal economy and improving the business environment.[381]

Second, the wording should be clear and procedures transparent. Clear wording and transparency should go a long way to reduce the potential for corruption. Whenever it applies, a regulation should clearly specify the responsibilities of administrators. For instance, the regulation concerning the issuance of a licence to conduct business activity should specify the areas that are excluded (rather than those which are included) and set the timetable that the administration has to observe in issuing the document. By infusing more predictability into the policy environment, these simple rules can positively influence economic activity.

Third, simplicity reduces the potential for rent extraction. Complex tax regulations, for instance, offer opportunities for tax evasion and provide incentives to elicit payments. For instance, eliminating various tax incentives (exemptions), and using the savings to reduce corporate or personal income tax rates, and simplifying the tax code do not only reduce the opportunities for extracting bribes but also improve the business climate because of their limiting impact on bureaucratic discretion. The reduction in tax rates minimizes administration and compliance costs and thus contributes to efficiency in the allocation of resources. One may thus expect higher foreign investment inflows as well as an increase in domestic business activity.

Broadening the tax base through lowering tax rates and eliminating special exemptions decreases opportunities for rent-seeking. It tends to make tax liabilities transparent; it lessens the compliance cost to taxpayers and makes tax calculations easier; and more importantly, it also reduces the discretionary power of tax inspectors.[382] In brief, tax simplification does not only limit the opportunity for extracting illicit payments but also increases economic efficiency.

Simplicity entails greater neutrality of applied measures and therefore reduces distortions. A good example of measures that simultaneously contribute to greater allocative efficiency and helps a country to take advantage of the opportunities offered by global markets is a uniform tariff structure.[383] The best foreign trade policy, especially for a small developing economy, is that of free trade. The second best option is a uniform tariff structure with, preferably, a low tariff rate. Both options – if enforced by special constitutional arrangements removing tariff policy from the realms of current policy – are also superior on other grounds. Because of administrative simplicity, they reduce (or almost completely eliminate with a zero tariff rate) the potential for *administrative corruption*. And, if the commitment to the principle of uniformity is legally binding, it also discourages firms from shopping around for protection through tariffs. In other words, the potential for *policy capture* by a protectionist lobby is then significantly curtailed (for a more detailed discussion, see box 5.3.1).

Hence, any plan to address the intrinsic problems related to corruption in tax and customs administration as well as in government procurement should first consider measures to reduce the potential for corruption both at the policy and implementation levels. Government procurement offers enormous opportunities for corruption. Note first that its focus lies in the shady area between the private and public sectors. Second, the market for government procurement is huge, often accounting for a sizeable proportion of a country's GDP. Soft or single tendering procedures under which no competitive bidding is initiated offer opportunities for the misappropriation of public funds and for raising the cost of public services to taxpayers. Despite some steps toward making the public procurement systems more transparent and, therefore, less prone to corruption, the results are far from satisfactory. The fact that not a single WTO member from among the post-communist countries has joined the WTO Agreement on Government Procurement seems to indicate the strength of entrenched domestic interests.[384]

As for tax and customs administration, a simple tax code with low rates and a wide tax base, and a relatively uniform tariff structure on imports, are ways to start an anti-corruption campaign in tax and customs administration.[385] This should be accompanied by the introduction of non-discretionary penalties for non-compliance. The elimination of administrative discretion in setting penalties removes one source of corruption.

---

*Development Challenge,* World Bank Policy Research Working Paper, No. 2215 (Washington, D.C.), October 1999.

[381] H. De Soto, *The Other Path. The Invisible Revolution in the Third World* (New York, Harper & Row, 1991); D. Kaufman and A. Kaliberda, op. cit.

[382] World Bank, "An anti-corruption strategy for revenue administration", *PREM Notes Public Sector,* No. 33 (Washington, D.C.), October 1999.

[383] A. Panagariya and D. Rodrik, "Political economy arguments for a uniform tariff", *International Economic Review,* Vol. 34, No. 3, 1993, pp. 685-703.

[384] The WTO Agreement on Government Procurement provides a set of rules for opening procurement to foreign competition as well as establishing detailed rules to enhance the transparency of tendering procedures. This agreement, although negotiated under GATT auspices is plurilateral, i.e. applies only to signatories. No post-communist country has yet acceded to that agreement, although some made commitments to join it as part of their recent accessions. The countries furthest along in the accession process are Estonia and Latvia. A related activity at the WTO is the effort by many members to launch negotiations on an Agreement on Transparency in Government Procurement. Such an agreement would apply to all members of the WTO and would address, among other things, issues relating to corruption and governance.

[385] Tax and customs administration has been notoriously corrupt in most post-communist countries, especially in the CIS economies. Anecdotal evidence from customs administrations provide a good illustration of the World Bank's *state capture* (policies are formulated explicitly to create opportunities for rents) and *administrative corruption* (complex regulations offer customs officials significant discretion in setting border charges). Despite meager salaries, positions in customs border control are highly sought after (illicit payments to obtain this position are sometimes equivalent to a total of 50 years of salary).

> **Box 5.3.1**
>
> **Good economics and a corruption reducing measure: the case for a uniform tariff structure**
>
> The introduction of a uniform tariff structure with, preferably, a low tariff rate is mainly informed by the following considerations:
>
> *It equalizes distortions associated with tariffs across sectors of the economy since it affords the same level of protection to all firms.* In other words, it offers the same protection, for instance, to a producer of computer chips and one of computers. This is very important in the case of a small transition economy facing readjustment to world markets. High tariffs on products of sectors which seemed to have had a comparative advantage (that is, were competitive) before the collapse of central planning would only halt the process of adjustment.
>
> *It is transparent and administratively simple.* This in turn reduces the administrative burden of customs by simplifying clearing procedures. A uniform tariff rate also helps to prevent losses in customs revenue due to the misclassification of imported goods. While it is not a universal remedy for corrupt practices, it certainly improves the administrative capacity to monitor customs revenues.
>
> *It reduces drastically the scope for unproductive lobbying activities for higher tariffs provided that a government is firmly committed to the principle of a uniform tariff rate.* This commitment can be achieved by a legally adopted requirement that the principle of a uniform tariff rate can be overturned only by a majority or, even more effectively, super majority (e.g. two-thirds majority) vote in a parliament. The reduction in lobbying activities is due to the fact that they are driven by the prospects of gains being bestowed upon a small group of firms. The uniform tariff rate gives the appearance of equal protection to all import-competing sectors of the economy.
>
> *It compels decision makers to take into account not only the interests of producers* (as is usually the case with the exception of some countries, Australia for example, which have set up institutions to conduct a full cost/benefit analysis of new foreign trade policy measures) *but also consumers and users of imports.* This is so because the change in a tariff rate affects all import-competing and export industries and, therefore, users of imports are likely to oppose lobbying efforts to change a tariff rate.
>
> *The adherence to the principle of a uniform rate tariff effectively removes tariff decisions from the realm of current politics, which is usually destructive of economic efficiency and the business climate.* Consider for instance the following: many countries are tempted to offer higher tariffs to attract foreign direct investment. Leaving aside the fact that international experience suggests that this is not effective in attracting high quality investment, this may lead to significant disagreements and fiscally irresponsible competition between regional authorities. Note also that the principle of a uniform tariff rate seems to be particularly well suited to the institutional arrangements in countries with weak administrative capacity and vulnerable to producers' lobbying. By establishing clear rules of the game, it reduces, if not entirely eliminates, the potential friction over the use of tariffs to protect specific industries or attract foreign investment.

The simplification and expansion of nondiscretionary rules alone will not uproot corruption. These have to be supplemented by measures addressing motives. These will be most effective if simultaneously pursued through four different channels: basic motivation (elite ethos combined with mission statements); supply-side elements (effective sanctions for bribe payers and publicity for penalties); positive incentives (autonomy, competitive base pay, non-arbitrary reward procedures); and negative incentives (effective sanctions for corruption).[386]

To be effective, these measures should be accompanied by reforms in organization and management as well as by the establishment of external mechanisms increasing the accountability of the respective branches of administration. The former includes the setting of performance standards that can be captured by an effective management information system, whereas the latter involves offering "voice" through independent surveys, citizen reviews and oversight.

Hence combating administrative corruption boils down to two, relatively straightforward propositions. As for laws and regulations, these must first set tough constraints on the supply of regulations, and second, make sure that they are transparent, simple, and that they limit the discretionary powers of officials. To be effective, they have to be accompanied by measures to enhance the transparency, accountability and capacity of public administration; to increase resistance to lobbying and outside political influence; and to develop and implement clear criteria for personnel management decisions in the civil service. In addition, adequate institutional capacity should be in place for financial management, including effective agencies for internal and external audit, the treasury system and for public procurement. The mass media and civil society can provide important external checks by increasing accountability and tracing corruption.

---

[386] For an extensive discussion of these major components and their implementation in World Bank-financed programmes in several countries see World Bank, "An anti-corruption strategy …", op. cit.

### (iii) NGOs and independent media

The level of consciousness about "the cancer of corruption" has dramatically expanded worldwide. So has the number of national anti-corruption programmes, often supported by international organizations. A large number of studies, sponsored mostly by international organizations and NGOs, have identified the major issues and the steps that anti-corruption campaigns should contain.[387] Their review would go beyond the format of this paper.

While we shall draw on a number of their observations, the approach that we take here differs subtly in its emphasis on the institutional underpinnings of governance and on explicitly putting governance first rather than corruption per se. In the absence of civil society and political accountability, even the best designed anti-corruption programmes are not worth more than the paper they are written on, simply because the privileged will do their best to undermine their implementation.

NGOs and free and open media can help to suppress the level of both types of corruption (*state capture* and *administrative* corruption) by uncovering and exposing abuses. Both can raise public awareness about corruption and the need to subvert it. Their full independence from the state is the necessary condition of their effectiveness in anti-corruption campaigns.

The media and many NGOs still remain within state's reach in most post-communist countries. This is so either because of funding arrangements (NGOs) or conflicts of interests created by ownership arrangements or corruption within both. State controls through issuing licenses or allocating frequencies have weakened the mass media's (especially television and radio) contribution to checking corruption. NGOs often remain weak unless supported by government. In fact, an illustration of the weaknesses of NGOs may be that they have "… proven to be most effective when the government treats [them] as a particular ally".[388] When NGOs are strong and independent, the government, whether it wants to or not, must listen to their *voice* rather than choose to treat them as an ally.

By the same token, the conventional measure of the robustness of civil society in terms of the density of non-governmental organizations may be highly misleading. Some NGOs in post-communist countries are so closely related to the government that they have almost become fused with it. Some may perform some analytical functions while offering incumbent politicians the chance of earning additional income, others simply offer employment opportunities to politicians who are temporarily out of office. The irony is that in the post-communist world, these NGOs – let us call them, for the lack of a better term, semi-official NGOs – are often the best financed, and the most praised by the media and public officials. These are a new variety of NGOs with strong political commitments.

But in some countries there has emerged another category of civic associations, which appear to be much more valuable at least in terms of strengthening democracy. They resemble the more traditional "civil society", as described by Putnam,[389] with a strong commitment to civic responsibility and community service. These are organizations founded by groups of citizens concerned with the state of local or regional affairs – we shall refer to them as spontaneous NGOs. People who form these associations are brought together not by political ambitions or the prospect of improving their standard of living, but by a sense of public duty and civic responsibility. These people do not receive any income for their activity. On the contrary, they materially support their associations by paying membership dues. This category of civic associations is a much better indicator of the maturity of civil society. In contrast to semi-official NGOs with their strong political commitments, they also contribute to the growth in social cohesion and the development of social capital.[390]

Unlike the semi-official NGOs, spontaneous NGOs often tend to be unpopular with political parties and public administrations, which they criticize for their shortcomings. Both central and local government tend to disregard their existence wherever possible. In some post-communist countries, semi-official NGOs give a cold shoulder to spontaneous NGOs, which comes as no surprise as the former benefit from state capture. Therefore, spontaneous NGOs are much more important for democracy and rooting out corruption (especially state capture) then their counterparts. Where they exist, reformers, both domestic and external, should actively seek their involvement in the reform process.

---

[387] The World Bank provides probably the most comprehensive discussion of the design of anti-corruption strategies under various categories defined in terms of two aspects of corruption (state capture and administrative corruption). World Bank, *Anti-corruption in Transition* …, op. cit. While the report focuses on post-communist countries, its strategy recommendations are fairly general and applicable to a large number of developing countries. M. Gonzales de Asis, "Coalition-building to fight corruption", World Bank Institute, paper prepared for the Anti-corruption Summit (Washington, D.C.), November 2000 (http://www.worldbank.org/wbi/governance), examines the best ways to build coalitions to fight corruption. T. Wolf and E. Gurgen, op. cit., summarize the principles underlying the IMF's effort to curb corruption. J. Braga de Macedo, "Globalization and institutional change: a development perspective", revised version of the paper presented at the General Assembly of the Pontificat Academy of Social Sciences on *Globalization and the Common Humanity: Ethical and Institutional Concerns* (Vatican City), 25-28 April 2001, discusses the efforts to subvert corruption undertaken by NGOs mainly in Latin America.

[388] World Bank, *Anti-corruption in Transition* …, op. cit., p. xxiii.

[389] R. Putnam, *Making Democracy Work: Civic Traditions in Modern Italy* (Princeton, Princeton University Press, 1993).

[390] Ibid.

### (iv) Concluding comments

The common denominator of the failure to downsize corruption in most post-communist countries is the continued penetration of the economic realm by politics. Its most visible symptom is the pursuit of material gain through political rather than business careers. Corruption is a severe hindrance to sustained growth and the effective management of public services. All countries appear to suffer from *state capture* and relatively high *administrative corruption*.

Corruption revealed in *state capture* presents a particularly hard problem. Developing a strategy to combat solely *administrative corruption* without addressing one of its sources, i.e. the lack of constraints on the supply of laws and regulations offering opportunities for rent-seeking (state capture), would be ineffective, if not counterproductive. A change of climate can only come from the top, and only this can start a needed virtuous circle of change. The root of state corruption is the absence of well-developed formal and informal checks and balances. Accountability is further constrained by underdeveloped expenditure controls, close personal links between leading politicians and businessmen and, in CIS countries, by the executive's dominance over other branches of government. Civil society, NGOs and the media are still feeling their way towards real influence. This, together with the pervasive "control" activities of government entities, weakens the responsiveness of government to public and business concerns. The civil service needs to become more professional and meritocratic, and better able to resist pressures from powerful interest groups.

While the achievements have been considerable in many post-communist countries, in most of them the capacity for effective policy remains unbalanced and accountability structures underdeveloped. In effect, there are shortcomings in both the supply and the demand for good public policy. On the supply side, the capacity to generate good policy options remains limited by the endowment of institutional structures and skills which, in the aftermath of the collapse of communism, are unsuited to the needs of the nation-state and a market-oriented economy.

On the demand side, the existing political arrangements (including electoral procedures) still suppress "voice". Without opening the political system to more intense competition, the chances for a significant reduction in the level of corruption are slim.

## 5.4 Corruption and development: policy convergence

Although corruption is as old as government and the distinction between private and public, until recently international organizations and national governments studiously avoided addressing this phenomenon; whenever international organizations did so, they would use the "less offensive" term of rent-seeking. This changed, however, around the mid-1990s, first with the establishment of Transparency International in 1993. Ever since, corruption has become one of the main themes on the agenda of all major international organizations. For instance, combating corruption was the main issue at the annual World Bank/IMF meeting in Hong Kong in 1997. Both organizations have spared no effort to raise consciousness about its negative impact on economic performance and both have encouraged and assisted countries' efforts to fight it. The Organization for Economic Cooperation and Development has also designed its own programmes to combat bribery in international business transactions by imposing on its member states the duty to penalize acts of corruption of foreign officials by business executives who are OECD citizens (OECD Convention of 1997) and developing codes of sound business practices.[391] Together with its national chapters, Transparency International has become highly visible working at the grass roots level to raise awareness of corruption and to remove its most evident symptoms.

No single factor can be pinned down as responsible for the emergence of interest in corruption. In fact, several developments have contributed to it. First of all, the end of the cold war made it possible to end the practice of buying the support of corrupt political leaders of developing countries for the western cause. Simultaneously, the demise of communism removed from the agenda the need to protect corrupt, centre-right governments in the west against the danger of a communist electoral victory. Italy is a case in point.[392]

The last two decades have also witnessed a convergence in thinking about what constitutes the right package of policies to enter the path of sustainable economic development.[393] While in the 1960s and the 1970s, it was assumed that developing countries should pursue policies widely divergent from those in highly developed economies, this view began to change following the world debt crisis of the early 1980s. The earlier emphasis on inward-orientation, financing development through inflation and the direct involvement of the state in economic activity gave way to foreign trade liberalization and macroeconomic stability. However, the experience of reforming economies in Latin America in the 1980s or the post-communist transition economies in the 1990s has demonstrated that macro-stability alone is not sufficient to trigger economic growth and attract foreign investors. Neither is a competitive exchange rate an automatic spur to export-led growth because, for instance, "… inefficiency and

---

[391] OECD, *Fighting Corruption in Developing Countries and Emerging Markets*, OECD Development Centre (Paris), 2000.

[392] L. Meldolesi, "Corruption, accountability, and democracy in Italy: an outline", in J. Tulchin and R. Espach (eds.), *Combating Corruption in Latin America* (Washington, D.C., Woodrow Wilson Center Press, 2000), pp. 71-87.

[393] J. Braga de Macedo, "Converging European …", op. cit.; J. Sachs and A. Warner, "Economic reform and the process of global integration", *Brookings Papers on Economic Activity: 1* (Washington, D.C.), 1995.

corruption paralysed the ports…".[394] It turns out that even the best-designed legal framework to guide private economic activity matters little if the capacity to implement it is lacking.

Thus, it became apparent that development could not be tackled without establishing sound institutions which could assure high quality governance together with macroeconomic stability and economic openness. The rediscovery of the importance of institutions for economic performance was bound to trigger interest in corruption, which is clearly a symptom of poor governance. Subsequent research has provided strong empirical evidence that whereas high corruption stifles economic growth. Mauro[395] shows that corruption decreases investment as a share of GDP. Correspondingly, the decrease in corruption significantly increases both the investment rate and the annual growth rate of per capita GDP.

Second, the world economy has changed dramatically over the last two decades. Integration into the production and marketing arrangements of the multinational corporations rather than the pursuit of an autarchic national development strategy appears to offer the most efficient way to take the advantage of the growth opportunities offered by the global economy. Managerial know-how and the technological requirements for successfully competing in international markets have become simply too great to meet through reliance on a country's own resources. Foreign participation – in the form of either "outsourcing" or direct investment – offers direct access to the global networks of a parent company as well as to its know-how. Technology has made possible the fragmentation of production processes, i.e. dividing the industry's value added chain into smaller functions that can be contracted out to independent suppliers.[396] This possibility of "dividing up the value chain" of production has resulted in – to borrow an apt phrase from Feenstra[397] – the integration of trade and the disintegration of production in the global economy.[398]

Globalization, fuelled by revolutionary changes in technology and information as well as by liberalization in international economic exchanges, has redefined the formula for economic success. The success of the so-called third tier of east Asian economies – Malaysia and Thailand – has hinged heavily on opening up to foreign investment and participating in the new division of labour based on fragmentation of production. Attracting long-term foreign capital to invest in upstream production activity is justifiably regarded as a necessary condition for tapping the opportunities offered by the global economy.[399] This, however, requires not only adequate physical infrastructure but also, above all, a "low corruption" environment that is friendly to private business activity.

Thus, two findings of empirical research, that high corruption depresses the magnitude of foreign direct investment[400] and lowers the quality in terms of research-and-development intensity[401] are not counter-intuitive. They provide strong ammunition to the argument about the importance of low levels of corruption for economic growth in contemporary global economy.

In a similar vein, the few empirical studies that have been undertaken to determine the effect of institutional factors including among others property rights, government regulation, political and civil rights, the rule of law, governance and public administration, legal reform and political risk on growth in transition economies, testify to their importance. While they have not tested explicitly the "corruption" variable, they nonetheless shed light on the importance of good, non-corrupt governance. For instance, an empirical study tracing the role of institutional factors in the growth of transition economies has found that property rights and political stability are particularly important for per capita GDP growth, while an indicator of credibility (a simple average of five indicators – predictability of rules, political stability, security of property rights, reliability of the judiciary and corruption) was highly significant for both FDI and per GDP capita growth.[402] Campos[403] has found that the rule of law was the most important institutional dimension (compared with accountability of the executive, the quality of bureaucracy, transparency and accountability, and the strength of civil society) in terms of its effect on per capita income and school enrolment. Havrylyshyn and van Rooden[404] found that while among institutional factors legal reform was the

---

[394] M. Naim, "Washington consensus or Washington confusion", *Foreign Policy*, Spring 2000, pp. 87-103.

[395] P. Mauro, "The effects of corruption on growth, investment and government expenditure: a cross-country analysis", in K. Elliot (ed.), *Corruption and the Global Economy* (Washington, D.C., Institute for International Economics, 1997).

[396] M. Borrus and J. Zysman, "Globalization with borders: the rise of Wintelism as the future of industrial competition", *Industry and Innovation*, Vol. 4, No. 2, December 1997.

[397] R. Feenstra, "Integration of trade and disintegration of production", *Journal of Economic Perspectives*, Vol. 12, No. 4, 1998, pp. 31-50.

[398] For an extensive discussion of fragmentation of production and trade see contributions in S. Arndt and H. Kierzkowski, "Introduction", in S. Arndt and H. Kierzkowski (eds.), *Fragmentation and International Trade* (Oxford and New York, Oxford University Press, 2000).

[399] Ibid.

[400] S.-J. Wei, "How taxing is corruption on international investors?", *Review of Economics and Statistics*, Vol. 82, No. 1, February 2000, pp. 1-11.

[401] B. Smarzynska and S.-J. Wei, op. cit.

[402] A. Brunetti, G. Kisunko and B. Weder, *Institutions in Transition: Reliability of Rules and Economic Performance in Former Socialist Countries*, World Bank Policy Research Working Paper, No. 1809 (Washington, D.C.), August 1997.

[403] N. Campos, *Context is Everything: Measuring Institutional Change in Transition Economies*, World Bank Policy Research Working Paper, No. 2269 (Washington, D.C.), January 2000.

[404] O. Havrylyshyn and R. van Rooden, *Institutions Matter in Transition, but so do Policies*, IMF Working Paper WP/00/70, (Washington, D.C.), March 2000.

most important,[405] their effect was not as strong as that of macroeconomic stability or structural reforms.

Last but not least, since both Latin American and post-communist transitions have involved not only a shift towards a new market-based economic regime but also democratization, these "double transitions" have changed the institutional factors underpinning corruption and might have temporarily increased it. With the authority of the party and state ebbing, an outburst of corruption accompanied the disintegration of communism. Transition has created huge opportunities for corruption especially in countries that adhered to partial reforms.[406] While reforms in Latin America called for less radical measures because initial conditions were less remote from the market economy, corruption there may not have increased. But its perception has probably increased as democratization, and legal investigations that had been suppressed under authoritarian regimes, have brought corruption into public debate.

## 5.5 Conclusion

The recent upsurge in interest in corruption stems from the discovery of the importance of institutions and governance in the context of the current global economic and political environment. The emphasis on good institutions and good governance illustrates a convergence in policies among countries seeking to exploit the opportunities created by the ongoing technological revolution and globalization. Its common denominator has been the rejection of authoritarianism and centralization, represented in the past by bureaucratic central planning, as conducive to effective industrialization and social stability. Two waves of democratization, witnessed over the last two decades in Latin America in the 1980s and in the former Soviet bloc in the late 1980s and early 1990s coincided with the dramatic change in economic policies. Because of initial conditions, the post-communist transition has brought to fore the issue of institutions and governance more prominently than the transitions in Latin America.

It has also brought to the fore the realization that good governance, macroeconomic stability and integration into global markets are necessary conditions for sustainable economic growth. The cost of doing nothing in terms of improving the quality of the policy environment and subverting corruption is extremely high. It spells economic devolution and expansion in poverty, simply because the growth opportunities offered by the second global economy remain untapped. Economies that are over-regulated and overprotected are unlikely to attract the high quality FDI that would assist domestic producers to integrate into global markets. Nor are they likely to increase the intensity of competition domestically or to trigger sustainable economic growth and improved efficiency in the allocation and use of resources.

If sustainable growth is the objective of a government, then the issue of establishing a good policy environment should move to the top of its reform agenda. How to improve governance? Our argument offers three recommendations. First, while the dominant body of literature suggests a combination of measures addressing motives and the organizational environment, this study argues that the best course of action is to establish conditions for competition not only in the economy but also in the polity. The main argument here is that the political system, including electoral procedures, matters and that democracies are not handicapped in creating good governance structures with a low incidence of corruption. They also seem to be better endowed to deal with corruption. While more research is needed, there is clearly a link between the quality of governance and electoral procedures. Unfortunately, the link has yet to receive more attention from international organizations and policy think tanks. Electoral procedures with proportional representation and very low or no thresholds and relatively large districts favour the emergence of a strong multiparty democracy at the national level. However, at the local level the majority formula seems to assure greater accountability of local government officials to their constituencies as well as their stronger involvement in local governance.

Second, economy-wide liberalization rather than active state involvement in sectoral management offers the best way to reduce the potential for corruption. But market freedom requires supervisory and regulatory vigilance. Following international practice in the development of regulatory institutions makes good economic sense and would subvert the sources of corrupt practices.

Third, the reliance on penalties to combat corruption may be a good solution for societies with sound governance structures robust enough to resist policy capture by private interests, but this would be a wrong approach for countries with a poor policy environment controlled by narrow private interests. Instead, their point of departure is to identify vulnerabilities to corruption and reduce the potential for corruption. Over-controlled economies provide an ideal breeding ground for corruption. Therefore, the initial effort should focus on reducing the potential for corruption. Simplicity and transparency usually go hand-in-hand with good economics and governance.

Privatization of the state or *res privata*, as reflected in the World Bank's *state capture* index, does not augur well for the improvement of governance. Most CIS countries seem to fall into the category of countries with corruption perpetuated from above. Under these circumstances, internationally supported programmes focusing on one particular policy domain appear to be a waste of resources. Instead, international efforts should focus on the development of a "good civil society".

---

[405] Ibid. Havrylyshyn and van Rooden tested nine institutional variables for 25 transition economies for the period 1991-1998. Their basic equation regresses GDP growth on inflation, a structural reform index (contemporaneous and lagged) and initial conditions. They found that adding an institutional variable added little explanatory power. Legal reform was found to be the most significant institutional factor. They concluded that good economic policies remain the dominant statistical determinant of growth.

[406] J. Hellman, op. cit.

# DISCUSSANTS' COMMENTS ON PROFESSORS A. AND B. KAMINSKIS' PAPER

## 5.A Mark Pieth

First I would like to thank you very much for the opportunity to comment on this excellent paper. As a lawyer and as someone involved in international policy development, I was particularly impressed by the authors' straightforward approach and readiness to approach some very touchy issues. We are given some explanations of why corruption might even increase after the start of transition to a market economy and moves towards democracy. You have, in your short summary, mentioned the example of Belarus before transition. We are also given a very clear account of what makes institutions ineffective, even if the impression of change, I think "electocracy" is the word used, has been created.

I would like to make two points. First, I have a comment as a non-economist and then I would like to add something to the issues that have been discussed this afternoon. I will, of course, abstain from taking a position on the methodology, since this is not my speciality, although I can't quite abstain from slightly overstepping my background with my first comment. If I share your overall approach, there is one topic I have been missing, not just in your talk, but also in the discourse of this afternoon. Neither the invisible hand of Adam Smith nor the separation of powers of the French Revolution, to quote those two ideas, will deal with the large number of destitute, impoverished and possibly homeless people generated by the transition to the market economy. It may be unfair to raise this issue at this point because it was not your topic of course, but I have looked in vain for an element of what I might call solidarity, although I don't know if you want to use that word. I'm not satisfied anyway and I would just like to mention that the logic of the argument that if the economy flourishes everybody will benefit ignores the element of distribution, of adequate distribution. Take your example of taxation: I fully agree that we need simple, straightforward tax systems that leave little room for dispute or discretion. As for tax reduction, I also agree that we need to have realistic rates of taxation because we must be able to actually enforce them. I think that is crucial. Beyond this, however, I insist that the state needs the resources to deal with issues such as health care, education, poverty relief, and so on. Of course, this comment would lead us further afield into the very complex area of tax harmonization and the issue of harmful tax competition in a much wider context; it would also lead us to matters that are sometimes misrepresented. It would certainly be wrong for the north and west to tell the south and east that a very simple concept of a market economy would really be the solution to their problems. I think there we have to be very careful and that is basically the direction of my comment.

I come to my second point, to the additional dimension. It has been mentioned at various points in the discussion that corruption has not only a demand side but also a supply side. I feel that the Kaminskis' paper lets off somewhat lightly those who traditionally pay the large bribes. Foreign companies especially appear to be the victims rather than the agents in corruption in your account. Now it is certainly correct that companies would rarely offer bribes without being solicited. However, they act in tune with the local customs and they have not shown themselves in the past to be very courageous in resisting them. We have heard of the example given by the representative of the United Kingdom of small- and medium-sized companies which have little margin for manoeuvre. My experience as the Chairman of the OECD Working Group on Bribery for the last 10 years is that in order to change the attitude of companies there is a need for clear, internationally agreed standards and for the threat of tough action by countries playing host to multinational enterprises. Using the authors' term, one has to create a "common bad" out of bribery, especially active bribery, and the contribution of IGOs such as the United Nations, the Council of Europe, the European Union, the Organization of American States and the OECD has been precisely to raise the costs of bribery to those potentially ready to bribe. The means for doing this are relatively classical: the criminal law, with sanctions including forfeiture of profits; civil law, providing for compensation; and administrative law, with disbarment from future contracts. The OECD has been especially effective in this area because it has been able to use the instruments of peer pressure to make countries change their laws and even to apply them. Somewhat cynically, I would say that being really serious about application is new in the area of international law. On the aspect of implementation and ratification, I just want to mention that in the last three years, and this is maybe a world record, we have been able to get 31 out of 34 participating countries to actually write and ratify laws on bribery and they account for 70 per cent of world exports and 90 per cent of all FDI.

Now, these countries will have to explain to examiners on site how they apply these rules. The principle is the same as that applied in your paper: use straightforward language and let people say what they think of each other. The charm of this approach, and I don't know if this hypothesis will pass the scrutiny of my learned colleagues, is that it is potentially effective independently of whether or not a country follows your suggestions. Even if a country does not improve its governance we think that a unilateral supply-side approach can make a difference. It is like turning off the tap: if large funds no longer flow into some of these economies it will be very difficult for local elites or someone who has been elected but controls, let us say, the media, to pay their local allies. They will not then be able to stay in power for such a long time. So the idea is that the supply-side approach stands alone, even if something also needs to be done on the demand side. Even if the latter doesn't work out, action on the supply side could still make sense.

There is another aspect to the supply side that I want to mention briefly. It does not function on the basis of government pressure alone, but actually puts into effect a model that political scientists call governance at a distance. It only works if the entities in the private sector actually pick up these issues and make them work. Such action is currently being developed, for example, in the banking sector. There has been the Wellsburg declaration of the 10 largest private banks, just to mention one example, and similar activities in project financing are just starting. Also in the very sensitive area of power generation, one has only to think of any dam built over the last 10 years and you will find the most interesting corruption cases cropping up. This is the kind of topic we are now starting to tackle by getting the actors in the private sector to monitor each other and to implement those rules that have been generated, for example, in the form of OECD instruments. So you have intergovernmental action, governmental action and private sector action on the supply side. And this is actually the point I want to emphasize: you must look at both sides, not only the demand side, but also the supply side. And, of course, you also have to look at the differences between north and south. The north is not just about supply, it is also very much about receiving as well. But we have to be very careful in searching for solutions because I believe that in combating corruption you can actually go across the north-south divide, or, if you want, the east-west rift insofar as it still exists. There is a lot of potential here, but with this I will have to stop.

## 5.B Jorge Braga de Macedo

I liked the paper very much and I am going to make seven points about it, including some critical remarks, but let me start by congratulating Professors A. and B. Kaminski for having been able to work so well worlds apart.

The first point I want to make is about method as we are busy at the OECD Development Centre working on this. If you look at chart 5.B.1, you see above the box that endowments (which may be natural or cultural) determine performance, and that there is a mutual interaction between productivity and performance (this is the work-horse of economics). Below the box there are also interactions involving globalization (openness to trade and capital movements) and governance. These interactions eventually change the endowments: institutions improve, institutions

**CHART 5.B.1**

**G&G links to national economic performance**

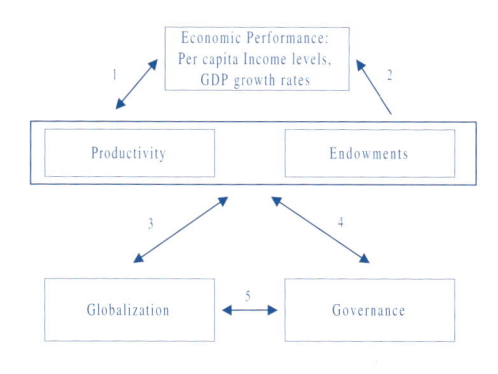

Source: F. Bonaglia, J. Braga de Macedo and M. Bussolo, *How Globalization Improves Governance*, CEPR Discussion Paper, No. 2992 (London), October 2001.

change. And this is why the Development Centre's work programme (available at www.oecd.org/dev) is about globalization and governance, or as we like to say G&G, not a brand of Italian shoes but rather a framework of interactions which helps determine which countries are going to grow faster (the fifth arrow in chart 5.B.1). You cannot analyse this G&G interaction without an interdisciplinary approach, so I very much welcome the words about interdisciplinary approaches that are implicit or explicit in the paper.

My second point concerns what they say about the heterogeneity of their sample. They mention that "Luxembourg countries" are in better shape than "Helsinki countries", which themselves are in better shape than the former Soviet Union and they suggest that this may simply have to do with the respective levels of income per capita. But that is why we have to be very careful about putting countries in particular boxes because what matters is the movement from one box to another. The words transition and development suggest precisely that. The movement, the path, now perhaps matters even more than endowments, which is also why credibility, which concerns how the future path is perceived, also matters. This of course is the flip side of what was said about Azerbaijan, which reflects the situation of many other countries, especially those where there are abundant natural resources.

Far from being a good thing for governance, natural resources create great pressure for immediate satisfaction and for corruption. If that hypothesis is true, then what can quantitative economic analysis do? And here, I want to appeal to banish the kinds of graph that are presented by A. and B. Kaminski. Let's just drop them, because they suggest a causality that is not there. When I want to suggest that there is a relationship between two variables, as is the case here, I use a histogram where it is obvious that there is no implication about what is the independent and what the dependent variable.[407] You throw in a regression and everybody gets confused, so here is my first critical remark. I recommend you drop the chart, and instead, if you want to use quantitative analysis, attempt to measure what you can call the independent effects of openness and income per capita on corruption, the dependent variable. Since you have an interaction, you have to use instruments (like Professor Wei's natural openness) or determinants of income that are not affected by corruption. By using those instruments – independent determinants of income and openness – you can identify corruption. Then you will find strikingly that globalization actually improves governance, that it actually lowers corruption.[408] It is quite consistent with what you said but your diagram in my view confuses things, not because the data are faulty, a point you make, but rather because the implicit statistical analysis in my view confuses the issue. That was my third point.

Now I want to attack directly the main point of your analysis, that which involves political analysis, which I found extremely refreshing. There is in the economics literature, and even in the work of some international organizations, the idea that the only way to get reform is to have a crisis. And you came close to saying that. Of course, given what happened to your native Poland, which is a striking example of a successful radical change, we could go along these lines. Well, maybe! Gradualism has some advantages over radical change, which have to do simply with the likelihood of reversal. If you have a striking change followed by reversal, you have got nowhere. Of course if gradualism also involves reversal you still have a problem, but I believe that in this regard you have to differentiate between macroeconomic stability, and the harder task of reform which involves governance, political institutions, NGOs, etc. Actually, with respect to financial stability, or to what I would call financial freedom (and I shall give an example of that in a moment), maybe shock therapy is needed.

I took note of what you said about electocracies. I think the Greeks used to call them tyrants: there were democracies, and then there were tyrannies, which were simply elections, so there is a very illustrious classical origin to your point. I also appreciate the quote from Juan Linz about parliamentary democracies and I think that this is the part of the paper that is most original.[409] Let me very quickly mention the example of my own country, Portugal, where we had a gradual regime change. Few people know that until 1989 it was forbidden to privatize in Portugal. After the first free elections had taken place in Hungary and Poland we still had a constitutional ban on privatization that required a two-thirds majority to amend. The amendment was passed in 1989, and that was the beginning of the regime change which was consolidated with entry into the ERM in the spring of 1992.[410]

But now, 10 years later, what is happening? A lot of the reforms that were supposed to follow adoption of the euro did not, and so you still have the same public administration, the same slow courts, bad hospitals, and so on. We call this the "euro hold-up". In other words, you introduce a reform but it escapes because the structural reforms do not follow. I would have given more attention to sustaining structural reforms and I would claim that it would be difficult to defend further radical change.

Fifth point: you draw on the literature on rent-seeking and you show that the behaviour of public and private actors is the key to the outcome of reform, because there are vested interests which resist any change that threatens their entitlements. There are very few parts of society that have a sense of the general interest. Paradoxically, private enterprise sometimes has more sense of the common good than the government simply because it is more forward looking.

This is why the fight against corruption requires a coalition and business must be part of it. The World Commission on Dams is a good example. This is a coalition involving governments and NGOs but also private firms. I think it is very important to take into account that corruption is bad for business and business is bad for corruption. In other words private firms, governments and NGOs are allies in this struggle.

Let me get back to the point I made earlier about natural resources. There are papers about the voracity effect, by Aaron Tornell and Philip Lane, very much along the lines of what you say. They have both theoretical and empirical studies showing that countries that discover oil or which increase the price of their main export product or get more aid or structural funds, see a fall in their growth rate, exactly along the lines you suggest. I don't need to elaborate but I think there you have a very important effect.

And that is why I am so excited about the empirical results that I obtained, which show that globalization per se improves governance, because you don't have to wait for income to rise to the level of countries with good governance. You just open up faster and that is of course exactly what you said with your uniform tariff example, which I found very persuasive. Apparently (as I heard from David Dollar when he visited Nova University in Lisbon some years ago), when a Minister goes to the World Bank and says "I am worried about corruption – what should I do?" they reply, "just liberalize trade". And the effect I picked up is exactly along the same lines. I should mention a very interesting example of your one-stop shopping idea of public procurement: such a scheme for electronic commerce

---

[407] J. Braga de Macedo, "Globalization and institutional change: a development perspective", April 2001, forthcoming contribution in the proceedings of the Pontifical Academy of Social Sciences.

[408] Ibid., and F. Bonaglia, J. Braga de Macedo and M. Bussolo, *How Globalization Improves Governance*, CEPR Discussion Paper, No. 2992 (London), October 2001.

[409] See J. Linz's views on post-revolutionary Portugal in J. Braga de Mecedo and S. Serfaty (eds.), *Portugal since the Revolution: Economic and Political Perspectives* (Boulder, CO, Westview Press, 1981).

[410] This regime change is described at length in F. Bonaglia et al., op. cit.

CHART 5.B.2

**European institutional architecture**

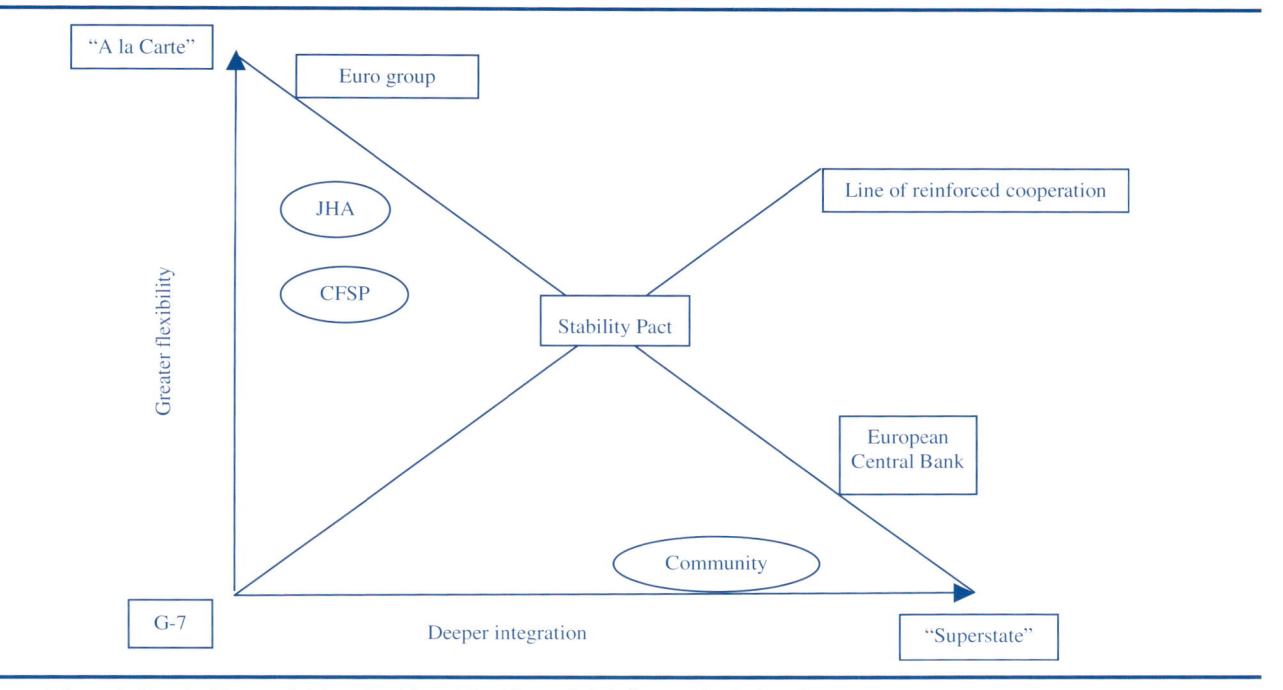

"A la Carte" — Euro group — JHA — CFSP — Stability Pact — Line of reinforced cooperation — European Central Bank — Community — G-7 — "Superstate"

Greater flexibility / Deeper integration

Source: J. Braga de Macedo, "The euro in international financial architecture", *Acta Oeconomica*, forthcoming.

in Chile is working very well and there are other examples in the Development Centre website already cited.

Let me very quickly make two last points. A case I would like to draw your attention to is Argentina. It is an example of what I call moving from *Amnesty International* to *Transparency International*, in other words a sequence in the kinds of struggle that matter for people but doing it via financial freedom. In the case of Argentina it is a story covering three decades. In 1983 you get democracy and civil society was interested in fighting for human rights; at that time no one was worried about corruption. Why? Because corruption did not exist because there was hyperinflation. Then Cavallo introduced convertibility in 1991. The introduction of financial freedom, far from being a reform for just a few rich people, allowed people to concentrate on corruption. Indeed, recently a very ambitious law has been introduced eliminating the impunity of elected officials in Argentina, very much along the lines of what Transparency International and other organizations have proposed. So here is a case of where in each decade reforms have been completed in sequence and sustained. As I mentioned above, for the specifics of the macroeconomic stability they used shock therapy. So, again, let's not confuse macrostability with structural reform.[411]

My last point is about peer pressure, which I think is important for two reasons. The first is that many countries are not used to it. Okay Belgium is used to it, the Netherlands, Italy and the founders of the European Community are used to it, maybe the large countries in Europe are used to it, certainly among themselves. But many countries are not. Even in the countries in transition you very often have instead a so-called "beauty contest". Who is going to show up first in the picture of those seeking accession to the European Union? That is not peer pressure, that is the opposite. The OECD is a peer pressure organization and it is very important to learn about peer pressure, much more so than worrying about top-down approaches. Unless you think that peer pressure only applies in OECD, which is inter-governmental and not in Europe, which is a far more ambitious undertaking. Look at chart 5.B.2. Basically, you have to balance an understanding of European integration, and therefore this idea of going towards a federal state, that is the horizontal axis, with more flexibility, which is on the vertical axis. If you combine the two you see that Europe is at heart a group of countries that have learned to work together. And if you take that view, the newcomers are going to find their way much faster than if they are seen as entering an already complete architecture. Instead, what you have is a "line of reinforced cooperation" and there will be chances for Hungary, Poland and the other countries to follow that line just as Portugal did. How many people 10 years ago would have thought Portugal would be a member of the euro zone and also part of the Schengen agreement, the two most dynamic parts of the European Union? Probably very few. Nevertheless there it is.

---

[411] Elaborated in the preface to J. Braga de Macedo, D. Cohen and H. Reisen (eds.), *Don't Fix, Don't Float*, OECD Development Centre Studies (Paris), August 2001.

# PART THREE

## STATISTICAL APPENDIX

For the user's convenience, as well as to lighten the text, the *Economic Survey of Europe* includes a set of appendix tables showing time series for the main economic indicators over a longer period. The data are presented in two sections, following the structure of the text: *Appendix A* provides macroeconomic indicators for the market economies in western Europe, North America and Japan for 1986-2000, *Appendix B* does the same for the east European countries, the Baltic states and the Commonwealth of Independent States for 1980-2000.

Re-estimated historical series are not yet available for all the transition economies, and longer time series could in some instances be obtained only by splicing older data with the new statistics (as explained in the notes to the tables). Historical series for Czechoslovakia, the former SFR of Yugoslavia and the Soviet Union can be found in previous issues of this *Survey*. For the economies of western Europe and North America data for the more recent years may also be subject to revision as more comprehensive benchmark figures become available.

Data were compiled from international and national statistical sources. Details on recent changes in national accounts methodology were provided in chapter 7 of the *Economic Survey of Europe, 2000 No. 1*. Aggregates are UNECE secretariat calculations, using PPPs obtained from the 1996 European Comparison Programme. Greece has become a member of the euro area at the beginning of 2001. In order to ensure continuity of time series and comparability with the text tables, Greece has been included in the euro area aggregates for all years shown in the appendix tables.

The figures for 2000 are based on data available at mid-October 2001.

## APPENDIX TABLE A.1

### Real GDP in western Europe, North America and Japan, 1986-2000
*(Percentage change over preceding year)*

|  | 1986 | 1987 | 1988 | 1989 | 1990 | 1991 | 1992 | 1993 | 1994 | 1995 | 1996 | 1997 | 1998 | 1999 | 2000 |
|---|---|---|---|---|---|---|---|---|---|---|---|---|---|---|---|
| France [a] | 2.4 | 2.5 | 4.6 | 4.2 | 2.6 | 1.0 | 1.5 | -0.9 | 2.1 | 1.7 | 1.1 | 1.9 | 3.4 | 2.9 | 3.1 |
| Germany [b] | 2.5 | 1.7 | 3.7 | 3.5 | 3.2 | 2.8 | 2.2 | -1.1 | 2.3 | 1.7 | 0.8 | 1.4 | 2.0 | 1.8 | 3.0 |
| Italy | 2.5 | 3.0 | 3.9 | 2.9 | 2.0 | 1.4 | 0.8 | -0.9 | 2.2 | 2.9 | 1.1 | 2.0 | 1.8 | 1.6 | 2.9 |
| Austria | 2.3 | 1.7 | 3.2 | 4.2 | 4.7 | 3.3 | 2.3 | 0.4 | 2.6 | 1.6 | 2.0 | 1.3 | 3.3 | 2.8 | 3.3 |
| Belgium | 1.8 | 2.7 | 4.6 | 3.6 | 2.8 | 1.9 | 1.6 | -1.5 | 3.0 | 2.6 | 1.2 | 3.4 | 2.4 | 2.7 | 4.0 |
| Finland | 2.5 | 4.2 | 4.7 | 5.1 | – | -6.3 | -3.3 | -1.1 | 4.0 | 3.8 | 4.0 | 6.3 | 5.3 | 4.0 | 5.7 |
| Greece | 0.5 | -2.3 | 4.3 | 3.8 | – | 3.1 | 0.7 | -1.6 | 2.0 | 2.1 | 2.4 | 3.5 | 3.0 | 3.4 | 4.1 |
| Ireland | -0.4 | 4.7 | 5.2 | 5.8 | 8.5 | 1.9 | 3.3 | 2.7 | 5.8 | 10.0 | 7.8 | 10.8 | 8.6 | 10.8 | 11.5 |
| Luxembourg | 7.8 | 2.3 | 10.4 | 9.8 | 2.2 | 6.1 | 4.5 | 8.7 | 4.2 | 3.8 | 2.9 | 7.3 | 5.0 | 7.5 | 8.5 |
| Netherlands | 2.8 | 1.4 | 3.1 | 5.0 | 4.1 | 2.5 | 1.7 | 0.9 | 2.6 | 2.9 | 3.0 | 3.8 | 4.3 | 3.7 | 3.5 |
| Portugal | 4.1 | 6.4 | 7.5 | 5.4 | 4.8 | 2.4 | 1.9 | -1.4 | 2.5 | 3.6 | 3.7 | 3.8 | 3.8 | 3.3 | 3.2 |
| Spain | 3.2 | 5.6 | 5.2 | 4.7 | 3.7 | 2.3 | 0.7 | -1.2 | 2.3 | 2.7 | 2.4 | 3.9 | 4.3 | 4.0 | 4.1 |
| *Euro area* [c] | 2.5 | 2.6 | 4.2 | 3.9 | 2.9 | 1.9 | 1.6 | -0.9 | 2.4 | 2.2 | 1.4 | 2.3 | 2.9 | 2.6 | 3.4 |
| United Kingdom | 3.9 | 4.5 | 5.2 | 2.2 | 0.8 | -1.4 | 0.2 | 2.5 | 4.7 | 2.9 | 2.6 | 3.4 | 3.0 | 2.1 | 2.9 |
| Denmark | 4.1 | -0.3 | -0.7 | 0.2 | 1.0 | 1.1 | 0.6 | – | 5.5 | 2.8 | 2.5 | 3.0 | 2.8 | 2.1 | 3.2 |
| Sweden | 2.7 | 3.3 | 2.6 | 2.7 | 1.1 | -1.1 | -1.7 | -1.8 | 4.1 | 3.7 | 1.1 | 2.1 | 3.6 | 4.1 | 3.6 |
| *European Union* [d] | 2.8 | 2.9 | 4.2 | 3.5 | 2.5 | 1.3 | 1.3 | -0.5 | 2.7 | 2.5 | 1.6 | 2.5 | 2.9 | 2.6 | 3.3 |
| Cyprus | 3.8 | 7.0 | 8.5 | 7.9 | 7.4 | 0.6 | 9.8 | 0.7 | 5.9 | 6.1 | 1.9 | 2.5 | 5.0 | 4.5 | 4.8 |
| Iceland | 6.3 | 8.5 | -0.1 | 0.3 | 1.2 | 0.7 | -3.3 | 0.6 | 4.5 | 0.1 | 5.2 | 4.8 | 4.5 | 4.3 | 3.6 |
| Israel [a] | 3.6 | 6.2 | 3.4 | 1.4 | 6.3 | 5.7 | 6.8 | 3.4 | 6.9 | 8.3 | 4.6 | 2.9 | 2.4 | 2.6 | 6.2 |
| Malta | 3.9 | 4.1 | 8.4 | 8.2 | 6.3 | 6.3 | 4.7 | 4.5 | 5.7 | 6.2 | 4.0 | 4.9 | 3.4 | 4.0 | 4.3 |
| Norway [a] | 3.6 | 2.0 | -0.1 | 0.9 | 2.0 | 3.1 | 3.3 | 2.7 | 5.5 | 3.8 | 4.9 | 4.7 | 2.4 | 1.1 | 2.3 |
| Switzerland | 1.6 | 0.7 | 3.1 | 4.3 | 3.7 | -0.8 | -0.1 | -0.5 | 0.5 | 0.5 | 0.3 | 1.7 | 2.3 | 1.5 | 3.4 |
| Turkey | 7.0 | 9.5 | 2.1 | 0.3 | 9.3 | 0.9 | 6.0 | 8.0 | -5.5 | 7.2 | 7.0 | 7.5 | 3.1 | -5.0 | 7.2 |
| **Western Europe** | 3.0 | 3.2 | 4.1 | 3.3 | 2.9 | 1.3 | 1.4 | 0.1 | 2.4 | 2.7 | 1.9 | 2.8 | 2.9 | 2.2 | 3.5 |
| Canada [a] | 2.4 | 4.2 | 4.9 | 2.6 | 0.2 | -2.1 | 0.9 | 2.4 | 4.7 | 2.8 | 1.6 | 4.3 | 3.9 | 5.1 | 4.4 |
| United States [a] | 3.4 | 3.4 | 4.2 | 3.5 | 1.8 | -0.5 | 3.0 | 2.7 | 4.0 | 2.7 | 3.6 | 4.4 | 4.3 | 4.1 | 4.1 |
| **North America** | 3.3 | 3.5 | 4.2 | 3.4 | 1.6 | -0.6 | 2.9 | 2.6 | 4.1 | 2.7 | 3.4 | 4.4 | 4.3 | 4.2 | 4.2 |
| Japan | 3.0 | 4.5 | 6.5 | 5.3 | 5.3 | 3.1 | 0.9 | 0.4 | 1.0 | 1.6 | 3.5 | 1.8 | -1.1 | 0.8 | 1.5 |
| **Total above** | 3.1 | 3.5 | 4.5 | 3.7 | 2.7 | 0.8 | 1.9 | 1.2 | 2.9 | 2.5 | 2.8 | 3.3 | 2.8 | 2.8 | 3.5 |
| *Memorandum items:* | | | | | | | | | | | | | | | |
| *4 major west European economies* [e] | 2.8 | 2.8 | 4.3 | 3.2 | 2.3 | 1.2 | 1.3 | -0.2 | 2.8 | 2.2 | 1.3 | 2.1 | 2.5 | 2.1 | 3.0 |
| *Western Europe and North America* | 3.1 | 3.3 | 4.1 | 3.4 | 2.2 | 0.3 | 2.1 | 1.4 | 3.3 | 2.7 | 2.7 | 3.6 | 3.6 | 3.2 | 3.8 |

*Source:* Eurostat, New Cronos Database; OECD, *National Accounts* (Paris), various issues; national statistics.

*Note:* All aggregates exclude Israel. Growth rates of regional aggregates have been calculated as weighted averages of growth rates in individual countries. Weights were derived from 1996 GDP data converted from national currency units into dollars using 1996 purchasing power parities. From 1992 onwards, data for the European Union and the Euro area are as reported by Eurostat; before 1992, growth rates are computed as for other regional aggregates. 1993 SNA/ESA95 definitions except for Iceland, Malta, Switzerland and Turkey.

[a] Annual changes are calculated from chained national currency series.

[b] Data before 1992 are estimates for the whole of Germany, based on statistics published by Deutsches Institut fur Wirtschaftforschung and by the East German Statistical Office.

[c] Twelve countries above.

[d] Fifteen countries above.

[e] France, Germany, Italy and the United Kingdom.

## APPENDIX TABLE A.2

**Real private consumption expenditure in western Europe, North America and Japan, 1986-2000**

*(Percentage change over preceding year)*

|  | 1986 | 1987 | 1988 | 1989 | 1990 | 1991 | 1992 | 1993 | 1994 | 1995 | 1996 | 1997 | 1998 | 1999 | 2000 |
|---|---|---|---|---|---|---|---|---|---|---|---|---|---|---|---|
| France [a] | 3.6 | 3.0 | 2.7 | 3.0 | 2.7 | 0.7 | 0.9 | -0.4 | 1.2 | 1.2 | 1.3 | 0.2 | 3.4 | 2.8 | 2.5 |
| Germany [b] | 3.5 | 3.4 | 2.7 | 2.8 | 5.4 | 5.6 | 2.7 | 0.1 | 1.0 | 2.0 | 1.0 | 0.6 | 1.8 | 3.1 | 1.5 |
| Italy | 4.0 | 3.8 | 4.0 | 3.7 | 2.1 | 2.9 | 1.9 | -3.7 | 1.5 | 1.7 | 1.2 | 3.2 | 3.1 | 2.3 | 2.9 |
| Austria | 2.2 | 3.0 | 3.2 | 4.3 | 4.5 | 2.5 | 3.0 | 0.8 | 2.4 | 2.6 | 3.2 | 1.4 | 2.9 | 2.3 | 2.7 |
| Belgium | 3.1 | 1.8 | 3.7 | 3.9 | 3.2 | 3.0 | 2.2 | -1.0 | 2.0 | 1.0 | 0.7 | 2.1 | 3.3 | 1.9 | 3.1 |
| Finland | 4.0 | 5.1 | 5.3 | 4.6 | -0.6 | -3.8 | -4.4 | -3.1 | 2.6 | 4.4 | 4.2 | 3.5 | 5.1 | 4.0 | 3.0 |
| Greece | -1.5 | 2.7 | 6.1 | 6.3 | 2.6 | 2.9 | 2.3 | -0.8 | 1.9 | 2.5 | 2.4 | 2.8 | 3.1 | 3.0 | 3.0 |
| Ireland | 2.0 | 3.3 | 4.5 | 6.5 | 1.4 | 1.8 | 2.9 | 3.0 | 4.4 | 4.4 | 6.3 | 7.3 | 7.3 | 8.2 | 9.9 |
| Luxembourg | 5.7 | 4.6 | 4.6 | 5.1 | 5.7 | 6.3 | -0.9 | 1.7 | 2.4 | 2.4 | 4.4 | 3.8 | 2.3 | 4.1 | 3.5 |
| Netherlands | 3.0 | 2.7 | 0.6 | 3.3 | 3.9 | 2.7 | 0.8 | 0.5 | 0.9 | 3.0 | 4.0 | 3.0 | 4.8 | 4.5 | 3.7 |
| Portugal | 5.7 | 5.3 | 6.8 | 3.1 | 6.0 | 3.6 | 3.6 | 0.9 | 2.2 | 4.1 | 3.1 | 3.1 | 5.9 | 5.0 | 2.8 |
| Spain | 3.3 | 5.8 | 4.9 | 5.7 | 3.6 | 2.9 | 2.2 | -2.2 | 0.9 | 1.6 | 2.2 | 3.1 | 4.5 | 4.7 | 4.0 |
| *Euro area* [c] | 3.4 | 3.6 | 3.4 | 3.6 | 3.6 | 3.1 | 2.0 | -0.9 | 1.3 | 1.8 | 1.6 | 1.6 | 3.1 | 3.2 | 2.5 |
| United Kingdom | 6.4 | 5.3 | 7.5 | 3.3 | 1.0 | -1.5 | 0.6 | 3.2 | 3.3 | 1.9 | 3.8 | 3.8 | 3.8 | 4.2 | 4.0 |
| Denmark | 5.9 | -2.2 | -2.1 | -0.1 | 0.1 | 1.6 | 1.9 | 0.5 | 6.5 | 1.2 | 2.5 | 2.9 | 3.6 | 0.5 | -0.1 |
| Sweden | 5.2 | 5.3 | 2.6 | 1.2 | -0.4 | 1.0 | -1.3 | -3.0 | 1.8 | 0.6 | 1.4 | 2.0 | 2.7 | 3.8 | 3.1 |
| *European Union* [d] | 4.0 | 3.8 | 3.9 | 3.4 | 3.0 | 2.3 | 1.7 | -0.4 | 1.7 | 1.9 | 1.9 | 1.9 | 3.2 | 3.3 | 2.8 |
| Cyprus | 1.7 | 5.5 | 10.5 | 6.9 | 9.0 | 9.9 | 3.2 | -4.8 | 5.0 | 10.3 | 3.5 | 4.0 | 8.4 | 3.1 | 5.9 |
| Iceland | 6.9 | 16.2 | -3.8 | -4.2 | 0.5 | 2.9 | -3.1 | -4.7 | 2.9 | 2.2 | 5.4 | 5.5 | 10.0 | 6.9 | 4.0 |
| Israel [a] | 15.1 | 8.9 | 4.5 | 0.4 | 5.6 | 7.2 | 8.0 | 7.3 | 9.5 | 5.8 | 5.4 | 4.2 | 3.6 | 3.2 | 6.6 |
| Malta | 1.4 | 0.5 | 9.0 | 9.2 | 3.8 | 3.8 | 4.3 | 0.8 | 2.3 | 10.5 | 7.1 | 1.6 | 2.5 | 5.9 | 5.8 |
| Norway [a] | 5.0 | -0.8 | -2.0 | -0.6 | 0.7 | 1.5 | 2.2 | 2.2 | 4.0 | 3.4 | 5.3 | 3.6 | 3.4 | 2.2 | 2.4 |
| Switzerland | 2.3 | 2.2 | 1.7 | 2.3 | 1.2 | 1.6 | 0.1 | -0.9 | 1.0 | 0.6 | 0.7 | 1.4 | 2.2 | 2.2 | 2.0 |
| Turkey | 5.8 | -0.3 | 1.2 | -1.0 | 13.1 | 2.7 | 3.2 | 8.6 | -5.4 | 4.8 | 8.5 | 8.4 | 0.6 | -3.1 | 6.4 |
| *Western Europe* | 4.0 | 3.5 | 3.7 | 3.2 | 3.4 | 2.3 | 1.7 | 0.1 | 1.4 | 2.0 | 2.3 | 2.4 | 3.1 | 3.0 | 3.0 |
| Canada [a] | 3.7 | 4.1 | 4.3 | 3.4 | 1.2 | -1.6 | 1.6 | 1.8 | 3.0 | 2.1 | 2.6 | 4.6 | 3.0 | 3.4 | 3.6 |
| United States [a] | 4.2 | 3.3 | 4.0 | 2.7 | 1.8 | -0.2 | 2.9 | 3.4 | 3.8 | 3.0 | 3.2 | 3.6 | 4.8 | 5.0 | 4.8 |
| *North America* | 4.2 | 3.4 | 4.1 | 2.7 | 1.8 | -0.3 | 2.8 | 3.2 | 3.7 | 2.9 | 3.1 | 3.6 | 4.6 | 4.9 | 4.7 |
| Japan | 3.2 | 4.1 | 5.1 | 4.7 | 4.4 | 2.7 | 2.6 | 1.8 | 2.6 | 1.4 | 2.4 | 0.8 | 0.1 | 1.2 | 0.5 |
| **Total above** | 4.0 | 3.6 | 4.1 | 3.2 | 2.9 | 1.3 | 2.3 | 1.7 | 2.6 | 2.3 | 2.7 | 2.7 | 3.3 | 3.5 | 3.3 |
| *Memorandum items:* | | | | | | | | | | | | | | | |
| 4 major west European economies [e] | 4.3 | 3.8 | 4.1 | 3.2 | 3.1 | 2.3 | 1.7 | -0.2 | 1.7 | 1.8 | 1.7 | 1.8 | 2.9 | 3.1 | 2.6 |
| Western Europe and North America | 4.1 | 3.4 | 3.9 | 2.9 | 2.6 | 1.0 | 2.2 | 1.6 | 2.6 | 2.5 | 2.7 | 3.0 | 3.9 | 3.9 | 3.9 |

*Source:* Eurostat, New Cronos Database; OECD, *National Accounts* (Paris), various issues; national statistics.

*Note:* See appendix table A.1.

[a] Annual changes are calculated from chained national currency series.

[b] Data before 1992 are estimates for the whole of Germany, based on statistics published by Deutsches Institut fur Wirtschaftforschung and by the East German Statistical Office.

[c] Twelve countries above.

[d] Fifteen countries above.

[e] France, Germany, Italy and the United Kingdom.

## APPENDIX TABLE A.3

**Real general government consumption expenditure in western Europe, North America and Japan, 1986-2000**
*(Percentage change over preceding year)*

|  | 1986 | 1987 | 1988 | 1989 | 1990 | 1991 | 1992 | 1993 | 1994 | 1995 | 1996 | 1997 | 1998 | 1999 | 2000 |
|---|---|---|---|---|---|---|---|---|---|---|---|---|---|---|---|
| France [a] | 2.4 | 2.2 | 3.2 | 1.6 | 2.5 | 2.7 | 3.8 | 4.6 | 0.7 | -0.1 | 2.3 | 2.1 | -0.1 | 2.0 | 2.2 |
| Germany [b] | 2.3 | 2.1 | 2.2 | -1.3 | 2.3 | 0.8 | 5.0 | 0.1 | 2.4 | 1.5 | 1.8 | 0.4 | 1.2 | 1.6 | 1.2 |
| Italy | 2.6 | 4.8 | 4.0 | 0.2 | 2.5 | 1.7 | 0.6 | -0.2 | -0.9 | -2.2 | 1.0 | 0.2 | 0.3 | 1.5 | 1.6 |
| Austria | 1.8 | 0.2 | 1.1 | 1.7 | 2.3 | 3.2 | 3.5 | 3.7 | 3.0 | 1.3 | 1.2 | -1.4 | 2.8 | 3.2 | 0.6 |
| Belgium | 1.3 | 2.7 | -0.7 | 1.1 | -0.3 | 3.6 | 1.5 | -0.1 | 1.4 | 1.2 | 2.4 | 0.1 | 1.4 | 3.4 | 2.0 |
| Finland | 3.4 | 4.4 | 1.9 | 2.2 | 4.0 | 2.1 | -2.4 | -4.2 | 0.3 | 2.0 | 2.5 | 4.1 | 1.7 | 1.9 | 0.7 |
| Greece | -1.1 | 0.2 | -5.5 | 5.4 | 0.6 | -1.5 | -3.0 | 2.6 | -1.1 | 5.6 | 0.9 | 3.0 | 1.7 | -0.1 | 0.8 |
| Ireland | 2.6 | -4.8 | -5.0 | -1.3 | 5.4 | 2.8 | 3.0 | -0.4 | 4.1 | 3.0 | 3.2 | 5.5 | 5.5 | 6.5 | 5.4 |
| Luxembourg | 2.7 | 4.7 | 4.9 | 3.9 | 3.1 | 3.9 | 1.5 | 3.7 | 2.0 | 2.2 | 4.4 | 2.1 | 2.8 | 12.8 | 4.9 |
| Netherlands | 2.4 | 2.8 | 1.9 | 2.0 | 2.3 | 3.0 | 2.8 | 1.6 | 1.5 | 1.3 | -0.4 | 3.2 | 3.6 | 2.8 | 1.9 |
| Portugal | 7.2 | 3.8 | 8.6 | 6.6 | 5.4 | 10.3 | 1.1 | 0.9 | 2.1 | 2.3 | 3.4 | 2.2 | 3.0 | 4.5 | 3.8 |
| Spain | 5.4 | 8.9 | 4.0 | 8.3 | 6.6 | 5.6 | 4.0 | 2.4 | -0.3 | 1.8 | 1.3 | 2.9 | 3.7 | 2.9 | 2.6 |
| **Euro area** [c] | 2.7 | 3.3 | 2.7 | 1.3 | 2.8 | 2.4 | 3.4 | 1.7 | 1.2 | 0.7 | 1.7 | 1.3 | 1.2 | 2.0 | 1.8 |
| United Kingdom | 1.5 | – | 0.2 | 1.0 | 2.2 | 3.0 | 0.7 | -0.7 | 1.0 | 1.7 | 1.2 | 0.1 | 1.5 | 2.8 | 1.6 |
| Denmark | 0.9 | 2.2 | -0.1 | -0.8 | -0.2 | 0.6 | 0.8 | 4.1 | 3.0 | 2.1 | 3.4 | 0.8 | 3.1 | 1.4 | 1.0 |
| Sweden | 1.8 | 1.2 | 1.1 | 3.0 | 2.5 | 3.4 | 0.2 | -0.1 | -0.9 | -0.6 | 0.9 | -1.2 | 3.2 | 1.7 | 0.1 |
| **European Union** [d] | 2.5 | 2.7 | 2.2 | 1.3 | 2.7 | 2.5 | 2.9 | 1.3 | 1.2 | 0.8 | 1.6 | 1.0 | 1.4 | 2.1 | 1.6 |
| Cyprus | 3.6 | 5.3 | 10.5 | 1.9 | 17.4 | 3.9 | 13.8 | -14.3 | 4.1 | 2.9 | 11.7 | 4.9 | 7.3 | -5.0 | 2.4 |
| Iceland | 7.3 | 6.5 | 4.7 | 3.0 | 4.4 | 3.1 | -0.7 | 2.3 | 4.0 | 1.8 | 1.2 | 2.5 | 3.4 | 4.9 | 3.7 |
| Israel [a] | -9.7 | 18.3 | -2.5 | -8.6 | 7.7 | 4.1 | 1.4 | 4.2 | -0.2 | 1.8 | 5.1 | 1.9 | 2.7 | 3.1 | 1.1 |
| Malta | 4.4 | 9.1 | 6.0 | 12.7 | 5.7 | 10.9 | 8.9 | 6.0 | 6.4 | 8.5 | 8.4 | -1.1 | -4.0 | -1.1 | 3.6 |
| Norway [a] | 1.9 | 4.6 | -0.1 | 1.9 | 4.9 | 4.3 | 5.3 | 2.2 | 1.4 | 0.3 | 2.8 | 1.9 | 3.8 | 3.3 | 1.4 |
| Switzerland | 3.4 | 1.7 | 4.5 | 5.4 | 5.4 | 3.5 | 0.7 | -0.1 | 2.0 | -0.1 | 2.0 | – | 0.7 | -0.4 | 0.2 |
| Turkey | 9.2 | 9.4 | -1.1 | 0.8 | 8.0 | 3.7 | 3.6 | 8.6 | -5.5 | 6.8 | 8.6 | 4.1 | 7.8 | 6.5 | 7.1 |
| **Western Europe** | 2.8 | 3.0 | 2.1 | 1.4 | 3.0 | 2.6 | 2.6 | 1.4 | 0.7 | 1.0 | 1.9 | 1.2 | 1.8 | 2.3 | 1.9 |
| Canada [a] | 1.8 | 1.4 | 4.5 | 2.7 | 3.5 | 2.9 | 0.9 | – | -1.3 | -0.6 | -1.4 | -0.8 | 1.8 | 2.6 | 2.2 |
| United States [a,e] | 5.4 | 3.0 | 1.2 | 2.8 | 3.3 | 1.2 | 0.5 | -0.8 | 0.1 | 0.4 | 1.1 | 2.4 | 1.9 | 3.3 | 2.7 |
| **North America** | 5.1 | 2.8 | 1.4 | 2.7 | 3.3 | 1.3 | 0.5 | -0.7 | – | 0.4 | 0.9 | 2.1 | 1.9 | 3.2 | 2.6 |
| Japan | 4.8 | 3.5 | 3.4 | 2.9 | 2.5 | 3.2 | 2.7 | 3.2 | 2.9 | 4.3 | 2.8 | 1.3 | 1.9 | 4.0 | 3.6 |
| **Total above** | 4.1 | 3.0 | 2.0 | 2.2 | 3.0 | 2.1 | 1.7 | 0.8 | 0.7 | 1.3 | 1.6 | 1.6 | 1.8 | 3.0 | 2.5 |
| *Memorandum items:* | | | | | | | | | | | | | | | |
| *4 major west European economies* [f] | 2.2 | 2.3 | 2.4 | 0.2 | 2.4 | 1.9 | 2.8 | 0.9 | 1.0 | 0.4 | 1.6 | 0.7 | 0.8 | 1.9 | 1.6 |
| *Western Europe and North America* | 4.0 | 2.9 | 1.8 | 2.1 | 3.2 | 1.9 | 1.6 | 0.3 | 0.3 | 0.7 | 1.4 | 1.6 | 1.8 | 2.8 | 2.3 |

*Source:* Eurostat, New Cronos Database; OECD, *National Accounts* (Paris), various issues; national statistics.

*Note:* See appendix table A.1.

[a] Annual changes are calculated from chained national currency series.

[b] Data before 1992 are estimates for the whole of Germany, based on statistics published by Deutsches Institut fur Wirtschaftforschung and by the East German Statistical Office.

[c] Twelve countries above.

[d] Fifteen countries above.

[e] Includes also government gross investment expenditure.

[f] France, Germany, Italy and the United Kingdom.

## APPENDIX TABLE A.4

**Real gross domestic fixed capital formation in western Europe, North America and Japan, 1986-2000**
*(Percentage change over preceding year)*

|  | 1986 | 1987 | 1988 | 1989 | 1990 | 1991 | 1992 | 1993 | 1994 | 1995 | 1996 | 1997 | 1998 | 1999 | 2000 |
|---|---|---|---|---|---|---|---|---|---|---|---|---|---|---|---|
| France[a] | 6.0 | 6.0 | 9.5 | 7.3 | 3.3 | -1.5 | -1.6 | -6.4 | 1.5 | 2.0 | – | -0.1 | 7.0 | 6.2 | 6.1 |
| Germany[b] | 3.6 | 2.5 | 4.6 | 5.6 | 9.0 | 9.8 | 4.5 | -4.5 | 4.0 | -0.7 | -0.8 | 0.6 | 3.0 | 4.2 | 2.3 |
| Italy | 2.3 | 4.2 | 6.7 | 4.2 | 4.0 | 1.0 | -1.4 | -10.9 | 0.1 | 6.0 | 3.6 | 2.1 | 4.3 | 4.6 | 6.1 |
| Austria | 2.4 | 4.4 | 6.8 | 4.1 | 6.2 | 6.6 | 0.6 | -0.9 | 4.6 | 1.3 | 2.2 | 1.0 | 2.7 | 3.2 | 3.4 |
| Belgium | 3.2 | 6.2 | 15.7 | 12.6 | 8.5 | -4.1 | 1.7 | -3.1 | -0.1 | 4.9 | 0.8 | 6.7 | 4.6 | 4.8 | 4.5 |
| Finland | 1.0 | 4.9 | 11.0 | 13.0 | -4.6 | -18.6 | -16.7 | -16.6 | -2.7 | 10.6 | 8.4 | 11.9 | 9.3 | 3.0 | 5.5 |
| Greece | 0.1 | -5.6 | 2.6 | 6.1 | 4.5 | 4.2 | -3.5 | -4.0 | -3.1 | 4.1 | 8.4 | 13.2 | 8.0 | 7.3 | 8.1 |
| Ireland | -2.8 | -1.1 | 5.2 | 10.1 | 13.4 | -6.2 | -1.8 | -3.5 | 12.0 | 12.8 | 16.5 | 17.9 | 16.5 | 14.0 | 7.0 |
| Luxembourg | 31.0 | 17.9 | 15.0 | 7.0 | 2.7 | 31.6 | -9.0 | 28.4 | -14.9 | 3.5 | -3.5 | 10.5 | 1.5 | 26.6 | 0.5 |
| Netherlands | 7.0 | 0.8 | 5.3 | 5.1 | 2.5 | 0.4 | 0.7 | -3.2 | 2.1 | 3.9 | 6.3 | 6.6 | 4.2 | 7.8 | 3.8 |
| Portugal | 10.9 | 18.0 | 14.8 | 4.2 | 8.2 | 3.5 | 4.8 | -5.8 | 3.5 | -1.6 | 6.3 | 14.4 | 9.1 | 6.6 | 5.2 |
| Spain | 9.9 | 14.0 | 13.9 | 13.6 | 6.6 | 1.6 | -4.4 | -10.5 | 2.5 | 8.2 | 2.1 | 5.0 | 9.7 | 8.9 | 5.9 |
| *Euro area*[c] | 4.6 | 5.0 | 7.8 | 6.8 | 5.8 | 2.9 | 0.5 | -6.1 | 2.4 | 2.5 | 1.3 | 2.5 | 5.2 | 5.5 | 4.5 |
| United Kingdom | 1.9 | 9.3 | 14.9 | 6.0 | -2.6 | -8.2 | -0.9 | 0.3 | 4.7 | 3.1 | 4.7 | 7.1 | 13.2 | 0.9 | 4.9 |
| Denmark | 16.7 | -0.9 | -6.2 | -0.8 | -2.1 | -3.3 | -2.0 | -4.0 | 7.6 | 11.6 | 4.0 | 10.9 | 7.6 | 1.5 | 9.9 |
| Sweden | 1.1 | 8.0 | 6.4 | 12.1 | 0.2 | -8.6 | -11.6 | -15.0 | 6.1 | 9.4 | 5.0 | -1.1 | 8.5 | 8.1 | 4.5 |
| *European Union*[d] | 4.3 | 5.7 | 8.6 | 6.7 | 4.3 | 0.8 | 0.1 | -5.6 | 2.7 | 2.9 | 1.8 | 3.1 | 6.2 | 4.9 | 4.6 |
| Cyprus | -7.1 | 4.5 | 10.6 | 20.0 | -2.8 | -1.6 | 16.2 | -12.8 | -2.5 | -1.7 | 7.4 | -4.5 | 8.0 | -1.2 | 1.0 |
| Iceland | -1.6 | 18.8 | -0.2 | -7.9 | 3.0 | 3.3 | -11.1 | -10.7 | 0.6 | -1.1 | 25.7 | 9.6 | 26.6 | -0.8 | 9.0 |
| Israel[a] | 7.4 | 6.1 | 1.6 | -2.2 | 25.3 | 41.9 | 5.2 | 5.3 | 8.4 | 6.6 | 7.7 | -2.4 | -5.6 | 0.4 | 0.4 |
| Malta | -8.7 | 30.7 | 6.1 | 1.0 | 17.9 | – | -0.2 | 11.1 | 8.5 | 17.8 | -8.4 | -4.5 | -3.4 | 0.8 | 18.7 |
| Norway[a] | 7.6 | 0.3 | -1.8 | -6.9 | -10.8 | -0.4 | -3.1 | 4.3 | 4.5 | 3.4 | 9.9 | 13.9 | 10.6 | -8.2 | -1.1 |
| Switzerland | 5.4 | 4.0 | 8.1 | 5.3 | 3.8 | -2.9 | -6.6 | -2.7 | 6.5 | 1.8 | -2.4 | 1.5 | 4.5 | 1.8 | 6.8 |
| Turkey | 8.4 | 45.1 | -1.0 | 2.2 | 15.9 | 0.4 | 6.4 | 26.4 | -16.0 | 9.1 | 14.1 | 14.8 | -3.9 | -16.0 | 16.5 |
| *Western Europe* | 4.6 | 7.4 | 8.0 | 6.3 | 4.6 | 0.6 | -0.2 | -4.0 | 1.9 | 3.6 | 2.9 | 4.1 | 6.2 | 3.6 | 5.3 |
| Canada[a] | 4.7 | 10.5 | 9.4 | 5.6 | -4.0 | -5.5 | -2.7 | -2.0 | 7.5 | -2.1 | 4.4 | 15.2 | 2.4 | 7.3 | 6.7 |
| United States[a,e] | .. | .. | 3.6 | 2.7 | -1.8 | -6.9 | 6.5 | 8.1 | 9.1 | 6.1 | 9.3 | 9.6 | 11.4 | 7.8 | 7.6 |
| *North America* | .. | .. | 4.1 | 3.0 | -2.0 | -6.8 | 5.7 | 7.3 | 9.0 | 5.4 | 8.9 | 10.0 | 10.6 | 7.8 | 7.5 |
| Japan | 5.1 | 9.4 | 12.0 | 8.6 | 8.8 | 2.2 | -2.5 | -3.1 | -1.4 | 0.3 | 6.8 | 1.0 | -4.0 | -0.9 | 0.6 |
| **Total above** | .. | .. | 7.0 | 5.2 | 2.5 | -2.3 | 1.9 | 0.9 | 4.4 | 3.8 | 6.1 | 6.1 | 6.5 | 4.7 | 5.5 |
| *Memorandum items:* | | | | | | | | | | | | | | | |
| 4 major west European economies[f] | 3.5 | 5.1 | 8.4 | 5.8 | 4.1 | 1.3 | 0.6 | -5.3 | 2.7 | 2.3 | 1.6 | 2.2 | 6.4 | 4.0 | 4.6 |
| Western Europe and North America | .. | .. | 6.1 | 4.6 | 1.3 | -3.1 | 2.8 | 1.6 | 5.5 | 4.5 | 5.9 | 7.1 | 8.4 | 5.7 | 6.4 |

*Source:* Eurostat, New Cronos Database; OECD, *National Accounts* (Paris), various issues; national statistics.

*Note:* See appendix table A.1.

[a] Annual changes are calculated from chained national currency series.

[b] Data before 1992 are estimates for the whole of Germany, based on statistics published by Deutsches Institut fur Wirtschaftforschung and by the East German Statistical Office.

[c] Twelve countries above.

[d] Fifteen countries above.

[e] Private sector only. See appendix table A.3.

[f] France, Germany, Italy and the United Kingdom.

## APPENDIX TABLE A.5

**Real total domestic expenditures in western Europe, North America and Japan, 1986-2000**

*(Percentage change over preceding year)*

|  | 1986 | 1987 | 1988 | 1989 | 1990 | 1991 | 1992 | 1993 | 1994 | 1995 | 1996 | 1997 | 1998 | 1999 | 2000 |
|---|---|---|---|---|---|---|---|---|---|---|---|---|---|---|---|
| France [a] | 3.5 | 3.2 | 4.7 | 3.9 | 2.8 | 0.5 | 0.8 | -1.6 | 2.1 | 1.6 | 0.7 | 0.7 | 4.0 | 3.1 | 3.3 |
| Germany [b] | 3.3 | 2.6 | 3.6 | 2.7 | 4.8 | 6.2 | 2.8 | -1.1 | 2.3 | 1.7 | 0.3 | 0.6 | 2.4 | 2.6 | 2.0 |
| Italy | 3.1 | 4.3 | 4.1 | 3.1 | 2.7 | 2.1 | 0.9 | -5.1 | 1.7 | 2.0 | 0.9 | 2.7 | 3.1 | 3.0 | 2.3 |
| Austria | 2.1 | 2.5 | 3.2 | 3.7 | 4.4 | 3.5 | 2.3 | 0.6 | 3.5 | 2.6 | 1.9 | 1.3 | 2.5 | 2.6 | 2.0 |
| Belgium | 2.6 | 3.5 | 4.8 | 4.3 | 2.9 | 1.7 | 1.8 | -1.5 | 2.1 | 1.9 | 0.9 | 2.6 | 3.9 | 2.1 | 3.4 |
| Finland | 2.7 | 5.7 | 6.6 | 7.0 | -0.5 | -7.9 | -5.7 | -5.5 | 3.2 | 3.2 | 4.1 | 4.7 | 4.8 | 2.7 | 3.3 |
| Greece | 0.4 | -2.7 | 5.9 | 5.3 | 2.2 | 3.5 | -0.5 | -1.0 | 1.1 | 3.5 | 3.3 | 3.6 | 4.7 | 2.9 | 3.8 |
| Ireland | 1.2 | -0.4 | 1.9 | 6.9 | 6.3 | 0.1 | -0.3 | 1.1 | 5.6 | 7.3 | 7.7 | 9.8 | 9.4 | 7.0 | 9.2 |
| Luxembourg | 7.9 | 5.3 | 6.5 | 8.0 | 3.2 | 8.1 | -1.2 | 8.8 | -0.2 | 3.0 | 2.7 | 5.5 | 2.4 | 11.3 | 3.1 |
| Netherlands | 3.8 | 1.4 | 2.2 | 5.0 | 3.4 | 2.2 | 1.2 | -1.6 | 2.0 | 3.5 | 2.8 | 3.9 | 4.8 | 4.2 | 3.0 |
| Portugal | 7.8 | 9.0 | 10.1 | 3.9 | 5.9 | 3.8 | 3.7 | -1.4 | 2.9 | 2.9 | 3.2 | 5.0 | 6.3 | 5.5 | 3.4 |
| Spain | 5.4 | 8.1 | 7.0 | 7.8 | 4.8 | 2.9 | 1.0 | -4.1 | 1.3 | 3.2 | 1.9 | 3.4 | 5.6 | 5.5 | 4.1 |
| *Euro area* [c] | 3.5 | 3.6 | 4.5 | 4.0 | 3.7 | 3.0 | 1.5 | -2.1 | 2.1 | 2.1 | 1.0 | 1.7 | 3.6 | 3.2 | 2.8 |
| United Kingdom | 4.6 | 4.8 | 7.8 | 2.9 | -0.2 | -2.3 | 0.9 | 2.2 | 3.7 | 2.0 | 3.1 | 3.9 | 5.1 | 3.4 | 3.4 |
| Denmark | 6.6 | -2.3 | -1.6 | -0.1 | -0.7 | -0.1 | 0.9 | -0.3 | 7.0 | 4.2 | 2.2 | 4.9 | 4.5 | -0.6 | 2.6 |
| Sweden | 2.9 | 4.3 | 3.0 | 4.0 | 0.7 | -1.6 | -1.8 | -4.7 | 3.1 | 2.0 | 0.7 | 0.9 | 4.2 | 3.4 | 2.7 |
| *European Union* [d] | 3.7 | 3.7 | 4.8 | 3.7 | 2.9 | 2.0 | 1.3 | -1.6 | 2.4 | 2.1 | 1.3 | 2.1 | 3.8 | 3.2 | 2.9 |
| Cyprus | -1.5 | 5.6 | 11.3 | 9.7 | 6.3 | 5.1 | 9.3 | -9.5 | 7.4 | 7.8 | 4.6 | 1.4 | 8.8 | 0.7 | 4.6 |
| Iceland | 4.7 | 15.5 | -0.9 | -4.2 | 1.6 | 4.2 | -4.5 | -4.2 | 2.5 | 2.0 | 7.4 | 5.7 | 12.3 | 4.6 | 5.4 |
| Israel [a] | 5.4 | 9.0 | 2.5 | -2.8 | 9.3 | 12.1 | 5.3 | 6.8 | 5.3 | 6.7 | 5.5 | 1.2 | 1.0 | 4.9 | 3.2 |
| Malta | – | 3.5 | 11.5 | 8.3 | 7.6 | 4.9 | 0.1 | 4.8 | 5.9 | 9.5 | 2.8 | -0.1 | -1.1 | 5.0 | 11.9 |
| Norway [a] | 7.4 | -0.8 | -3.0 | -2.0 | -0.5 | 0.8 | 1.5 | 3.1 | 4.0 | 4.3 | 4.2 | 6.4 | 5.4 | -0.7 | 2.3 |
| Switzerland | 4.5 | 2.0 | 2.6 | 4.1 | 3.9 | -0.6 | -2.7 | -1.0 | 2.7 | 1.8 | 0.4 | 1.3 | 4.3 | 1.4 | 3.1 |
| Turkey | 7.0 | 8.9 | -1.3 | 1.5 | 14.6 | -0.6 | 5.6 | 14.2 | -12.5 | 11.4 | 7.6 | 9.0 | 0.6 | -4.0 | 9.4 |
| **Western Europe** | 3.9 | 3.9 | 4.4 | 3.6 | 3.5 | 1.8 | 1.4 | -0.8 | 1.8 | 2.6 | 1.8 | 2.6 | 3.8 | 2.8 | 3.2 |
| Canada [a] | 3.3 | 4.9 | 5.4 | 4.0 | -0.3 | -1.9 | 0.5 | 1.6 | 3.2 | 1.8 | 1.2 | 6.1 | 2.3 | 4.0 | 4.5 |
| United States [a] | 3.8 | 3.7 | 3.2 | 2.9 | 1.4 | -1.1 | 3.1 | 3.2 | 4.4 | 2.4 | 3.7 | 4.7 | 5.5 | 5.0 | 4.8 |
| **North America** | 3.7 | 3.8 | 3.4 | 3.0 | 1.3 | -1.1 | 2.9 | 3.0 | 4.3 | 2.4 | 3.5 | 4.8 | 5.2 | 4.9 | 4.8 |
| Japan | 3.8 | 5.3 | 7.3 | 5.6 | 5.3 | 2.7 | 0.6 | 0.3 | 1.2 | 2.1 | 4.0 | 0.9 | -1.5 | 0.9 | 1.1 |
| **Total above** | 3.8 | 4.1 | 4.4 | 3.6 | 2.8 | 0.7 | 1.9 | 1.0 | 2.8 | 2.4 | 2.8 | 3.3 | 3.6 | 3.4 | 3.5 |
| *Memorandum items:* | | | | | | | | | | | | | | | |
| 4 major west European economies [e] | 3.6 | 3.6 | 4.9 | 3.1 | 2.8 | 2.1 | 1.5 | -1.4 | 2.4 | 1.8 | 1.1 | 1.8 | 3.5 | 3.0 | 2.7 |
| Western Europe and North America | 3.8 | 3.8 | 3.9 | 3.3 | 2.4 | 0.3 | 2.1 | 1.1 | 3.0 | 2.5 | 2.6 | 3.7 | 4.5 | 3.8 | 4.0 |

*Source:* Eurostat, New Cronos Database; OECD, *National Accounts* (Paris), various issues; national statistics.

*Note:* See appendix table A.1.

[a] Annual changes are calculated from chained national currency series.

[b] Data before 1992 are estimates for the whole of Germany, based on statistics published by Deutsches Institut fur Wirtschaftforschung and by the East German Statistical Office.

[c] Twelve countries above.

[d] Fifteen countries above.

[e] France, Germany, Italy and the United Kingdom.

## APPENDIX TABLE A.6

**Real exports of goods and services in western Europe, North America and Japan, 1986-2000**
*(Percentage change over preceding year)*

|  | 1986 | 1987 | 1988 | 1989 | 1990 | 1991 | 1992 | 1993 | 1994 | 1995 | 1996 | 1997 | 1998 | 1999 | 2000 |
|---|---|---|---|---|---|---|---|---|---|---|---|---|---|---|---|
| France [a] | -0.4 | 3.4 | 8.7 | 10.0 | 4.8 | 5.9 | 5.4 | – | 7.7 | 7.7 | 3.5 | 11.8 | 8.3 | 4.0 | 12.6 |
| Germany [b] | -0.5 | 0.5 | 5.1 | 9.8 | 0.5 | -2.9 | -0.8 | -5.5 | 7.6 | 5.7 | 5.1 | 11.2 | 6.8 | 5.6 | 13.2 |
| Italy | 0.8 | 4.5 | 5.1 | 7.8 | 7.5 | -1.4 | 7.3 | 9.0 | 9.8 | 12.6 | 0.6 | 6.4 | 3.6 | – | 10.2 |
| Austria | -2.3 | 3.1 | 10.2 | 9.7 | 7.8 | 5.2 | 1.5 | -1.4 | 5.6 | 6.7 | 6.2 | 9.9 | 5.5 | 7.6 | 10.3 |
| Belgium | 2.8 | 5.0 | 9.6 | 8.3 | 4.6 | 3.1 | 3.7 | -0.4 | 8.4 | 5.7 | 1.2 | 6.7 | 4.4 | 5.2 | 11.8 |
| Finland | 0.7 | 2.9 | 3.5 | 1.6 | 1.2 | -7.3 | 10.3 | 16.7 | 13.1 | 8.6 | 5.8 | 14.1 | 8.9 | 6.8 | 18.1 |
| Greece | 16.8 | 5.9 | -2.1 | 1.9 | -3.5 | 4.1 | 10.0 | -2.6 | 7.4 | 3.0 | 3.5 | 18.2 | 5.9 | 6.5 | 12.3 |
| Ireland | 2.9 | 13.7 | 9.0 | 10.3 | 8.7 | 5.7 | 13.9 | 9.7 | 15.1 | 20.0 | 12.2 | 17.4 | 21.4 | 15.7 | 17.8 |
| Luxembourg | 3.3 | 4.4 | 11.7 | 8.1 | 3.4 | 6.7 | 4.8 | 2.8 | 4.4 | 4.4 | 4.0 | 10.5 | 9.9 | 7.9 | 14.3 |
| Netherlands | 1.8 | 4.1 | 8.9 | 7.9 | 5.1 | 5.6 | 2.4 | 5.7 | 9.7 | 8.8 | 4.6 | 8.8 | 7.4 | 5.4 | 9.5 |
| Portugal | 6.8 | 11.2 | 8.2 | 13.0 | 10.0 | 2.6 | 5.0 | -3.6 | 8.7 | 13.6 | 7.1 | 7.2 | 7.7 | 3.4 | 7.0 |
| Spain | 1.9 | 6.3 | 5.1 | 3.0 | 3.2 | 7.9 | 7.4 | 8.5 | 16.7 | 10.0 | 10.4 | 15.3 | 8.3 | 6.6 | 10.8 |
| *Euro area* [c] | 0.8 | 3.5 | 6.3 | 8.3 | 3.9 | 1.6 | 3.2 | 0.8 | 8.9 | 7.9 | 4.3 | 10.6 | 7.2 | 5.2 | 12.1 |
| United Kingdom | 4.3 | 6.1 | 0.6 | 4.5 | 5.4 | -0.1 | 4.3 | 4.4 | 9.2 | 9.0 | 8.2 | 8.3 | 3.0 | 5.4 | 10.2 |
| Denmark | 0.5 | 4.5 | 7.6 | 4.2 | 6.2 | 6.1 | -0.9 | -1.5 | 7.0 | 2.9 | 4.3 | 4.1 | 2.4 | 9.7 | 11.6 |
| Sweden | 3.4 | 4.3 | 2.8 | 3.2 | 1.8 | -1.9 | 2.2 | 8.3 | 14.1 | 11.3 | 3.5 | 13.7 | 8.4 | 5.9 | 9.8 |
| *European Union* [d] | 1.4 | 3.9 | 5.3 | 7.5 | 4.1 | 1.3 | 3.2 | 1.4 | 9.1 | 5.8 | 4.6 | 12.8 | 6.5 | 5.2 | 11.6 |
| Cyprus | -1.7 | 13.7 | 13.5 | 16.8 | 7.9 | -8.4 | 18.7 | -1.3 | 7.9 | 4.6 | 4.1 | 0.8 | -2.4 | 6.5 | 9.8 |
| Iceland | 5.9 | 3.3 | -3.6 | 2.9 | – | -5.9 | -1.9 | 7.0 | 9.9 | -2.1 | 9.9 | 5.7 | 2.2 | 5.5 | 5.1 |
| Israel [a] | 5.6 | 10.2 | -1.5 | 4.0 | 2.0 | -2.6 | 14.1 | 9.9 | 12.8 | 8.4 | 6.6 | 7.7 | 6.3 | 11.6 | 23.9 |
| Malta | 7.0 | 12.6 | 6.1 | 10.7 | 13.3 | 7.5 | 9.7 | 5.3 | 7.1 | 5.4 | -5.9 | 4.0 | 8.1 | 8.1 | 5.6 |
| Norway [a] | 2.2 | 1.1 | 6.4 | 11.0 | 8.6 | 6.1 | 5.2 | 3.2 | 8.7 | 4.3 | 9.3 | 6.1 | 0.3 | 2.8 | 2.7 |
| Switzerland | -0.4 | 2.3 | 6.5 | 6.6 | 2.1 | -2.1 | 3.0 | 1.5 | 1.8 | 1.6 | 2.5 | 8.6 | 5.0 | 5.9 | 9.5 |
| Turkey | -5.1 | 26.4 | 18.4 | -0.3 | 2.5 | 3.7 | 11.0 | 7.7 | 15.2 | 8.0 | 22.0 | 19.1 | 12.0 | -7.0 | 19.3 |
| **Western Europe** | 1.1 | 4.9 | 6.0 | 7.2 | 4.1 | 1.4 | 4.3 | 2.4 | 9.4 | 8.3 | 5.7 | 10.5 | 6.3 | 4.1 | 11.7 |
| Canada [a] | 4.3 | 2.9 | 8.9 | 1.0 | 4.7 | 1.8 | 7.2 | 10.8 | 12.7 | 8.5 | 5.6 | 8.3 | 8.9 | 9.9 | 7.6 |
| United States [a] | 7.4 | 11.2 | 16.1 | 11.8 | 8.7 | 6.5 | 6.2 | 3.3 | 8.9 | 10.3 | 8.2 | 12.3 | 2.1 | 3.2 | 9.5 |
| **North America** | 7.1 | 10.5 | 15.5 | 10.9 | 8.4 | 6.1 | 6.2 | 4.0 | 9.3 | 10.1 | 7.9 | 11.9 | 2.7 | 3.8 | 9.3 |
| Japan | -5.5 | -0.5 | 5.9 | 9.1 | 7.0 | 4.1 | 3.9 | -0.1 | 3.5 | 4.1 | 6.5 | 11.2 | -2.3 | 1.4 | 12.1 |
| **Total above** | 2.6 | 6.4 | 10.0 | 9.1 | 6.4 | 3.8 | 5.1 | 2.7 | 8.4 | 8.4 | 6.8 | 11.2 | 3.4 | 3.6 | 10.8 |
| *Memorandum items:* | | | | | | | | | | | | | | | |
| *4 major west European economies* [e] | 0.8 | 3.3 | 4.9 | 8.2 | 4.1 | – | 3.6 | 1.2 | 8.5 | 8.4 | 4.4 | 9.6 | 5.6 | 3.9 | 11.8 |
| *Western Europe and North America* | 4.1 | 7.7 | 10.7 | 9.1 | 6.2 | 3.8 | 5.3 | 3.2 | 9.3 | 9.2 | 6.8 | 11.2 | 4.5 | 4.0 | 10.5 |

*Source:* Eurostat, New Cronos Database; OECD, *National Accounts* (Paris), various issues; national statistics.

*Note:* See appendix table A.1. Data on national account basis.

[a] Annual changes are calculated from chained national currency series.

[b] Data before 1992 are estimates for the whole of Germany, based on statistics published by Deutsches Institut fur Wirtschaftforschung and by the East German Statistical Office.

[c] Twelve countries above.

[d] Fifteen countries above.

[e] France, Germany, Italy and the United Kingdom.

## APPENDIX TABLE A.7

**Real imports of goods and services in western Europe, North America and Japan, 1986-2000**

*(Percentage change over preceding year)*

|  | 1986 | 1987 | 1988 | 1989 | 1990 | 1991 | 1992 | 1993 | 1994 | 1995 | 1996 | 1997 | 1998 | 1999 | 2000 |
|---|---|---|---|---|---|---|---|---|---|---|---|---|---|---|---|
| France [a] | 6.5 | 7.7 | 8.8 | 8.0 | 5.5 | 3.1 | 1.8 | -3.7 | 8.2 | 8.0 | 1.6 | 6.9 | 11.6 | 4.7 | 14.2 |
| Germany [b] | 2.5 | 4.2 | 4.9 | 7.8 | 5.2 | 9.6 | 1.5 | -5.5 | 7.4 | 5.6 | 3.1 | 8.3 | 8.9 | 8.5 | 10.0 |
| Italy | 4.0 | 12.2 | 5.9 | 8.9 | 11.5 | 2.3 | 7.4 | -10.9 | 8.1 | 9.7 | -0.3 | 10.1 | 9.0 | 5.1 | 8.3 |
| Austria | -2.9 | 5.4 | 10.4 | 8.0 | 6.9 | 5.8 | 1.4 | -1.1 | 8.2 | 9.3 | 5.8 | 9.7 | 3.7 | 7.1 | 7.5 |
| Belgium | 4.5 | 6.7 | 10.4 | 9.6 | 4.8 | 2.8 | 4.1 | -0.4 | 7.2 | 5.0 | 0.8 | 5.7 | 6.5 | 4.5 | 11.4 |
| Finland | 1.5 | 9.2 | 10.9 | 9.0 | -0.8 | -13.5 | 0.6 | 1.3 | 12.8 | 7.8 | 6.4 | 11.3 | 8.5 | 4.0 | 15.7 |
| Greece | 13.9 | 2.1 | 7.3 | 10.5 | 8.4 | 5.8 | 1.1 | 0.6 | 1.5 | 8.9 | 7.0 | 13.9 | 11.3 | 3.9 | 8.7 |
| Ireland | 5.6 | 6.2 | 4.9 | 13.5 | 5.1 | 2.4 | 8.2 | 7.5 | 15.5 | 16.4 | 12.5 | 16.8 | 25.8 | 11.9 | 16.6 |
| Luxembourg | 3.8 | 7.5 | 8.2 | 6.6 | 4.5 | 9.0 | -0.8 | 2.8 | -0.1 | 3.8 | 4.0 | 9.3 | 8.3 | 11.2 | 10.5 |
| Netherlands | 3.5 | 4.2 | 6.9 | 8.2 | 3.6 | 5.1 | 1.4 | 0.7 | 9.4 | 10.6 | 4.4 | 9.5 | 8.5 | 6.3 | 9.4 |
| Portugal | 16.9 | 23.1 | 18.0 | 6.1 | 14.0 | 7.3 | 10.7 | -3.3 | 9.0 | 9.4 | 4.9 | 10.0 | 13.8 | 8.9 | 6.5 |
| Spain | 14.4 | 20.1 | 14.4 | 17.3 | 7.8 | 9.0 | 6.9 | -5.2 | 11.3 | 11.0 | 8.0 | 13.3 | 13.4 | 11.9 | 10.4 |
| *Euro area* [c] | 5.4 | 8.8 | 7.8 | 9.2 | 6.9 | 5.5 | 3.1 | -4.0 | 8.2 | 7.7 | 3.2 | 9.1 | 9.9 | 7.1 | 10.8 |
| United Kingdom | 6.9 | 7.9 | 12.8 | 7.4 | 0.5 | -4.5 | 6.8 | 3.3 | 5.7 | 5.4 | 9.6 | 9.7 | 9.6 | 8.9 | 10.7 |
| Denmark | 9.7 | -2.8 | 5.3 | 4.1 | 1.2 | 3.0 | -0.4 | -2.7 | 12.3 | 7.3 | 3.5 | 10.0 | 7.4 | 2.2 | 10.8 |
| Sweden | 3.8 | 7.6 | 4.5 | 7.7 | 0.7 | -4.9 | 1.5 | -2.2 | 12.2 | 7.2 | 3.0 | 12.5 | 11.2 | 4.3 | 9.7 |
| *European Union* [d] | 5.7 | 8.4 | 8.5 | 8.8 | 5.7 | 3.7 | 3.4 | -3.0 | 8.0 | 4.8 | 3.8 | 12.1 | 9.7 | 7.1 | 10.6 |
| Cyprus | -11.0 | 5.5 | 13.4 | 20.4 | 5.8 | 2.2 | 18.2 | -18.1 | 8.2 | 11.5 | 6.7 | -0.5 | 7.5 | -3.3 | 8.5 |
| Iceland | 0.9 | 23.3 | -4.6 | -10.3 | 1.0 | 5.3 | -5.9 | -7.7 | 4.2 | 4.0 | 16.7 | 8.5 | 23.3 | 6.1 | 9.3 |
| Israel [a] | 11.2 | 19.6 | -2.8 | -5.0 | 9.5 | 16.0 | 8.8 | 14.1 | 10.9 | 4.5 | 8.2 | 1.9 | 1.8 | 14.8 | 12.2 |
| Malta | 0.1 | 12.3 | 11.1 | 11.1 | 15.7 | 5.4 | 3.0 | 5.9 | 7.5 | 10.0 | -5.9 | -1.7 | 2.5 | 9.1 | 13.9 |
| Norway [a] | 11.8 | -6.5 | -2.4 | 2.2 | 2.5 | 0.2 | 0.7 | 4.4 | 4.9 | 5.6 | 8.0 | 11.3 | 8.0 | -1.6 | 2.5 |
| Switzerland | 8.1 | 6.2 | 5.2 | 5.9 | 2.6 | -1.6 | -4.2 | 0.1 | 7.9 | 5.1 | 2.7 | 7.6 | 9.6 | 5.5 | 8.6 |
| Turkey | -3.5 | 23.0 | -4.5 | 6.9 | 33.0 | -5.2 | 10.9 | 35.8 | -21.9 | 29.6 | 20.5 | 22.4 | 2.3 | -3.7 | 25.4 |
| **Western Europe** | 5.4 | 8.8 | 7.6 | 8.6 | 6.8 | 3.1 | 4.1 | -1.6 | 6.6 | 8.6 | 4.8 | 10.0 | 9.5 | 6.4 | 11.1 |
| Canada [a] | 7.2 | 5.3 | 13.5 | 5.9 | 2.0 | 2.5 | 4.7 | 7.4 | 8.0 | 5.7 | 5.1 | 14.2 | 4.9 | 7.3 | 8.1 |
| United States [a] | 8.4 | 6.1 | 3.8 | 4.0 | 3.8 | -0.5 | 6.6 | 9.1 | 12.0 | 8.2 | 8.6 | 13.7 | 11.8 | 10.5 | 13.4 |
| **North America** | 8.3 | 6.0 | 4.6 | 4.1 | 3.7 | -0.3 | 6.5 | 8.9 | 11.6 | 8.0 | 8.3 | 13.7 | 11.2 | 10.2 | 12.9 |
| Japan | 3.2 | 11.3 | 19.5 | 15.7 | 7.0 | -1.1 | -0.7 | -1.4 | 7.8 | 12.8 | 13.2 | 1.2 | -6.8 | 3.0 | 9.9 |
| **Total above** | 6.3 | 8.0 | 8.2 | 7.8 | 5.5 | 1.0 | 4.4 | 2.9 | 8.9 | 9.0 | 7.6 | 10.2 | 7.6 | 7.5 | 11.7 |
| *Memorandum items:* | | | | | | | | | | | | | | | |
| *4 major west European economies* [e] | 4.7 | 7.6 | 7.7 | 8.0 | 5.7 | 3.4 | 4.0 | -4.4 | 7.4 | 7.0 | 3.4 | 8.7 | 9.7 | 6.9 | 10.7 |
| *Western Europe and North America* | 6.8 | 7.4 | 6.1 | 6.3 | 5.3 | 1.4 | 5.3 | 3.7 | 9.1 | 8.3 | 6.6 | 11.8 | 10.4 | 8.3 | 12.0 |

*Source:* Eurostat, New Cronos Database; OECD, *National Accounts* (Paris), various issues; national statistics.

*Note:* See appendix table A.1. Data on national account basis.

[a] Annual changes are calculated from chained national currency series.

[b] Data before 1992 are estimates for the whole of Germany, based on statistics published by Deutsches Institut fur Wirtschaftforschung and by the East German Statistical Office.

[c] Twelve countries above.

[d] Fifteen countries above.

[e] France, Germany, Italy and the United Kingdom.

APPENDIX TABLE A.8

**Industrial output in western Europe, North America and Japan, 1986-2000**
*(Percentage change over preceding year)*

|  | 1986 | 1987 | 1988 | 1989 | 1990 | 1991 | 1992 | 1993 | 1994 | 1995 | 1996 | 1997 | 1998 | 1999 | 2000 |
|---|---|---|---|---|---|---|---|---|---|---|---|---|---|---|---|
| France | 0.6 | 1.2 | 4.6 | 3.7 | 3.1 | -0.3 | -1.0 | -3.8 | 4.2 | 2.5 | 0.9 | 3.8 | 5.1 | 2.0 | 3.4 |
| Germany[a] | 1.8 | 0.4 | 3.6 | 4.9 | 5.2 | 3.5 | -2.4 | -7.9 | 3.2 | 0.8 | 0.8 | 3.7 | 4.1 | 1.5 | 6.3 |
| Italy | 4.1 | 2.6 | 6.9 | 3.9 | 6.3 | -0.4 | -1.3 | -2.1 | 6.2 | 5.0 | -1.9 | 3.8 | 1.1 | – | 4.8 |
| Austria | 1.2 | 1.0 | 4.4 | 5.8 | 6.8 | 1.9 | -1.2 | -1.5 | 4.0 | 4.9 | 1.0 | 6.4 | 8.2 | 6.0 | 9.2 |
| Belgium | 0.8 | 2.1 | 5.8 | 3.4 | 1.5 | -1.9 | -0.4 | -5.1 | 2.1 | 6.5 | 0.5 | 4.7 | 3.4 | 0.9 | 5.5 |
| Finland | 1.8 | 5.0 | 3.2 | 3.6 | -0.6 | -8.7 | 1.3 | 5.6 | 11.3 | 7.3 | 3.6 | 10.1 | 7.4 | 6.1 | 11.1 |
| Greece | -0.3 | -1.2 | 5.1 | 1.8 | -2.5 | -1.0 | -1.1 | -2.9 | 1.3 | 1.8 | 1.2 | 1.3 | 7.1 | 3.9 | 0.5 |
| Ireland | 2.1 | 8.9 | 10.7 | 11.6 | 4.7 | 3.3 | 9.1 | 5.6 | 11.9 | 20.5 | 8.1 | 17.5 | 19.8 | 14.8 | 15.4 |
| Luxembourg | 1.9 | -0.6 | 8.7 | 7.8 | 2.6 | 0.4 | -0.8 | -4.3 | 5.9 | 2.0 | 0.1 | 5.8 | -0.1 | 11.5 | 4.3 |
| Netherlands | 0.2 | 1.1 | 0.1 | 5.1 | 2.4 | 1.8 | -0.2 | -1.1 | 4.9 | 4.6 | 2.4 | 0.2 | 2.4 | 2.2 | 2.9 |
| Portugal | 7.3 | 4.4 | 3.8 | 6.7 | 9.0 | – | -2.3 | -5.2 | -0.2 | 11.6 | 5.3 | 2.6 | 5.7 | 3.0 | 0.5 |
| Spain | 3.3 | 4.6 | 3.1 | 5.1 | -0.3 | -0.7 | -3.1 | -4.7 | 7.7 | 4.8 | -1.3 | 6.9 | 5.5 | 2.6 | 4.4 |
| *Euro area*[b] | 2.2 | 1.8 | 4.5 | 4.6 | 4.2 | 0.9 | -1.5 | -4.4 | 4.7 | 3.6 | 0.4 | 4.2 | 4.1 | 1.8 | 5.1 |
| United Kingdom | 1.4 | 4.1 | 5.2 | 2.1 | – | -3.3 | 0.4 | 2.1 | 5.2 | 1.8 | 1.0 | 1.0 | 0.8 | 0.5 | 1.5 |
| Denmark | 7.8 | -3.6 | 2.5 | 2.4 | 2.4 | – | 3.5 | -2.2 | 10.3 | 4.2 | 2.0 | 5.9 | 1.9 | 2.7 | 6.2 |
| Sweden | 0.2 | 2.8 | 2.9 | 2.9 | 0.3 | -5.2 | -1.7 | -0.4 | 11.5 | 9.8 | 1.0 | 6.5 | 4.2 | 3.0 | 8.4 |
| *European Union*[c] | 2.2 | 2.1 | 4.5 | 4.1 | 3.5 | 0.1 | -1.2 | -3.4 | 5.0 | 3.5 | 0.5 | 3.9 | 3.6 | 1.7 | 4.7 |
| Israel | 3.6 | 4.9 | -3.1 | -1.6 | 8.0 | 6.8 | 8.2 | 6.9 | 7.4 | 8.4 | 5.4 | 1.8 | 2.8 | 1.4 | 10.1 |
| Norway | 4.1 | 6.6 | 2.9 | 9.3 | 2.5 | 2.5 | 5.6 | 3.6 | 7.0 | 5.9 | 5.4 | 3.4 | -0.6 | -0.2 | 2.9 |
| Switzerland | 3.8 | 1.2 | 7.8 | 1.5 | 4.8 | 0.5 | -1.0 | -1.8 | 4.3 | 2.0 | – | 4.6 | 3.6 | 3.6 | 8.2 |
| Turkey | 11.7 | 10.5 | 1.6 | 3.6 | 9.5 | 2.7 | 5.0 | 8.0 | -6.2 | 12.7 | 7.5 | 10.7 | 1.2 | -3.7 | 6.1 |
| **Western Europe** | 2.5 | 2.4 | 4.4 | 4.1 | 3.7 | 0.3 | -0.9 | -2.9 | 4.4 | 3.8 | 0.9 | 4.2 | 3.5 | 1.5 | 4.9 |
| Canada | -0.6 | 4.4 | 6.1 | -0.4 | -2.8 | -3.8 | 1.1 | 4.5 | 6.5 | 4.5 | 1.4 | 4.4 | 2.4 | 4.4 | 5.6 |
| United States | 1.2 | 4.6 | 4.5 | 1.8 | -0.2 | -2.0 | 3.1 | 3.5 | 5.4 | 4.8 | 4.6 | 6.8 | 4.9 | 4.2 | 5.6 |
| **North America** | 1.0 | 4.6 | 4.7 | 1.6 | -0.4 | -2.1 | 3.0 | 3.6 | 5.5 | 4.8 | 4.3 | 6.7 | 4.7 | 4.2 | 5.6 |
| Japan | -0.2 | 3.4 | 9.4 | 5.8 | 4.2 | 1.9 | -5.7 | -3.5 | 1.3 | 3.3 | 2.3 | 3.5 | -6.5 | 0.8 | 5.7 |
| **Total above** | 1.4 | 3.4 | 5.4 | 3.5 | 2.2 | -0.3 | -0.5 | -0.6 | 4.2 | 4.1 | 2.5 | 5.1 | 2.2 | 2.5 | 5.3 |
| *Memorandum items:* | | | | | | | | | | | | | | | |
| 4 major west European economies[d] | 2.0 | 1.8 | 4.8 | 3.9 | 4.0 | 0.5 | -1.4 | -3.9 | 4.5 | 2.3 | 0.2 | 3.2 | 3.0 | 1.1 | 4.4 |
| Western Europe and North America | 1.8 | 3.4 | 4.5 | 2.9 | 1.8 | -0.8 | 0.8 | 0.1 | 4.9 | 4.3 | 2.5 | 5.4 | 4.1 | 2.9 | 5.3 |

*Source:* National statistics; OECD, *Main Economic Indicators* (Paris), various issues; UNECE secretariat estimates.

*Note:* Growth rates of regional aggregates have been calculated as weighted averages of growth rates in individual countries. Weights were derived from 1995 gross value added originating in industry converted from national currency units into dollars using 1995 GDP purchasing power parities.

[a] West Germany, 1986-1991.
[b] Twelve countries above.
[c] Fifteen countries above.
[d] France, Germany, Italy and the United Kingdom.

## APPENDIX TABLE A.9

**Total employment in western Europe, North America and Japan, 1986-2000**

*(Percentage change over preceding year)*

|  | 1986 | 1987 | 1988 | 1989 | 1990 | 1991 | 1992 | 1993 | 1994 | 1995 | 1996 | 1997 | 1998 | 1999 | 2000 |
|---|---|---|---|---|---|---|---|---|---|---|---|---|---|---|---|
| France | 0.4 | 0.8 | 0.9 | 1.7 | 1.0 | 0.1 | -0.5 | -1.2 | -0.1 | 0.8 | 0.3 | 0.3 | 1.2 | 1.8 | 1.9 |
| Germany [a] | 1.4 | 0.7 | 0.8 | 1.5 | 3.0 | 2.5 | -1.5 | -1.4 | -0.2 | 0.2 | -0.3 | -0.2 | 0.9 | 1.1 | 1.5 |
| Italy | 0.7 | 0.2 | 1.1 | 0.7 | 1.6 | 1.9 | -0.5 | -2.5 | -1.5 | -0.1 | 0.6 | 0.4 | 1.0 | 1.2 | 1.4 |
| Austria | 0.4 | – | 0.6 | 1.3 | 1.6 | 1.4 | 0.2 | -0.6 | -0.2 | – | -0.6 | 0.5 | 0.8 | 1.4 | 0.9 |
| Belgium | 0.6 | 0.6 | 1.7 | 1.2 | 0.9 | 0.1 | -0.5 | -0.8 | -0.4 | 0.7 | 0.4 | 0.8 | 1.2 | 1.3 | 1.3 |
| Finland | -0.4 | 0.5 | 1.0 | 0.9 | -0.5 | -5.7 | -7.2 | -6.2 | -1.1 | 1.6 | 1.4 | 3.3 | 2.1 | 2.1 | 1.5 |
| Greece | 0.4 | -0.1 | 1.6 | 0.4 | 1.3 | -2.3 | 1.5 | 0.9 | 1.9 | 0.9 | -0.4 | -0.3 | 3.4 | -0.7 | 1.2 |
| Ireland | 0.5 | 0.6 | 0.3 | -0.1 | 3.3 | -0.1 | 3.1 | 1.5 | 3.2 | 5.4 | 3.8 | 5.6 | 5.0 | 6.4 | 5.0 |
| Luxembourg | 2.5 | 2.7 | -7.0 | 1.5 | 1.6 | 1.3 | 0.2 | -0.5 | 0.7 | 0.9 | 2.3 | 3.2 | 4.4 | 5.1 | 5.2 |
| Netherlands | 2.5 | 1.6 | 2.3 | 1.8 | 3.0 | 2.6 | 1.6 | 0.7 | -0.1 | 2.4 | 2.3 | 3.2 | 3.0 | 2.8 | 2.7 |
| Portugal | 2.7 | 2.3 | 2.2 | 1.9 | 1.7 | 2.8 | -1.6 | -2.0 | -1.0 | -0.2 | 2.3 | 1.7 | 2.7 | 1.8 | 1.5 |
| Spain | 2.2 | 3.1 | 2.9 | 4.1 | 2.6 | 0.2 | -1.9 | -4.3 | -0.9 | 1.8 | 1.3 | 2.8 | 3.7 | 3.5 | 3.1 |
| *Euro area* [b] | 1.1 | 0.9 | 1.3 | 1.6 | 2.0 | 1.3 | -0.9 | -1.7 | -0.4 | 0.7 | 0.5 | 0.8 | 1.7 | 1.7 | 1.8 |
| United Kingdom [c] | 0.2 | 2.0 | 3.6 | 3.2 | 1.0 | -3.2 | -2.6 | -1.3 | 0.6 | 1.1 | 1.0 | 2.0 | 1.2 | 1.0 | 0.6 |
| Denmark | 2.3 | 0.4 | -0.7 | -0.7 | -0.7 | -0.6 | -0.8 | -1.5 | 1.4 | 0.5 | 0.7 | 1.2 | 1.2 | 1.1 | 0.9 |
| Sweden | 0.6 | 0.8 | 1.4 | 1.5 | 0.9 | -1.5 | -4.4 | -5.1 | -0.8 | 1.3 | -0.6 | -1.1 | 1.2 | 2.3 | 2.0 |
| *European Union* [d] | 1.0 | 1.1 | 1.6 | 1.8 | 1.8 | 0.4 | -1.3 | -1.8 | -0.2 | 0.7 | 0.6 | 1.0 | 1.6 | 1.6 | 1.6 |
| Cyprus [e] | 1.1 | 3.1 | 4.7 | 3.9 | 2.8 | 0.6 | 4.5 | -0.1 | 2.3 | 3.4 | 1.0 | -0.2 | 1.2 | 0.8 | .. |
| Iceland [e] | 3.2 | 5.8 | -3.0 | -1.5 | -1.1 | -0.1 | -1.4 | -0.8 | 0.5 | 0.9 | 3.1 | 1.0 | 3.4 | 2.7 | 2.0 |
| Israel | 1.4 | 2.6 | 3.5 | 0.5 | 2.1 | 6.1 | 4.2 | 6.1 | 6.9 | 5.2 | 2.4 | 1.4 | 1.6 | 3.1 | 4.0 |
| Malta [f] | 2.1 | 5.9 | 2.5 | 0.9 | 0.8 | 2.5 | 1.0 | 0.5 | -1.3 | 3.3 | 1.0 | 0.5 | 0.4 | 0.6 | 2.1 |
| Norway | 3.3 | 2.0 | -0.5 | -2.8 | -0.8 | -0.7 | -0.3 | 0.2 | 1.3 | 2.1 | 2.1 | 2.9 | 2.3 | 0.7 | 0.5 |
| Switzerland | 2.3 | 2.5 | 2.6 | 2.7 | 3.2 | 1.8 | -1.6 | -0.7 | -0.3 | 0.3 | 0.3 | -0.3 | 1.2 | 0.6 | 1.4 |
| Turkey | 1.9 | 2.3 | 1.5 | 2.6 | 1.7 | 1.7 | 0.2 | 0.2 | 2.8 | 3.7 | 2.0 | -2.5 | 2.8 | 2.2 | 2.7 |
| *Western Europe* | 1.1 | 1.3 | 1.6 | 1.9 | 1.8 | 0.6 | -1.1 | -1.4 | 0.2 | 1.1 | 0.8 | 0.6 | 1.7 | 1.6 | 1.7 |
| Canada | 3.1 | 2.8 | 3.2 | 2.2 | 0.8 | -1.8 | -0.7 | 0.8 | 2.0 | 1.9 | 0.8 | 2.3 | 2.6 | 2.8 | 2.6 |
| United States | 2.3 | 2.6 | 2.2 | 2.1 | 1.2 | -0.9 | 0.7 | 1.5 | 2.3 | 1.5 | 1.4 | 2.2 | 1.5 | 1.5 | 1.3 |
| *North America* | 2.4 | 2.6 | 2.3 | 2.1 | 1.2 | -1.0 | 0.5 | 1.4 | 2.3 | 1.5 | 1.4 | 2.3 | 1.6 | 1.7 | 1.4 |
| Japan | 0.5 | 0.4 | 1.2 | 1.5 | 1.7 | 2.0 | 1.1 | 0.4 | 0.1 | 0.2 | 0.5 | 1.1 | -0.7 | -0.8 | -0.5 |
| **Total above** | 1.4 | 1.6 | 1.8 | 1.9 | 1.5 | 0.3 | -0.2 | -0.1 | 0.9 | 1.1 | 0.9 | 1.3 | 1.2 | 1.2 | 1.3 |
| *Memorandum items:* | | | | | | | | | | | | | | | |
| 4 major west European economies [g] | 0.7 | 0.9 | 1.6 | 1.8 | 1.8 | 0.4 | -1.3 | -1.6 | -0.2 | 0.5 | 0.3 | 0.6 | 1.1 | 1.2 | 1.3 |
| Western Europe and North America | 1.6 | 1.8 | 1.9 | 2.0 | 1.5 | -0.1 | -0.4 | -0.2 | 1.1 | 1.3 | 1.0 | 1.3 | 1.6 | 1.6 | 1.6 |

*Source:* OECD, *Main Economic Indicators* and *Quarterly Labour Force Statistics* (Paris), latest issues; Eurostat, New Cronos Database; national statistics.

*Note:* Total employment is defined as the number of persons engaged in some productive activity within resident production units (national accounts concept). The labour force survey concept (based on resident household surveys) is used for Canada, Israel, Turkey, the United Kingdom and the United States; Austria (up to 1987); Portugal (up to 1990); Ireland, the Netherlands and Spain (up to 1994). All aggregates exclude Israel.

[a] West Germany, 1986-1991.

[b] Twelve countries above.

[c] Number of jobs.

[d] Fifteen countries above.

[e] Full-time equivalent.

[f] Full-time occupied at the end of the year.

[g] France, Germany, Italy and the United Kingdom.

## APPENDIX TABLE A.10

**Standardized unemployment rates [a] in western Europe, North America and Japan, 1986-2000**
*(Per cent of civilian labour force)*

|  | 1986 | 1987 | 1988 | 1989 | 1990 | 1991 | 1992 | 1993 | 1994 | 1995 | 1996 | 1997 | 1998 | 1999 | 2000 |
|---|---|---|---|---|---|---|---|---|---|---|---|---|---|---|---|
| France | 10.3 | 10.5 | 10.0 | 9.4 | 9.0 | 9.5 | 10.4 | 11.7 | 12.3 | 11.7 | 12.4 | 12.3 | 11.8 | 11.2 | 9.5 |
| Germany [b] | 6.5 | 6.3 | 6.2 | 5.6 | 4.8 | 6.0 | 6.6 | 7.9 | 8.4 | 8.2 | 8.9 | 9.9 | 9.3 | 8.6 | 8.1 |
| Italy | 9.0 | 9.8 | 9.8 | 9.8 | 9.0 | 8.6 | 8.8 | 10.2 | 11.1 | 11.6 | 11.7 | 11.7 | 11.8 | 11.3 | 10.5 |
| Austria | 3.1 | 3.8 | 3.6 | 3.1 | 3.2 | 3.5 | 3.6 | 4.0 | 3.8 | 3.9 | 4.3 | 4.4 | 4.5 | 4.0 | 3.7 |
| Belgium | 10.3 | 10.1 | 9.0 | 7.5 | 6.7 | 6.6 | 7.2 | 8.8 | 10.0 | 9.9 | 9.7 | 9.4 | 9.5 | 8.8 | 7.0 |
| Finland | 6.7 | 4.9 | 4.2 | 3.1 | 3.2 | 6.6 | 11.7 | 16.3 | 16.6 | 15.4 | 14.6 | 12.7 | 11.4 | 10.2 | 9.8 |
| Greece | 6.6 | 6.7 | 6.8 | 6.7 | 6.4 | 7.0 | 7.9 | 8.6 | 8.9 | 9.2 | 9.6 | 9.8 | 10.9 | 11.6 | 11.1 |
| Ireland | 16.8 | 16.6 | 16.2 | 14.7 | 13.4 | 14.7 | 15.4 | 15.6 | 14.3 | 12.3 | 11.7 | 9.9 | 7.5 | 5.6 | 4.2 |
| Luxembourg | 2.6 | 2.5 | 2.0 | 1.8 | 1.7 | 1.7 | 2.1 | 2.6 | 3.2 | 2.9 | 3.0 | 2.7 | 2.7 | 2.4 | 2.2 |
| Netherlands | 8.3 | 8.1 | 7.6 | 6.9 | 6.2 | 5.8 | 5.6 | 6.5 | 7.1 | 6.9 | 6.3 | 5.2 | 4.0 | 3.4 | 3.0 |
| Portugal | 8.8 | 7.3 | 5.9 | 5.2 | 4.8 | 4.2 | 4.3 | 5.7 | 6.9 | 7.3 | 7.3 | 6.8 | 5.2 | 4.5 | 4.1 |
| Spain | 21.2 | 20.6 | 19.5 | 17.2 | 16.3 | 16.4 | 18.4 | 22.7 | 24.2 | 22.9 | 22.2 | 20.8 | 18.8 | 15.9 | 14.1 |
| *Euro area* [c] | 9.9 | 9.9 | 9.5 | 8.8 | 8.1 | 8.2 | 9.1 | 10.8 | 11.5 | 11.2 | 11.5 | 11.5 | 10.8 | 10.0 | 8.9 |
| United Kingdom | 11.6 | 10.6 | 8.7 | 7.3 | 7.1 | 8.8 | 10.0 | 10.5 | 9.6 | 8.7 | 8.2 | 7.0 | 6.3 | 6.1 | 5.6 |
| Denmark | 5.4 | 5.4 | 6.1 | 7.3 | 7.7 | 8.4 | 9.2 | 10.2 | 8.2 | 7.2 | 6.8 | 5.6 | 5.2 | 5.2 | 4.7 |
| Sweden | 2.7 | 2.2 | 1.8 | 1.5 | 1.7 | 3.1 | 5.6 | 9.1 | 9.4 | 8.8 | 9.6 | 9.9 | 8.3 | 7.2 | 5.9 |
| *European Union* [d] | 9.9 | 9.7 | 9.5 | 8.7 | 8.1 | 8.4 | 9.1 | 10.7 | 11.1 | 10.7 | 10.8 | 10.6 | 9.9 | 9.1 | 8.2 |
| Cyprus [e] | 3.7 | 3.4 | 2.8 | 2.3 | 1.8 | 3.0 | 1.8 | 2.6 | 2.7 | 2.6 | 3.1 | 3.4 | 3.4 | 3.6 | .. |
| Iceland | 0.6 | 0.5 | 0.6 | 1.6 | 1.8 | 2.6 | 4.3 | 5.3 | 5.4 | 4.9 | 3.8 | 3.9 | 2.7 | 2.1 | 2.3 |
| Israel [f] | 7.1 | 6.1 | 6.4 | 8.9 | 9.6 | 10.6 | 11.2 | 10.0 | 7.8 | 6.9 | 6.7 | 7.7 | 8.5 | 8.9 | 8.8 |
| Malta [g] | 6.9 | 4.4 | 4.0 | 3.7 | 3.8 | 3.6 | 4.0 | 4.5 | 4.1 | 3.8 | 4.4 | 5.0 | 5.1 | 5.3 | 4.5 |
| Norway | 2.0 | 2.1 | 3.2 | 5.0 | 5.3 | 5.6 | 6.0 | 6.1 | 5.5 | 5.0 | 4.9 | 4.1 | 3.3 | 3.3 | 3.5 |
| Switzerland | 0.8 | 0.7 | 0.6 | 0.5 | 0.5 | 2.0 | 3.1 | 4.0 | 3.8 | 3.5 | 3.9 | 4.2 | 3.5 | 3.0 | 2.7 |
| Turkey [f] | 7.9 | 8.3 | 8.4 | 8.7 | 8.2 | 7.9 | 8.1 | 7.8 | 8.1 | 6.9 | 6.0 | 6.7 | 6.8 | 7.7 | 6.6 |
| *Western Europe* | 0.9 | 1.0 | 9.1 | 8.5 | 7.9 | 8.2 | 8.8 | 10.2 | 10.5 | 10.0 | 10.0 | 9.9 | 9.3 | 8.7 | 8.0 |
| Canada | 9.6 | 8.8 | 7.8 | 7.6 | 8.1 | 10.3 | 11.2 | 11.4 | 10.4 | 9.4 | 9.6 | 9.1 | 8.3 | 7.6 | 6.8 |
| United States | 7.0 | 6.2 | 5.5 | 5.3 | 5.6 | 6.7 | 7.4 | 6.8 | 6.1 | 5.6 | 5.4 | 4.9 | 4.5 | 4.2 | 4.0 |
| *North America* | 7.3 | 6.5 | 5.7 | 5.5 | 5.9 | 7.1 | 7.8 | 7.3 | 6.5 | 6.0 | 5.8 | 5.3 | 4.9 | 4.5 | 4.3 |
| Japan | 2.8 | 2.9 | 2.5 | 2.3 | 2.1 | 2.1 | 2.2 | 2.5 | 2.9 | 3.1 | 3.4 | 3.4 | 4.1 | 4.7 | 4.7 |
| **Total above** | 3.5 | 3.3 | 6.8 | 6.4 | 6.2 | 6.8 | 7.4 | 7.9 | 7.8 | 7.4 | 7.4 | 7.2 | 6.9 | 6.6 | 6.0 |
| *Memorandum items:* |  |  |  |  |  |  |  |  |  |  |  |  |  |  |  |
| 4 major west European economies [h] | 9.3 | 9.2 | 8.6 | 7.9 | 7.3 | 7.3 | 8.7 | 9.8 | 10.1 | 9.7 | 10.0 | 10.1 | 9.6 | 9.1 | 8.3 |
| Western Europe and North America | 3.6 | 3.3 | 7.6 | 7.2 | 7.0 | 7.7 | 8.4 | 8.9 | 8.8 | 8.3 | 8.2 | 7.9 | 7.4 | 6.9 | 6.3 |

*Source:* OECD, *Main Economic Indicators* and *Quarterly Labour Force Statistics* (Paris), latest issues; Eurostat, New Cronos Database; national statistics.

*Note:* All aggregates exclude Israel. Comparisons with previous years are limited by changes in methodology in Austria (1993), Iceland (1991), Israel (1995), Norway (1989), Switzerland (1991) and the United States (1994).

[a] Eurostat-OECD definition except for Austria (1986-1992), Cyprus, Iceland (1986-1990), Israel, Malta, Switzerland (1986-1990) and Turkey.

[b] West Germany, 1986-1991.

[c] Twelve countries above.

[d] Fifteen countries above.

[e] Registered unemployment rate, average of monthly data.

[f] Definitions comply with ILO guidelines but do not follow the Eurostat-OECD standards.

[g] Registered unemployment rate at the end of the year.

[h] France, Germany, Italy and the United Kingdom.

## APPENDIX TABLE A.11

**Consumer prices in western Europe, North America and Japan, 1986-2000**
*(Percentage change over previous year)*

|  | 1986 | 1987 | 1988 | 1989 | 1990 | 1991 | 1992 | 1993 | 1994 | 1995 | 1996 | 1997 | 1998 | 1999 | 2000 |
|---|---|---|---|---|---|---|---|---|---|---|---|---|---|---|---|
| France | 2.5 | 3.3 | 2.7 | 3.5 | 3.5 | 3.2 | 2.4 | 2.1 | 1.7 | 1.8 | 2.0 | 1.2 | 0.8 | 0.5 | 1.7 |
| Germany[a] | -0.1 | 0.3 | 1.3 | 2.8 | 2.7 | 4.0 | 5.1 | 4.4 | 2.8 | 1.7 | 1.4 | 1.9 | 0.9 | 0.6 | 1.9 |
| Italy | 5.8 | 4.8 | 5.1 | 6.3 | 6.5 | 6.3 | 5.3 | 4.6 | 4.1 | 5.2 | 4.0 | 2.0 | 2.0 | 1.6 | 2.6 |
| Austria | 1.7 | 1.5 | 1.9 | 2.6 | 3.3 | 3.3 | 4.0 | 3.6 | 3.0 | 2.2 | 1.5 | 1.3 | 0.9 | 0.6 | 2.3 |
| Belgium | 1.3 | 1.6 | 1.2 | 3.1 | 3.4 | 3.2 | 2.4 | 2.8 | 2.4 | 1.5 | 2.1 | 1.6 | 0.9 | 1.1 | 2.5 |
| Finland | 2.9 | 4.1 | 5.1 | 6.6 | 6.1 | 4.3 | 2.9 | 2.2 | 1.1 | 0.8 | 0.6 | 1.2 | 1.4 | 1.2 | 3.4 |
| Greece | 23.0 | 16.4 | 13.5 | 13.7 | 20.4 | 19.5 | 15.9 | 14.4 | 10.9 | 8.9 | 8.2 | 5.5 | 4.8 | 2.6 | 3.2 |
| Ireland | 3.8 | 3.1 | 2.1 | 4.1 | 3.3 | 3.2 | 3.1 | 1.4 | 2.4 | 2.5 | 1.7 | 1.4 | 2.4 | 1.6 | 5.6 |
| Luxembourg | 0.3 | -0.1 | 1.4 | 3.4 | 3.3 | 3.1 | 3.2 | 3.6 | 2.2 | 1.9 | 1.3 | 1.4 | 1.0 | 1.0 | 3.2 |
| Netherlands | 0.1 | -0.7 | 0.7 | 1.1 | 2.4 | 3.2 | 3.2 | 2.6 | 2.8 | 1.9 | 2.0 | 2.2 | 2.0 | 2.2 | 2.5 |
| Portugal | 11.8 | 9.4 | 9.7 | 12.6 | 13.4 | 10.5 | 9.5 | 6.7 | 5.4 | 4.2 | 3.1 | 2.3 | 2.8 | 2.3 | 2.9 |
| Spain | 8.8 | 5.2 | 4.8 | 6.8 | 6.7 | 5.9 | 5.9 | 4.6 | 4.7 | 4.7 | 3.6 | 2.0 | 1.8 | 2.3 | 3.4 |
| *Euro area*[b] | 3.7 | 3.1 | 3.3 | 4.6 | 4.9 | 5.0 | 4.8 | 4.1 | 3.3 | 3.0 | 2.6 | 1.9 | 1.4 | 1.2 | 2.4 |
| United Kingdom | 3.4 | 4.2 | 4.9 | 7.8 | 9.5 | 5.9 | 3.7 | 1.6 | 2.5 | 3.4 | 2.5 | 3.1 | 3.4 | 1.6 | 2.9 |
| Denmark | 3.7 | 4.0 | 4.5 | 4.8 | 2.6 | 2.4 | 2.1 | 1.3 | 2.0 | 2.1 | 2.1 | 2.2 | 1.8 | 2.5 | 2.9 |
| Sweden | 4.2 | 4.2 | 6.1 | 6.6 | 10.4 | 9.7 | 2.6 | 4.7 | 2.4 | 2.9 | 0.8 | 0.9 | 0.4 | 0.3 | 1.3 |
| *European Union*[c] | 3.7 | 3.3 | 3.6 | 5.1 | 5.7 | 5.2 | 4.5 | 3.7 | 3.1 | 3.1 | 2.5 | 2.1 | 1.7 | 1.3 | 2.4 |
| Cyprus | 1.2 | 2.8 | 3.4 | 3.8 | 4.5 | 5.0 | 6.5 | 4.9 | 4.7 | 2.6 | 2.9 | 3.6 | 2.2 | 1.7 | 4.3 |
| Iceland | 21.2 | 17.8 | 25.7 | 20.8 | 15.9 | 6.8 | 3.7 | 4.1 | 1.5 | 1.7 | 2.3 | 1.8 | 1.7 | 3.4 | 5.0 |
| Israel | 48.1 | 19.9 | 16.3 | 20.2 | 17.2 | 19.0 | 12.0 | 11.0 | 12.3 | 10.1 | 11.3 | 9.0 | 5.4 | 5.2 | 1.1 |
| Malta | 2.0 | 0.5 | 0.9 | 0.9 | 3.0 | 2.5 | 1.6 | 4.1 | 4.1 | 4.0 | 2.5 | 3.2 | 2.2 | 2.1 | 2.3 |
| Norway | 7.2 | 8.7 | 6.7 | 4.5 | 4.1 | 3.4 | 2.3 | 2.3 | 1.4 | 2.4 | 1.3 | 2.6 | 2.3 | 2.3 | 3.1 |
| Switzerland | 0.8 | 1.4 | 1.9 | 3.1 | 5.4 | 5.9 | 4.0 | 3.3 | 0.9 | 1.8 | 0.8 | 0.5 | – | 0.8 | 1.6 |
| Turkey | 34.3 | 37.8 | 69.4 | 63.3 | 60.3 | 65.8 | 70.1 | 66.2 | 105.2 | 89.1 | 80.4 | 85.7 | 84.6 | 64.9 | 54.9 |
| *Western Europe* | 3.6 | 3.3 | 3.6 | 5.1 | 5.7 | 5.2 | 4.5 | 3.6 | 3.0 | 3.0 | 2.5 | 2.0 | 1.7 | 1.3 | 2.4 |
| Canada | 4.1 | 4.4 | 4.0 | 5.0 | 4.8 | 5.6 | 1.5 | 1.8 | 0.2 | 2.2 | 1.6 | 1.6 | 0.9 | 1.7 | 2.7 |
| United States | 1.9 | 3.6 | 4.1 | 4.8 | 5.4 | 4.2 | 3.0 | 3.0 | 2.6 | 2.8 | 3.0 | 2.3 | 1.6 | 2.2 | 3.4 |
| *North America* | 2.0 | 3.7 | 4.1 | 4.8 | 5.4 | 4.3 | 2.9 | 2.9 | 2.4 | 2.8 | 2.9 | 2.2 | 1.5 | 2.2 | 3.3 |
| Japan | 0.6 | 0.1 | 0.7 | 2.3 | 3.1 | 3.3 | 1.7 | 1.2 | 0.7 | -0.1 | 0.1 | 1.7 | 0.6 | -0.3 | -0.6 |
| **Total above** | 2.4 | 3.0 | 3.5 | 4.6 | 5.2 | 4.5 | 3.3 | 3.0 | 2.4 | 2.5 | 2.3 | 2.1 | 1.5 | 1.5 | 2.4 |
| *Memorandum items:* | | | | | | | | | | | | | | | |
| *4 major west European economies*[d] | 2.6 | 2.8 | 3.3 | 4.9 | 5.3 | 4.8 | 4.2 | 3.3 | 2.8 | 2.9 | 2.4 | 2.0 | 1.7 | 1.0 | 2.3 |
| **Western Europe and North America** | 2.7 | 3.5 | 3.9 | 4.9 | 5.5 | 4.7 | 3.6 | 3.2 | 2.7 | 2.9 | 2.7 | 2.2 | 1.6 | 1.8 | 2.9 |

*Source:* National statistics.

*Note:* All aggregates exclude Israel and Turkey. Growth rates of regional aggregates have been calculated as weighted averages of growth rates in individual countries. Weights were derived from 1996 private final consumption expenditure converted from national currency units into dollars using 1996 purchasing power parities.

[a] West Germany, 1986-1991.

[b] Twelve countries above.

[c] Fifteen countries above.

[d] France, Germany, Italy and the United Kingdom.

## APPENDIX TABLE B.1

**Real GDP/NMP in eastern Europe, the Baltic states and the CIS, 1980, 1987-2000**

*(Indices, 1989=100)*

|  | 1980 | 1987 | 1988 | 1989 | 1990 | 1991 | 1992 | 1993 | 1994 | 1995 | 1996 | 1997 | 1998 | 1999 | 2000 |
|---|---|---|---|---|---|---|---|---|---|---|---|---|---|---|---|
| **Eastern Europe** | 88.7 | 99.4 | 100.8 | 100.0 | 93.2 | 82.9 | 79.3 | 79.0 | 82.1 | 86.9 | 90.2 | 92.1 | 94.0 | 95.4 | 99.0 |
| Albania | 79.4 | 92.4 | 91.0 | 100.0 | 90.0 | 64.8 | 60.1 | 65.9 | 71.4 | 80.9 | 88.2 | 82.0 | 88.6 | 95.0 | 102.4 |
| Bosnia and Herzegovina | .. | .. | .. | .. | .. | .. | .. | .. | .. | .. | .. | .. | .. | .. | .. |
| Bulgaria | 76.2 | 99.3 | 101.9 | 100.0 | 90.9 | 83.3 | 77.2 | 76.1 | 77.5 | 79.7 | 71.6 | 66.6 | 68.9 | 70.6 | 74.7 |
| Croatia [a] | 99.0 | 102.5 | 101.6 | 100.0 | 92.9 | 73.3 | 64.7 | 59.5 | 63.0 | 67.3 | 71.3 | 76.2 | 78.1 | 77.8 | 80.7 |
| Czech Republic | .. | 93.7 | 95.7 | 100.0 | 98.8 | 87.3 | 86.9 | 86.9 | 88.9 | 94.1 | 98.2 | 97.4 | 96.3 | 95.9 | 98.7 |
| Hungary | 86.3 | 99.4 | 99.3 | 100.0 | 96.5 | 85.0 | 82.4 | 81.9 | 84.4 | 85.6 | 86.8 | 90.7 | 95.1 | 99.1 | 104.2 |
| Poland | 91.1 | 95.9 | 99.8 | 100.0 | 88.4 | 82.2 | 84.4 | 87.6 | 92.1 | 98.6 | 104.5 | 111.7 | 117.1 | 121.8 | 126.7 |
| Romania | 88.5 | 106.7 | 106.2 | 100.0 | 94.4 | 82.2 | 75.0 | 76.2 | 79.2 | 84.8 | 88.2 | 82.8 | 78.8 | 77.0 | 78.3 |
| Slovakia | .. | 97.1 | 99.0 | 100.0 | 97.5 | 83.3 | 77.9 | 75.1 | 78.7 | 84.0 | 89.3 | 94.8 | 98.7 | 100.6 | 102.8 |
| Slovenia | 98.9 | 103.5 | 100.5 | 100.0 | 91.9 | 83.7 | 79.1 | 81.4 | 85.7 | 89.3 | 92.4 | 96.6 | 100.3 | 105.5 | 110.4 |
| The former Yugoslav Republic of Macedonia | 93.3 | 101.4 | 98.1 | 100.0 | 89.8 | 84.3 | 78.7 | 72.8 | 71.6 | 70.8 | 71.6 | 72.6 | 75.1 | 78.3 | 81.7 |
| Yugoslavia [a] | 95.7 | 100.2 | 98.8 | 100.0 | 92.1 | 81.4 | 58.7 | 40.6 | 41.7 | 44.2 | 46.8 | 50.3 | 51.5 | 42.4 | 45.4 |
| **Baltic states** | 67.8 | 89.0 | 96.0 | 100.0 | 97.8 | 89.9 | 67.9 | 58.2 | 55.2 | 56.5 | 58.8 | 63.7 | 66.7 | 65.6 | 69.1 |
| Estonia | 74.5 | 89.2 | 93.8 | 100.0 | 91.9 | 82.7 | 71.0 | 65.0 | 63.7 | 66.6 | 69.2 | 76.5 | 80.3 | 79.8 | 85.3 |
| Latvia | 68.5 | 89.0 | 93.6 | 100.0 | 102.9 | 92.2 | 60.1 | 51.1 | 51.5 | 51.0 | 52.7 | 57.3 | 59.5 | 60.1 | 64.1 |
| Lithuania | 64.7 | 88.9 | 98.4 | 100.0 | 96.7 | 91.2 | 71.8 | 60.2 | 54.3 | 56.1 | 58.7 | 63.0 | 66.2 | 63.6 | 66.1 |
| **CIS** [b] | 77.5 | 93.9 | 98.1 | 100.0 | 96.8 | 90.9 | 78.0 | 70.4 | 60.3 | 56.9 | 55.0 | 55.6 | 54.0 | 56.4 | 60.8 |
| Armenia | 73.5 | 94.5 | 92.2 | 100.0 | 94.5 | 83.4 | 48.6 | 44.3 | 46.7 | 49.9 | 52.8 | 54.6 | 58.6 | 60.5 | 64.1 |
| Azerbaijan | 79.6 | 105.1 | 109.7 | 100.0 | 88.3 | 87.7 | 67.9 | 52.2 | 41.9 | 37.0 | 37.4 | 39.6 | 43.6 | 46.8 | 52.0 |
| Belarus | 65.7 | 91.3 | 92.4 | 100.0 | 98.1 | 96.9 | 87.6 | 81.0 | 70.8 | 63.4 | 65.2 | 72.6 | 78.7 | 81.4 | 86.1 |
| Georgia | 79.4 | 96.8 | 103.6 | 100.0 | 84.9 | 67.0 | 36.9 | 26.1 | 23.4 | 24.0 | 26.7 | 29.5 | 30.3 | 31.2 | 31.8 |
| Kazakhstan | 87.0 | 92.1 | 100.1 | 100.0 | 99.0 | 88.2 | 83.5 | 75.8 | 66.2 | 60.8 | 61.1 | 62.1 | 60.9 | 62.6 | 68.7 |
| Kyrgyzstan | 69.1 | 84.7 | 95.6 | 100.0 | 104.8 | 96.5 | 83.2 | 70.3 | 56.2 | 53.1 | 56.9 | 62.5 | 63.9 | 66.2 | 69.5 |
| Republic of Moldova | 72.1 | 90.3 | 91.9 | 100.0 | 97.6 | 80.5 | 57.2 | 56.5 | 39.0 | 38.5 | 36.2 | 36.8 | 34.4 | 33.2 | 33.9 |
| Russian Federation | 78.1 | 94.2 | 98.4 | 100.0 | 97.0 | 92.2 | 78.8 | 71.9 | 62.8 | 60.2 | 58.2 | 58.7 | 55.8 | 58.8 | 63.7 |
| Tajikistan | 80.8 | 93.9 | 106.9 | 100.0 | 100.2 | 91.7 | 62.1 | 52.0 | 40.9 | 35.8 | 29.8 | 30.3 | 32.0 | 33.1 | 35.9 |
| Turkmenistan | 80.7 | 97.1 | 107.5 | 100.0 | 101.8 | 97.0 | 82.5 | 83.7 | 69.2 | 64.2 | 68.5 | 60.7 | 63.8 | 74.0 | 87.0 |
| Ukraine | 75.0 | 93.4 | 95.2 | 100.0 | 96.4 | 88.0 | 79.3 | 68.0 | 52.5 | 46.1 | 41.5 | 40.2 | 39.4 | 39.3 | 41.6 |
| Uzbekistan | 76.0 | 88.4 | 97.0 | 100.0 | 99.2 | 98.7 | 87.7 | 85.7 | 81.2 | 80.5 | 81.9 | 86.1 | 89.9 | 93.9 | 97.6 |
| **Total above** | 80.3 | 95.2 | 98.7 | 100.0 | 95.9 | 88.8 | 78.2 | 72.4 | 65.9 | 64.8 | 64.3 | 65.3 | 64.7 | 66.8 | 71.0 |
| *Memorandum items:* | | | | | | | | | | | | | | | |
| CETE-5 | 88.6 | 97.3 | 99.6 | 100.0 | 93.3 | 84.1 | 83.8 | 85.0 | 88.5 | 93.5 | 97.9 | 102.4 | 105.8 | 109.0 | 113.2 |
| SETE-7 | 88.8 | 103.1 | 102.9 | 100.0 | 93.1 | 80.8 | 71.4 | 68.5 | 70.9 | 75.3 | 76.7 | 74.1 | 73.2 | 71.5 | 74.0 |
| Former GDR | .. | .. | .. | 100.0 | 84.5 | 68.3 | 73.3 | 80.1 | 87.9 | 91.9 | 94.9 | 96.7 | 97.7 | 99.0 | 100.1 |

*Source:* UNECE Common Database, derived from national and CIS statistics.

*Note:* Data for the east European countries are based on a GDP measure, except where otherwise mentioned. For the countries of the former Soviet Union, NMP data for 1980-1990 were chain-linked to GDP data from 1990. Country indices were aggregated with previous year PPP-based weights obtained from the European Comparison Programme for 1996.

[a] Gross material product (1980-1989 for Croatia).

[b] Net material product for 1980-1990 (until 1992 in the case of Turkmenistan).

## APPENDIX TABLE B.2
### Real total consumption expenditure in eastern Europe, the Baltic states and the CIS, 1980, 1987-2000
*(Indices, 1989=100 or earliest year available thereafter)*

|  | 1980 | 1987 | 1988 | 1989 | 1990 | 1991 | 1992 | 1993 | 1994 | 1995 | 1996 | 1997 | 1998 | 1999 | 2000 |
|---|---|---|---|---|---|---|---|---|---|---|---|---|---|---|---|
| Bulgaria | .. | .. | .. | 100.0 | 100.6 | 92.3 | 89.4 | 86.2 | 82.3 | 80.7 | 75.3 | 64.0 | 68.8 | 72.0 | 75.3 |
| Croatia | .. | .. | .. | .. | .. | 100.0 | 87.2 | 85.3 | 92.0 | 106.6 | 106.5 | 117.1 | 118.2 | 116.2 | 119.3 |
| Czech Republic | .. | 91.3 | 93.1 | 100.0 | 104.9 | 85.5 | 88.4 | 90.2 | 94.5 | 97.2 | 103.7 | 104.3 | 102.1 | 103.5 | 104.6 |
| Hungary | 92.2 | 104.9 | 102.0 | 100.0 | 97.3 | 92.2 | 92.8 | 97.9 | 95.6 | 89.3 | 86.1 | 88.1 | 91.7 | 95.6 | 98.6 |
| Poland | 108.0 | 111.8 | 114.7 | 100.0 | 88.3 | 94.9 | 98.2 | 103.0 | 107.0 | 110.5 | 118.4 | 125.6 | 130.8 | 136.6 | 139.8 |
| Romania | 83.9 | 88.7 | 90.6 | 100.0 | 108.9 | 96.0 | 90.7 | 91.8 | 95.3 | 105.5 | 112.9 | 108.1 | 109.3 | 104.4 | 106.2 |
| Slovakia | .. | 89.2 | 92.1 | 100.0 | 103.3 | 76.9 | 75.6 | 74.2 | 72.2 | 74.4 | 83.0 | 87.2 | 91.8 | 89.8 | 87.4 |
| Slovenia | .. | .. | .. | .. | 100.0 | 91.6 | 88.8 | 99.1 | 102.6 | 110.2 | 112.7 | 116.3 | 120.9 | 127.7 | 129.5 |
| The former Yugoslav Republic of Macedonia | .. | .. | .. | .. | 100.0 | 93.9 | 84.2 | 89.7 | 95.9 | 94.3 | 96.5 | 98.5 | 101.8 | 105.6 | .. |
| Estonia | .. | .. | .. | .. | .. | .. | .. | 100.0 | 102.1 | 110.9 | 116.7 | 124.7 | 131.1 | 131.9 | 139.7 |
| Latvia | .. | .. | .. | .. | 100.0 | 76.7 | 49.2 | 46.5 | 47.4 | 47.0 | 50.8 | 52.7 | 56.0 | 58.2 | 60.4 |
| Lithuania | .. | .. | .. | .. | .. | .. | .. | .. | .. | 100.0 | 108.2 | 116.4 | 125.4 | 122.4 | 126.8 |
| Armenia | .. | .. | .. | .. | 100.0 | 97.4 | 84.9 | 66.4 | 68.9 | 74.5 | 76.8 | 81.7 | 85.4 | 86.6 | 92.0 |
| Azerbaijan | .. | .. | .. | .. | .. | .. | 100.0 | 77.4 | 62.2 | 60.4 | 65.3 | 72.2 | 80.4 | .. | .. |
| Belarus | .. | .. | .. | .. | 100.0 | 93.5 | 84.1 | 78.8 | 70.0 | 63.4 | 65.7 | 72.3 | 80.9 | 87.7 | .. |
| Georgia | .. | .. | .. | .. | 100.0 | 79.2 | 77.1 | 45.4 | 42.4 | 46.1 | .. | .. | .. | .. | .. |
| Kazakhstan | .. | .. | .. | .. | 100.0 | 96.8 | 96.2 | 84.9 | 67.7 | 55.0 | 51.3 | 51.8 | 50.4 | 53.6 | .. |
| Kyrgyzstan | .. | .. | .. | .. | 100.0 | 83.5 | 72.8 | 64.3 | 51.8 | 43.4 | 46.2 | 42.4 | 48.8 | 49.3 | .. |
| Republic of Moldova | .. | .. | .. | .. | .. | .. | .. | 100.0 | 82.6 | 90.3 | 99.7 | 111.5 | 109.3 | 92.0 | 100.5 |
| Russian Federation | .. | .. | .. | .. | 100.0 | 93.9 | 89.0 | 88.1 | 85.8 | 83.3 | 80.7 | 82.7 | 81.5 | 79.5 | 85.1 |
| Ukraine | .. | .. | .. | .. | 100.0 | 94.7 | 88.7 | 72.1 | 65.1 | 62.7 | 57.5 | 56.4 | 56.4 | 55.7 | .. |

*Source:* UNECE Common Database, derived from national and CIS statistics.

## APPENDIX TABLE B.3
### Real gross fixed capital formation in eastern Europe, the Baltic states and the CIS, 1980, 1987-2000
*(Indices, 1989=100 or earliest year available thereafter)*

|  | 1980 | 1987 | 1988 | 1989 | 1990 | 1991 | 1992 | 1993 | 1994 | 1995 | 1996 | 1997 | 1998 | 1999 | 2000 |
|---|---|---|---|---|---|---|---|---|---|---|---|---|---|---|---|
| Bulgaria | .. | .. | .. | .. | 100.0 | 80.0 | 74.1 | 61.2 | 61.9 | 71.8 | 56.6 | 43.0 | 57.2 | 71.7 | 77.6 |
| Croatia | .. | .. | .. | .. | .. | 100.0 | 88.5 | 94.5 | 93.6 | 108.2 | 148.8 | 183.5 | 189.1 | 187.0 | 180.5 |
| Czech Republic | .. | 93.4 | 99.4 | 100.0 | 97.9 | 71.1 | 82.8 | 83.0 | 90.5 | 108.5 | 117.3 | 113.9 | 114.0 | 113.3 | 118.1 |
| Hungary | 114.7 | 110.7 | 100.6 | 100.0 | 92.9 | 83.1 | 81.0 | 82.6 | 92.9 | 88.9 | 94.8 | 103.6 | 117.3 | 124.2 | 132.5 |
| Poland | 124.6 | 116.5 | 126.5 | 100.0 | 75.2 | 71.9 | 73.6 | 75.7 | 82.6 | 96.2 | 115.2 | 140.1 | 160.0 | 170.9 | 176.3 |
| Romania | 163.7 | 161.0 | 157.6 | 100.0 | 64.4 | 44.0 | 48.9 | 52.9 | 63.9 | 68.3 | 72.2 | 73.4 | 69.2 | 66.3 | 70.0 |
| Slovakia | .. | .. | .. | .. | 100.0 | 74.8 | 71.5 | 67.7 | 64.3 | 67.7 | 89.4 | 100.1 | 111.2 | 90.3 | 89.7 |
| Slovenia | .. | .. | .. | .. | 100.0 | 88.5 | 77.1 | 85.4 | 97.4 | 113.8 | 123.9 | 138.2 | 153.9 | 183.2 | 183.6 |
| The former Yugoslav Republic of Macedonia | .. | .. | .. | .. | 100.0 | 95.8 | 79.9 | 73.6 | 67.3 | 74.1 | 79.0 | 75.6 | 73.6 | 72.6 | .. |
| Estonia | .. | .. | .. | .. | .. | .. | .. | 100.0 | 106.2 | 110.5 | 123.1 | 144.6 | 161.0 | 137.5 | 140.2 |
| Latvia | .. | .. | .. | .. | 100.0 | 36.1 | 25.7 | 21.6 | 21.8 | 23.7 | 29.0 | 35.0 | 50.4 | 48.4 | 53.6 |
| Lithuania | .. | .. | .. | .. | .. | .. | .. | .. | .. | 100.0 | 117.6 | 143.6 | 157.8 | 147.9 | 142.2 |
| Armenia | .. | .. | .. | .. | 100.0 | 67.0 | 8.6 | 7.9 | 11.5 | 9.5 | 10.5 | 10.7 | 12.0 | 12.1 | 13.5 |
| Azerbaijan | .. | .. | .. | .. | .. | .. | 100.0 | 61.0 | 115.3 | 94.5 | 199.8 | 333.7 | 483.9 | .. | .. |
| Belarus | .. | .. | .. | .. | 100.0 | 104.2 | 84.8 | 78.3 | 67.6 | 47.6 | 46.1 | 56.2 | 61.8 | 59.3 | .. |
| Georgia | .. | .. | .. | .. | 100.0 | 67.3 | 49.2 | 18.5 | 133.4 | 219.9 | .. | .. | .. | .. | .. |
| Kazakhstan | .. | .. | .. | .. | 100.0 | 74.2 | 61.9 | 44.2 | 39.2 | 24.4 | 18.5 | 19.2 | 16.1 | 16.2 | .. |
| Kyrgyzstan | .. | .. | .. | .. | 100.0 | 89.4 | 63.2 | 49.4 | 35.1 | 56.4 | 49.1 | 34.6 | 34.0 | 43.6 | .. |
| Republic of Moldova | .. | .. | .. | .. | .. | .. | .. | 100.0 | 56.5 | 50.8 | 63.8 | 60.4 | 66.0 | 50.8 | 53.4 |
| Russian Federation | .. | .. | .. | .. | 100.0 | 84.5 | 49.4 | 36.7 | 27.1 | 24.6 | 20.4 | 18.9 | 17.1 | 17.9 | 20.7 |
| Ukraine | .. | .. | .. | .. | 100.0 | 79.1 | 67.4 | 46.8 | 27.6 | 19.1 | 14.8 | 15.1 | 15.5 | 15.5 | .. |

*Source:* UNECE Common Database, derived from national and CIS statistics.

APPENDIX TABLE B.4

**Real gross industrial output in eastern Europe, the Baltic states and the CIS, 1980, 1987-2000**
*(Indices, 1989=100)*

|  | 1980 | 1987 | 1988 | 1989 | 1990 | 1991 | 1992 | 1993 | 1994 | 1995 | 1996 | 1997 | 1998 | 1999 | 2000 |
|---|---|---|---|---|---|---|---|---|---|---|---|---|---|---|---|
| **Eastern Europe** | 82.8 | 98.5 | 100.6 | 100.0 | 85.9 | 70.2 | 63.0 | 61.5 | 65.5 | 70.3 | 73.9 | 77.5 | 78.6 | 78.3 | 84.9 |
| Albania | 77.0 | 93.3 | 95.2 | 100.0 | 86.7 | 50.4 | 35.2 | 31.7 | 25.8 | 23.9 | 18.1 | 18.6 | 22.7 | 26.3 | 29.4 |
| Bosnia and Herzegovina | 106.0 | 101.1 | 98.1 | 100.0 | 101.8 | 76.9 | 25.5 | 2.0 | 1.7 | 2.8 | 5.2 | 7.0 | 8.7 | 9.6 | 10.5 |
| Bulgaria | 71.3 | 98.0 | 101.1 | 100.0 | 83.2 | 66.4 | 54.2 | 48.8 | 54.0 | 56.4 | 59.3 | 53.4 | 49.1 | 44.6 | 47.1 |
| Croatia | 88.7 | 102.0 | 100.6 | 100.0 | 88.7 | 63.4 | 54.2 | 51.0 | 49.6 | 49.7 | 51.3 | 54.8 | 56.8 | 56.0 | 57.0 |
| Czech Republic | 81.5 | 96.5 | 98.5 | 100.0 | 96.6 | 75.7 | 69.8 | 66.1 | 67.4 | 73.3 | 74.8 | 78.1 | 79.4 | 76.9 | 81.1 |
| Hungary | 92.9 | 106.4 | 105.3 | 100.0 | 90.7 | 74.0 | 66.8 | 69.5 | 76.2 | 79.7 | 82.4 | 91.5 | 103.0 | 113.7 | 134.5 |
| Poland | 86.3 | 95.5 | 100.5 | 100.0 | 75.8 | 69.7 | 71.7 | 76.3 | 85.5 | 93.8 | 101.6 | 113.3 | 117.3 | 121.5 | 129.7 |
| Romania | 76.9 | 99.2 | 101.9 | 100.0 | 81.9 | 63.3 | 49.4 | 50.1 | 51.7 | 56.6 | 60.1 | 55.8 | 48.1 | 44.3 | 47.9 |
| Slovakia | 76.7 | 98.6 | 100.8 | 100.0 | 94.0 | 75.9 | 68.6 | 66.1 | 69.3 | 75.1 | 76.9 | 77.9 | 80.9 | 78.4 | 85.4 |
| Slovenia | 90.3 | 101.6 | 98.9 | 100.0 | 89.5 | 78.4 | 68.1 | 66.1 | 70.4 | 71.8 | 72.5 | 73.2 | 75.9 | 75.6 | 80.2 |
| The former Yugoslav Republic of Macedonia | 72.1 | 97.3 | 95.6 | 100.0 | 89.4 | 74.0 | 62.3 | 53.7 | 48.0 | 42.9 | 44.3 | 45.0 | 47.0 | 45.8 | 47.4 |
| Yugoslavia | 80.0 | 97.6 | 98.4 | 100.0 | 88.0 | 72.5 | 57.0 | 35.7 | 36.2 | 37.6 | 40.4 | 44.2 | 45.8 | 35.2 | 39.1 |
| **Baltic states** | 72.1 | 92.7 | 96.6 | 100.0 | 98.6 | 95.1 | 64.7 | 44.4 | 36.3 | 37.1 | 38.9 | 42.1 | 44.6 | 41.0 | 44.1 |
| Estonia | 78.5 | 96.3 | 99.3 | 100.0 | 100.0 | 92.8 | 59.8 | 48.6 | 47.1 | 48.0 | 49.4 | 56.6 | 59.0 | 57.0 | 64.2 |
| Latvia | 72.5 | 93.7 | 97.1 | 100.0 | 100.8 | 100.2 | 65.6 | 44.6 | 40.1 | 38.7 | 40.8 | 46.4 | 47.9 | 45.3 | 46.7 |
| Lithuania | 70.0 | 91.2 | 95.6 | 100.0 | 97.4 | 94.0 | 65.8 | 43.2 | 31.7 | 33.4 | 35.0 | 36.2 | 39.2 | 34.8 | 37.2 |
| **CIS** | 73.4 | 94.5 | 98.2 | 100.0 | 99.9 | 93.1 | 78.2 | 68.4 | 53.6 | 50.7 | 49.0 | 50.2 | 48.7 | 53.2 | 59.4 |
| Armenia | 76.3 | 110.3 | 109.1 | 100.0 | 92.5 | 85.4 | 44.2 | 39.5 | 41.6 | 42.2 | 42.8 | 43.2 | 42.3 | 44.5 | 47.4 |
| Azerbaijan | 76.1 | 96.1 | 99.4 | 100.0 | 93.7 | 85.4 | 59.4 | 47.7 | 35.9 | 28.2 | 26.3 | 26.4 | 27.0 | 28.0 | 29.9 |
| Belarus | 61.1 | 89.9 | 95.6 | 100.0 | 102.1 | 101.1 | 91.8 | 83.2 | 71.0 | 62.7 | 64.9 | 77.1 | 86.7 | 95.6 | 103.0 |
| Georgia | 70.6 | 96.2 | 99.3 | 100.0 | 94.3 | 73.0 | 39.6 | 25.0 | 15.2 | 13.2 | 14.1 | 15.2 | 15.0 | 16.1 | 17.1 |
| Kazakhstan | 72.4 | 94.1 | 97.6 | 100.0 | 99.2 | 98.3 | 84.7 | 72.2 | 51.9 | 47.7 | 47.8 | 49.7 | 48.5 | 49.8 | 57.5 |
| Kyrgyzstan | 66.7 | 89.0 | 95.1 | 100.0 | 99.4 | 99.1 | 73.5 | 56.3 | 35.5 | 26.7 | 27.8 | 38.8 | 40.9 | 39.1 | 41.4 |
| Republic of Moldova | 68.7 | 91.6 | 94.6 | 100.0 | 103.2 | 91.7 | 66.9 | 67.1 | 48.5 | 46.6 | 43.6 | 43.6 | 37.0 | 32.7 | 33.5 |
| Russian Federation | 74.4 | 95.0 | 98.6 | 100.0 | 99.9 | 91.9 | 75.4 | 64.7 | 51.2 | 49.5 | 47.5 | 48.5 | 46.0 | 51.0 | 57.1 |
| Tajikistan | 72.9 | 93.1 | 98.2 | 100.0 | 101.2 | 97.6 | 73.9 | 68.1 | 50.8 | 43.9 | 33.4 | 32.7 | 35.4 | 37.4 | 41.2 |
| Turkmenistan | 75.4 | 92.9 | 96.9 | 100.0 | 103.2 | 108.2 | 92.0 | 95.7 | 72.1 | 67.5 | 79.5 | 53.8 | 54.0 | 62.0 | 79.8 |
| Ukraine | 72.6 | 93.4 | 97.3 | 100.0 | 99.9 | 95.1 | 89.0 | 81.9 | 59.5 | 52.4 | 49.7 | 49.6 | 49.1 | 51.2 | 57.8 |
| Uzbekistan | 68.5 | 93.4 | 96.5 | 100.0 | 101.8 | 103.3 | 96.4 | 99.9 | 101.5 | 101.6 | 104.2 | 108.5 | 112.4 | 119.2 | 126.9 |
| **Total above** | 76.5 | 95.8 | 99.0 | 100.0 | 95.2 | 85.4 | 72.7 | 65.5 | 57.2 | 57.0 | 57.1 | 59.2 | 58.7 | 61.4 | 67.6 |
| *Memorandum items:* | | | | | | | | | | | | | | | |
| **CETE-5** | 85.4 | 98.3 | 100.7 | 100.0 | 87.0 | 73.3 | 69.8 | 70.6 | 76.3 | 82.4 | 86.5 | 93.7 | 97.8 | 100.2 | 108.9 |
| **SETE-7** | 78.6 | 98.8 | 100.6 | 100.0 | 84.1 | 65.4 | 52.0 | 47.0 | 48.3 | 51.1 | 53.9 | 51.8 | 48.0 | 43.5 | 46.7 |
| Former GDR | 75.2 | 94.7 | 97.7 | 100.0 | 72.7 | 37.0 | 34.7 | 34.9 | 38.1 | 39.9 | 41.7 | 44.6 | 48.2 | 51.7 | 57.2 |

*Source:* UNECE Common Database, derived from national and CIS statistics.

*Note:* For the countries of the former Soviet Union, data for 1980-1990 were chain-linked to national or CIS data from 1990. Country indices were aggregated with previous year PPP-based weights on the basis of data obtained from the European Comparison Programme for 1996.

## APPENDIX TABLE B.5

**Total employment in eastern Europe, the Baltic states and the CIS, 1980, 1987-2000**

*(Indices, 1989=100)*

|  | 1980 | 1987 | 1988 | 1989 | 1990 | 1991 | 1992 | 1993 | 1994 | 1995 | 1996 | 1997 | 1998 | 1999 | 2000 |
|---|---|---|---|---|---|---|---|---|---|---|---|---|---|---|---|
| **Eastern Europe** | 96.7 | 100.0 | 99.9 | 100.0 | 97.1 | 90.7 | 85.4 | 82.7 | 82.6 | 82.3 | 82.9 | 83.0 | 83.3 | 81.0 | .. |
| Albania | 77.9 | 95.9 | 97.6 | 100.0 | 99.2 | 97.5 | 76.0 | 72.7 | 80.7 | 80.4 | 77.5 | 76.9 | 75.4 | 74.0 | 74.2 |
| Bosnia and Herzegovina | .. | .. | .. | 100.0 | 97.1 | 58.1 | 22.1 | 9.9 | 9.1 | 10.1 | 22.5 | 34.4 | 36.4 | 37.6 | 38.0 |
| Bulgaria | 100.0 | 102.8 | 102.4 | 100.0 | 93.9 | 81.6 | 75.0 | 73.8 | 74.3 | 75.2 | 75.3 | 72.3 | 72.2 | 70.7 | 67.4 |
| Croatia | 87.4 | 100.6 | 100.4 | 100.0 | 97.1 | 89.2 | 79.3 | 76.6 | 74.8 | 73.9 | 74.5 | 73.9 | 78.8 | 78.5 | 77.0 |
| Czech Republic | 95.3 | 98.9 | 99.4 | 100.0 | 99.1 | 93.6 | 91.2 | 89.7 | 90.4 | 92.8 | 93.4 | 91.6 | 90.1 | 86.9 | 84.9 |
| Hungary [a] | 104.2 | 101.7 | 100.7 | 100.0 | 96.7 | 86.7 | 78.1 | 73.2 | 71.8 | 70.4 | 69.8 | 69.8 | 70.7 | 72.9 | 73.6 |
| Poland | 102.0 | 100.8 | 100.1 | 100.0 | 95.8 | 90.1 | 86.3 | 84.3 | 85.1 | 86.7 | 88.3 | 90.8 | 92.9 | 90.4 | 90.0 |
| Romania [b] | 94.6 | 97.9 | 98.7 | 100.0 | 99.0 | 98.5 | 95.5 | 91.9 | 91.5 | 86.7 | 85.7 | 82.4 | 80.5 | 76.9 | .. |
| Slovakia [c] | 90.8 | 99.0 | 99.8 | 100.0 | 98.2 | 85.9 | 86.8 | 87.7 | 84.0 | 85.7 | 88.6 | 88.1 | 87.8 | 85.1 | 83.9 |
| Slovenia [d] | 84.0 | 101.9 | 101.3 | 100.0 | 96.1 | 88.7 | 83.8 | 81.3 | 79.3 | 79.1 | 78.7 | 78.6 | 78.7 | 80.2 | 81.2 |
| The former Yugoslav Republic of Macedonia | 81.2 | 99.9 | 99.7 | 100.0 | 98.2 | 90.7 | 86.4 | 81.5 | 76.6 | 69.0 | 65.8 | 61.8 | 60.1 | 61.1 | 60.3 |
| Yugoslavia | 83.4 | 99.0 | 99.8 | 100.0 | 97.0 | 94.1 | 90.9 | 88.3 | 86.5 | 85.3 | 84.8 | 89.9 | 89.7 | 82.4 | 80.2 |
| **Baltic states** [e] | 94.9 | 99.4 | 99.6 | 100.0 | 98.5 | 98.9 | 94.4 | 89.0 | 83.2 | 80.6 | 80.1 | 80.9 | 80.5 | 79.5 | 77.9 |
| Estonia | .. | 97.6 | 97.6 | 100.0 | 98.6 | 96.4 | 91.4 | 84.5 | 82.7 | 78.3 | 77.0 | 77.4 | 76.4 | 73.3 | 72.6 |
| Latvia | 97.0 | 100.4 | 100.5 | 100.0 | 100.1 | 99.3 | 92.0 | 85.6 | 77.0 | 74.3 | 72.3 | 73.7 | 74.1 | 73.8 | 73.8 |
| Lithuania | 93.4 | 99.5 | 99.8 | 100.0 | 97.3 | 99.7 | 97.5 | 93.4 | 88.0 | 86.4 | 87.2 | 87.7 | 87.0 | 86.6 | 83.3 |
| **CIS** | 93.8 | 98.4 | 98.8 | 100.0 | 100.2 | 98.9 | 96.6 | 94.2 | 91.4 | 90.5 | 89.7 | 88.5 | 87.5 | .. | .. |
| Armenia | 86.6 | 99.5 | 101.4 | 100.0 | 102.4 | 105.0 | 99.2 | 97.0 | 93.5 | 92.8 | 90.2 | 86.2 | 84.0 | 81.6 | 80.3 |
| Azerbaijan | 62.7 | 74.7 | 75.2 | 100.0 | 100.9 | 101.7 | 101.4 | 101.2 | 98.9 | 98.4 | 100.5 | 100.7 | 100.9 | 100.9 | 100.9 |
| Belarus | 95.4 | 99.0 | 99.5 | 100.0 | 99.1 | 96.6 | 94.1 | 92.9 | 90.4 | 84.8 | 84.0 | 84.1 | 85.0 | 85.5 | 85.4 |
| Georgia | 92.7 | 101.0 | 101.1 | 100.0 | 102.3 | 93.3 | 73.5 | 66.4 | 64.8 | 79.0 | 75.4 | 82.7 | 84.6 | .. | .. |
| Kazakhstan | 86.2 | 94.7 | 96.0 | 100.0 | 101.3 | 100.1 | 98.3 | 89.9 | 85.4 | 85.0 | 84.6 | 84.0 | 79.5 | 79.2 | 80.5 |
| Kyrgyzstan | 81.9 | 97.9 | 98.7 | 100.0 | 100.5 | 99.6 | 105.6 | 96.6 | 94.6 | 94.4 | 95.0 | 97.1 | 98.0 | 101.5 | 101.7 |
| Republic of Moldova [f] | 97.3 | 99.7 | 98.9 | 100.0 | 99.1 | 99.0 | 98.0 | 80.7 | 80.4 | 80.0 | 79.4 | 78.7 | 78.5 | 71.5 | 72.5 |
| Russian Federation | 96.9 | 99.7 | 99.9 | 100.0 | 99.6 | 97.7 | 95.3 | 93.7 | 90.6 | 87.9 | 87.2 | 85.5 | 84.2 | 84.6 | 85.1 |
| Tajikistan | 76.7 | 94.5 | 96.9 | 100.0 | 103.2 | 104.9 | 101.6 | 98.7 | 98.7 | 98.6 | 92.1 | 95.3 | 95.6 | 92.5 | 92.9 |
| Turkmenistan | 79.8 | 98.5 | 101.2 | 100.0 | 103.4 | 107.0 | 110.5 | 114.0 | 118.5 | 122.5 | 124.7 | 127.2 | 128.8 | 133.6 | .. |
| Ukraine | 99.6 | 100.3 | 99.9 | 100.0 | 99.9 | 98.3 | 96.3 | 94.1 | 90.5 | 93.3 | 91.3 | 88.8 | 87.9 | 85.8 | 83.6 |
| Uzbekistan | 75.4 | 93.2 | 95.9 | 100.0 | 104.2 | 109.2 | 108.7 | 108.5 | 109.9 | 110.8 | 112.3 | 113.8 | 115.4 | 116.5 | 117.8 |
| **Total above** | 94.6 | 98.8 | 99.1 | 100.0 | 99.3 | 96.6 | 93.4 | 90.9 | 88.8 | 88.0 | 87.6 | 86.8 | 86.2 | .. | .. |
| *Memorandum items:* | | | | | | | | | | | | | | | |
| **CETE-5** | 99.7 | 100.5 | 100.1 | 100.0 | 96.7 | 89.8 | 85.8 | 83.6 | 83.5 | 84.7 | 85.8 | 86.8 | 87.9 | 86.1 | 85.6 |
| **SETE-7** | 92.3 | 99.2 | 99.7 | 100.0 | 97.6 | 91.9 | 84.9 | 81.5 | 81.4 | 79.0 | 78.8 | 77.7 | 77.0 | 74.1 | .. |

*Source:* UNECE Common Database, derived from national and CIS statistics.

[a] End of year, up to 1992; since 1992, annual average.

[b] End of year.

[c] End of year up to 1993; since 1993 annual average.

[d] Self-employed excluded until 1987.

[e] Excluding Estonia until 1985.

[f] Excluding Transdniestria since 1993.

APPENDIX TABLE B.6

**Employment in industry in eastern Europe, the Baltic states and the CIS, 1989-2000**
*(Indices, 1989=100)*

|  | 1989 | 1990 | 1991 | 1992 | 1993 | 1994 | 1995 | 1996 | 1997 | 1998 | 1999 | 2000 |
|---|---|---|---|---|---|---|---|---|---|---|---|---|
| **Eastern Europe**[a] | 100.0 | 95.6 | 86.7 | 76.7 | 71.9 | 69.5 | 68.6 | 68.3 | 66.2 | 65.1 | 60.9 | .. |
| Albania | .. | .. | .. | .. | .. | .. | .. | .. | .. | .. | .. | .. |
| Bosnia and Herzegovina | 100.0 | 98.3 | 60.3 | 23.3 | 8.3 | 9.2 | 18.7 | 22.0 | 26.8 | 28.9 | 28.9 | 29.0 |
| Bulgaria | 100.0 | 91.0 | 74.7 | 64.8 | 59.5 | 57.3 | 56.0 | 55.4 | 53.0 | 50.8 | 46.1 | 41.2 |
| Croatia | 100.0 | 102.4 | 84.8 | 70.3 | 70.4 | 67.3 | 59.5 | 59.0 | 56.6 | 60.0 | 58.3 | 57.1 |
| Czech Republic | 100.0 | 95.8 | 92.2 | 85.1 | 80.9 | 76.6 | 77.0 | 76.4 | 76.1 | 75.8 | 73.4 | 71.3 |
| Hungary | 100.0 | 97.0 | 85.4 | 77.3 | 69.1 | 66.0 | 62.5 | 61.9 | 63.0 | 65.9 | 66.4 | 65.6 |
| Poland | 100.0 | 93.7 | 86.1 | 76.7 | 74.4 | 73.8 | 76.2 | 75.6 | 75.8 | 75.0 | 69.5 | 67.8 |
| Romania[b] | 100.0 | 96.5 | 91.6 | 79.5 | 73.0 | 69.4 | 65.4 | 66.0 | 59.0 | 55.8 | 49.5 | .. |
| Slovakia | 100.0 | 95.7 | 88.2 | 78.7 | 74.0 | 71.7 | 71.6 | 71.6 | 70.2 | 67.3 | 65.4 | 63.3 |
| Slovenia | 100.0 | 95.1 | 85.2 | 84.6 | 78.1 | 75.1 | 72.2 | 71.5 | 68.5 | 67.8 | 66.7 | 66.3 |
| The former Yugoslav Republic of Macedonia | 100.0 | 95.3 | 87.3 | 81.6 | 77.5 | 72.9 | 63.1 | 59.0 | 54.4 | 52.5 | 55.4 | 52.9 |
| Yugoslavia | 100.0 | 100.9 | 92.1 | 87.2 | 85.0 | 82.7 | 80.6 | 78.8 | 76.3 | 73.8 | 66.3 | .. |
| **Baltic states** | 100.0 | 96.8 | 95.5 | 87.0 | 74.2 | 63.0 | 60.8 | 58.0 | 57.7 | 55.8 | 54.0 | 54.2 |
| Estonia | 100.0 | 96.9 | 93.0 | 85.8 | 73.6 | 70.7 | 76.4 | 73.0 | 68.8 | 67.5 | 63.4 | 65.6 |
| Latvia | 100.0 | 97.0 | 92.1 | 81.4 | 69.0 | 56.3 | 53.1 | 50.1 | 51.9 | 47.6 | 45.7 | 46.7 |
| Lithuania | 100.0 | 96.7 | 99.0 | 91.4 | 78.0 | 64.5 | 59.5 | 57.0 | 57.1 | 56.5 | 55.7 | 54.6 |
| **CIS** | 100.0 | 98.2 | 96.7 | 91.8 | 88.3 | 79.7 | 73.3 | 69.2 | 63.8 | 62.6 | .. | .. |
| Armenia | 100.0 | 102.6 | 95.0 | 84.0 | 75.2 | 73.7 | 62.8 | 52.9 | 47.5 | 43.4 | 40.5 | 37.3 |
| Azerbaijan | 100.0 | 97.1 | 94.9 | 88.9 | 81.1 | 77.4 | 72.9 | 58.6 | 50.1 | 52.0 | 53.6 | 51.6 |
| Belarus | 100.0 | 98.6 | 96.9 | 92.1 | 88.5 | 84.4 | 75.2 | 74.4 | 74.5 | 75.5 | 76.2 | 75.9 |
| Georgia | 100.0 | 104.2 | 92.6 | 66.0 | 56.5 | 51.6 | 46.8 | 33.6 | 25.8 | 26.4 | .. | .. |
| Kazakhstan | 100.0 | 98.5 | 99.9 | 96.2 | 83.6 | 76.9 | 69.6 | 66.9 | 59.0 | 57.8 | 57.9 | 54.7 |
| Kyrgyzstan | 100.0 | 99.9 | 92.7 | 89.5 | 80.5 | 72.0 | 61.2 | 54.6 | 51.2 | 50.1 | 47.4 | 44.0 |
| Republic of Moldova[c] | 100.0 | 102.4 | 95.1 | 93.1 | 55.0 | 52.1 | 44.7 | 43.9 | 42.9 | 40.8 | 35.9 | 37.5 |
| Russian Federation | 100.0 | 97.7 | 96.1 | 91.3 | 89.8 | 80.7 | 74.1 | 70.4 | 64.5 | 63.0 | 62.6 | 64.2 |
| Tajikistan | 100.0 | 102.5 | 100.8 | 98.2 | 86.1 | 81.9 | 71.9 | 71.1 | 62.5 | 60.1 | 52.3 | 47.7 |
| Turkmenistan | 100.0 | 104.2 | 100.8 | 101.0 | 110.7 | 110.5 | 115.4 | 119.9 | 132.9 | 150.2 | 152.9 | .. |
| Ukraine | 100.0 | 98.1 | 97.3 | 92.7 | 87.9 | 78.3 | 72.2 | 66.8 | 61.2 | 59.3 | 54.4 | 51.3 |
| Uzbekistan | 100.0 | 101.5 | 102.5 | 101.5 | 103.2 | 90.1 | 92.3 | 93.5 | 93.7 | 94.1 | 94.9 | .. |
| **Total above**[a] | 100.0 | 97.4 | 93.5 | 86.9 | 82.8 | 76.1 | 71.5 | 68.7 | 64.5 | 63.2 | .. | .. |
| *Memorandum items:* | | | | | | | | | | | | |
| CETE-5 | 100.0 | 94.9 | 87.5 | 79.1 | 75.1 | 73.0 | 73.6 | 73.1 | 73.1 | 72.8 | 69.4 | 67.8 |
| SETE-7[a] | 100.0 | 96.5 | 85.9 | 73.8 | 68.2 | 65.3 | 62.5 | 62.4 | 58.1 | 56.0 | 50.7 | .. |

*Source:* UNECE Common Database, derived from national and CIS statistics.

[a] Excluding Albania.

[b] End of year.

[c] Excluding Transdniestria since 1993.

## APPENDIX TABLE B.7

**Registered unemployment in eastern Europe, the Baltic states and the CIS, 1990-2000**

*(Per cent of labour force, end of period)*

|  | 1990 | 1991 | 1992 | 1993 | 1994 | 1995 | 1996 | 1997 | 1998 | 1999 | 2000 |
|---|---|---|---|---|---|---|---|---|---|---|---|
| **Eastern Europe** | .. | 9.6 | 12.4 | 14.0 | 13.6 | 12.5 | 11.7 | 11.9 | 12.6 | 14.6 | 15.2 |
| Albania | 9.5 | 9.2 | 27.0 | 22.0 | 18.0 | 12.9 | 12.3 | 14.9 | 17.6 | 18.2 | 16.9 |
| Bosnia and Herzegovina | .. | .. | .. | .. | .. | .. | .. | 39.0 | 38.7 | 39.0 | 39.4 |
| Bulgaria | 1.8 | 11.1 | 15.3 | 16.4 | 12.8 | 11.1 | 12.5 | 13.7 | 12.2 | 16.0 | 17.9 |
| Croatia | .. | 14.1 | 17.8 | 16.6 | 17.3 | 17.6 | 15.9 | 17.6 | 18.6 | 20.8 | 22.6 |
| Czech Republic | 0.7 | 4.1 | 2.6 | 3.5 | 3.2 | 2.9 | 3.5 | 5.2 | 7.5 | 9.4 | 8.8 |
| Hungary | 1.7 | 7.4 | 12.3 | 12.1 | 10.9 | 10.4 | 10.5 | 10.4 | 9.1 | 9.6 | 8.9 |
| Poland | 6.5 | 12.2 | 14.3 | 16.4 | 16.0 | 14.9 | 13.2 | 10.3 | 10.4 | 13.1 | 15.1 |
| Romania | 1.3 | 3.0 | 8.2 | 10.4 | 10.9 | 9.5 | 6.6 | 8.8 | 10.3 | 11.5 | 10.5 |
| Slovakia | 1.6 | 11.8 | 10.4 | 14.4 | 14.8 | 13.1 | 12.8 | 12.5 | 15.6 | 19.2 | 17.9 |
| Slovenia | .. | 10.1 | 13.3 | 15.5 | 14.2 | 14.5 | 14.4 | 14.8 | 14.6 | 13.0 | 12.0 |
| The former Yugoslav Republic of Macedonia | .. | 24.5 | 26.2 | 27.7 | 30.0 | 36.6 | 38.8 | 41.7 | 41.4 | 43.8 | 44.9 |
| Yugoslavia [a] | .. | 21.0 | 24.6 | 24.0 | 23.9 | 24.7 | 26.1 | 25.6 | 27.2 | 27.4 | 26.6 |
| **Baltic states** | .. | .. | 2.1 | 4.5 | 5.3 | 6.6 | 6.4 | 6.3 | 7.3 | 9.1 | 10.0 |
| Estonia [b] | .. | .. | 1.6 | 5.0 | 5.1 | 5.0 | 5.6 | 4.6 | 5.1 | 6.7 | 7.3 |
| Latvia | .. | .. | 2.3 | 5.8 | 6.5 | 6.6 | 7.2 | 7.0 | 9.2 | 9.1 | 7.8 |
| Lithuania | .. | .. | 3.5 | 3.4 | 4.5 | 7.3 | 6.2 | 6.7 | 6.9 | 10.0 | 12.6 |
| **CIS** | .. | .. | 2.7 | 3.6 | 4.4 | 5.8 | 6.6 | 7.6 | 9.0 | 8.3 | 7.0 |
| Armenia | .. | .. | 3.5 | 6.3 | 6.0 | 8.1 | 9.7 | 11.0 | 8.9 | 11.5 | 10.9 |
| Azerbaijan | .. | .. | 0.2 | 0.7 | 0.9 | 1.1 | 1.1 | 1.3 | 1.4 | 1.2 | 1.2 |
| Belarus | .. | .. | 0.5 | 1.3 | 2.1 | 2.7 | 4.0 | 2.8 | 2.3 | 2.0 | 2.1 |
| Georgia | .. | .. | 0.3 | 2.0 | 3.8 | 3.4 | 3.2 | 8.0 | 4.2 | 5.6 | .. |
| Kazakhstan | .. | .. | 0.4 | 0.6 | 1.0 | 2.1 | 4.1 | 3.9 | 3.7 | 3.9 | 3.7 |
| Kyrgyzstan | .. | .. | 0.1 | 0.2 | 0.8 | 3.0 | 4.5 | 3.1 | 3.1 | 3.0 | 3.1 |
| Republic of Moldova | .. | .. | 0.7 | 0.7 | 1.0 | 1.4 | 1.5 | 1.7 | 1.9 | 2.1 | 1.8 |
| Russian Federation [c] | .. | .. | 5.2 | 6.1 | 7.8 | 9.0 | 10.0 | 11.2 | 13.3 | 12.2 | 9.8 |
| Tajikistan | .. | .. | 0.4 | 1.1 | 1.8 | 1.8 | 2.4 | 2.8 | 2.9 | 3.1 | 3.0 |
| Turkmenistan | .. | .. | .. | .. | .. | .. | .. | .. | .. | .. | .. |
| Ukraine | .. | .. | 0.3 | 0.4 | 0.3 | 0.6 | 1.5 | 2.8 | 4.3 | 4.3 | 4.2 |
| Uzbekistan | .. | .. | 0.1 | 0.2 | 0.3 | 0.3 | 0.3 | 0.3 | 0.4 | 0.5 | 0.6 |
| *Memorandum items:* | | | | | | | | | | | |
| **CETE-5** | .. | 9.7 | 11.3 | 13.3 | 12.9 | 12.0 | 11.2 | 9.8 | 10.2 | 12.5 | 13.4 |
| **SETE-7** | .. | 9.3 | 14.2 | 15.1 | 14.6 | 13.7 | 12.5 | 14.3 | 15.4 | 16.5 | 17.8 |
| Russian Federation [d] | .. | 0.1 | 0.8 | 1.1 | 2.1 | 3.2 | 3.4 | 2.8 | 2.7 | 1.7 | 1.4 |
| Former-GDR | .. | .. | 13.5 | 15.4 | 13.5 | 14.9 | 15.9 | 19.4 | 17.4 | 17.7 | 17.2 |

*Source:* UNECE Common Database, derived from national and CIS statistics.

*Note:* Aggregates for eastern European countries till 1997 exclude Bosnia and Herzegovina, that for CIS excludes Turkmenistan.

[a] Since 1999, excluding Kosovo and Metohia.

[b] Job seekers till October 2000, thereafter – registered unemployed as percentage of the labour force.

[d] Based on Russian Federation Goskomstat's monthly estimates according to the ILO definition, i.e. including all persons not having employment but actively seeking work.

[d] Registered unemployment.

## APPENDIX TABLE B.8

**Consumer prices in eastern Europe, the Baltic states and the CIS, 1990-2000**

*(Annual average, percentage change over preceding year)*

| | 1990 | 1991 | 1992 | 1993 | 1994 | 1995 | 1996 | 1997 | 1998 | 1999 | 2000 |
|---|---|---|---|---|---|---|---|---|---|---|---|
| Albania | .. | 35.5 | 193.1 | 85.0 | 21.5 | 8.0 | 12.7 | 33.1 | 20.3 | - 0.1 | – |
| Bosnia and Herzegovina | 594.0 | 116.2 | 64 218.3 | 38 825.1 | 553.5 | - 12.1 | - 21.2 | 11.8 | 4.9 | - 0.6 | 1.7 |
| Bulgaria | 23.8 | 338.5 | 91.3 | 72.9 | 96.2 | 62.0 | 121.7 | 1 058.3 | 18.7 | 2.6 | 10.2 |
| Croatia [a] | 609.5 | 123.0 | 663.6 | 1 516.6 | 97.5 | 2.0 | 3.6 | 3.7 | 5.9 | 4.3 | 6.4 |
| Czech Republic | 9.9 | 56.7 | 11.1 | 20.8 | 10.0 | 9.1 | 8.9 | 8.4 | 10.6 | 2.1 | 3.9 |
| Hungary | 28.9 | 35.0 | 23.0 | 22.6 | 19.1 | 28.5 | 23.6 | 18.4 | 14.2 | 10.1 | 9.9 |
| Poland | 585.8 | 70.3 | 45.3 | 36.9 | 33.2 | 28.1 | 19.8 | 15.1 | 11.7 | 7.4 | 10.2 |
| Romania | 5.1 | 170.2 | 210.7 | 256.2 | 137.1 | 32.2 | 38.8 | 154.9 | 59.3 | 45.9 | 45.7 |
| Slovakia | 10.4 | 61.2 | 10.2 | 23.1 | 13.4 | 10.0 | 6.1 | 6.1 | 6.7 | 10.5 | 12.0 |
| Slovenia | 551.6 | 115.0 | 207.3 | 31.7 | 21.0 | 13.5 | 9.9 | 8.4 | 8.1 | 6.3 | 9.0 |
| The former Yugoslav Republic of Macedonia [a] | 608.4 | 114.9 | 1 505.5 | 353.1 | 121.0 | 16.9 | 4.1 | 3.8 | 1.1 | - 1.4 | 10.1 |
| Yugoslavia | 580.0 | 122.0 | 8 926.0 | 2.2E+14 | 7.9E+10 | 71.8 | 90.5 | 23.2 | 30.4 | 44.1 | 77.5 |
| Estonia | 18.0 | 202.0 | 1 078.2 | 89.6 | 47.9 | 28.9 | 23.1 | 11.1 | 10.6 | 3.5 | 3.8 |
| Latvia | 10.9 | 172.2 | 951.2 | 109.1 | 35.7 | 25.0 | 17.7 | 8.5 | 4.7 | 2.4 | 2.8 |
| Lithuania | 9.1 | 216.4 | 1 020.5 | 410.1 | 72.0 | 39.5 | 24.7 | 8.8 | 5.1 | 0.8 | 1.0 |
| Armenia | 6.9 | 174.1 | 728.7 | 3 731.8 | 4 964.0 | 175.5 | 18.7 | 13.8 | 8.7 | 0.7 | - 0.8 |
| Azerbaijan | 6.1 | 106.6 | 912.6 | 1 129.7 | 1 663.9 | 411.5 | 19.8 | 3.6 | - 0.8 | - 8.6 | 1.8 |
| Belarus | 4.7 | 94.1 | 971.2 | 1 190.9 | 2 219.6 | 709.3 | 52.7 | 63.9 | 73.2 | 293.7 | 168.9 |
| Georgia | 4.2 | 78.7 | 1 176.9 | 4 084.9 | 22 286.1 | 261.4 | 39.4 | 7.1 | 3.5 | 19.3 | 4.2 |
| Kazakhstan | 5.6 | 114.5 | 1 504.3 | 1 662.7 | 1 880.1 | 176.3 | 39.2 | 17.5 | 7.3 | 8.4 | 13.4 |
| Kyrgyzstan | 5.5 | 113.9 | 854.6 | 1 208.7 | 278.1 | 42.9 | 31.3 | 23.4 | 10.3 | 35.7 | 18.7 |
| Republic of Moldova | 5.7 | 114.4 | 1 308.0 | 1 751.0 | 486.4 | 29.9 | 23.5 | 11.8 | 7.7 | 39.3 | 31.3 |
| Russian Federation | 5.2 | 160.0 | 1 528.7 | 875.0 | 309.0 | 197.4 | 47.8 | 14.7 | 27.8 | 85.7 | 20.8 |
| Tajikistan | 5.9 | 112.9 | 822.0 | 2 884.8 | 350.3 | 682.1 | 422.4 | 85.4 | 43.1 | 27.5 | 32.9 |
| Turkmenistan | 5.7 | 88.5 | 483.2 | 3 128.4 | 2 562.1 | 1 105.3 | 714.0 | 83.7 | 16.8 | .. | .. |
| Ukraine | 5.4 | 94.0 | 1 485.8 | 4 734.9 | 891.2 | 376.7 | 80.2 | 15.9 | 10.6 | 22.7 | 28.2 |
| Uzbekistan | 5.8 | 97.3 | 414.5 | 1 231.8 | 1 550.0 | 76.5 | 54.0 | 58.8 | 17.7 | 29.0 | 24.9 |

*Source:* UNECE Common Database, derived from national statistics.

*Note:* From 1992 onwards indices derived from monthly data except for Armenia, Georgia, Hungary, Slovenia, Yugoslavia (from 1993); Turkmenistan (from 1995); Uzbekistan (from 1996).

[a] Retail prices.

## APPENDIX TABLE B.9

**Producer price indices in eastern Europe, the Baltic states and the CIS, 1990-2000**

*(Annual average, percentage change over preceding year)*

|  | 1990 | 1991 | 1992 | 1993 | 1994 | 1995 | 1996 | 1997 | 1998 | 1999 | 2000 |
|---|---|---|---|---|---|---|---|---|---|---|---|
| Albania | .. | .. | .. | .. | .. | .. | .. | .. | .. | .. | .. |
| Bosnia and Herzegovina | .. | 129.5 | 70 374.7 | 10 967.6 | 1 184.8 | 68.7 | -4.8 | 3.2 | 3.6 | 4.3 | 0.9 |
| Bulgaria | 14.7 | 296.4 | 56.1 | 28.3 | 59.1 | 48.9 | 129.7 | 888.1 | 20.0 | 3.2 | 16.9 |
| Croatia | 455.3 | 146.3 | 826.0 | 1 510.4 | 77.7 | 0.8 | 1.3 | 3.7 | -1.5 | 2.5 | 9.6 |
| Czech Republic | 2.5 | 70.3 | 10.8 | 9.3 | 5.4 | 7.8 | 4.9 | 5.1 | 4.9 | 1.1 | 5.1 |
| Hungary | 22.0 | 32.6 | 12.3 | 14.1 | 12.3 | 28.5 | 22.3 | 20.9 | 11.4 | 5.0 | 11.4 |
| Poland | 622.4 | 40.9 | 28.0 | 32.6 | 31.0 | 26.0 | 13.4 | 12.2 | 7.2 | 5.7 | 7.8 |
| Romania | 26.9 | 220.1 | 184.8 | 165.0 | 140.7 | 35.3 | 50.0 | 144.0 | 33.2 | 41.2 | 53.4 |
| Slovakia | 5.2 | 68.9 | 5.3 | 17.0 | 10.0 | 9.1 | 4.0 | 4.6 | 3.3 | 3.7 | 9.8 |
| Slovenia | 390.4 | 124.1 | 215.7 | 23.3 | 17.8 | 12.4 | 6.7 | 6.1 | 6.0 | 2.2 | 7.7 |
| The former Yugoslav Republic of Macedonia | 394.0 | 112.0 | 2 193.5 | 258.6 | 88.7 | 4.7 | -0.1 | 3.5 | 4.0 | -0.2 | 9.0 |
| Yugoslavia | 468.0 | 124.0 | 8 993.0 | 1.4E+13 | 7.9E+10 | 75.7 | 88.8 | 20.6 | 25.9 | 43.3 | 105.4 |
| Estonia | 19.3 | 208.4 | 1 208.0 | 75.2 | 36.1 | 25.5 | 14.7 | 8.8 | 4.1 | -1.3 | 4.9 |
| Latvia | .. | 192.0 | 1 310.0 | 117.1 | 17.0 | 12.0 | 13.8 | 4.3 | 2.0 | -4.0 | 0.8 |
| Lithuania | .. | 148.2 | 1 510.0 | 391.7 | 44.7 | 28.7 | 16.5 | 5.1 | -4.6 | 3.3 | 17.8 |
| Armenia | .. | 120.0 | 947.0 | 892.0 | 4 394.4 | 187.8 | 36.3 | 21.7 | 6.0 | 4.1 | -0.4 |
| Azerbaijan | .. | 179.5 | 7 453.6 | 1 974.0 | 3 971.6 | 1 340.1 | 70.6 | 11.4 | -5.5 | -1.3 | 9.4 |
| Belarus | 2.1 | 151.1 | 1 939.2 | 1 536.3 | 3 362.1 | 538.6 | 35.7 | 89.4 | 72.8 | 355.7 | 185.6 |
| Georgia | .. | .. | .. | .. | .. | .. | .. | .. | 2.2 | 15.5 | 5.8 |
| Kazakhstan | .. | 193.0 | 2 465.1 | 1 042.8 | 2 918.5 | 139.7 | 23.8 | 15.6 | 0.8 | 19.0 | 38.0 |
| Kyrgyzstan | .. | 160.0 | 1 664.0 | 831.0 | 72.9 | 34.0 | 29.1 | 34.8 | 12.7 | 51.7 | 31.7 |
| Republic of Moldova | .. | 130.0 | 1 210.9 | 1 078.5 | 711.7 | 52.2 | 30.2 | 14.9 | 9.7 | 47.1 | 33.6 |
| Russian Federation | 3.9 | 240.0 | 3 280.0 | 900.0 | 340.0 | 237.6 | 50.8 | 15.0 | 7.0 | 59.1 | 46.5 |
| Tajikistan | .. | 163.0 | 1 316.5 | 1 080.0 | 665.5 | 351.7 | 340.7 | 103.7 | 28.4 | 45.6 | 39.0 |
| Turkmenistan | .. | 211.0 | 994.0 | 1 610.0 | 911.0 | 296.5 | 2 974.9 | 260.6 | -30.5 | .. | .. |
| Ukraine | .. | 163.4 | 4 128.5 | 9 667.5 | 1 134.5 | 488.9 | 52.1 | 7.8 | 13.2 | 31.1 | 20.8 |
| Uzbekistan | .. | 147.0 | 1 296.0 | 1 119.0 | 2 162.6 | 792.5 | 128.5 | 53.9 | 40.0 | 38.0 | 61.1 |

*Source:* UNECE Common Database, derived from national statistics.

*Note:* From 1994 onwards indices derived from monthly data except: Bosnia and Herzegovina, Croatia, Czech Republic, Poland, The former Yugoslav Republic of Macedonia (from 1992); Hungary, Romania, Slovakia, Slovenia (from 1993); Turkmenistan, Yugoslavia (from 1995).

APPENDIX TABLE B.10

**Nominal gross wages in industry in eastern Europe, the Baltic states and the CIS, 1990-2000**

*(Annual average, percentage change over preceding year)*[a]

|  | 1990 | 1991 | 1992 | 1993 | 1994 | 1995 | 1996 | 1997 | 1998 | 1999 | 2000 |
|---|---|---|---|---|---|---|---|---|---|---|---|
| Albania [b] | .. | .. | .. | 69.5 | 34.5 | 29.3 | .. | .. | 20.4 | 10.4 | 17.7 |
| Bosnia and Herzegovina [c] | .. | .. | .. | .. | .. | .. | 253.6 | 50.8 | 31.1 | 15.4 | 12.1 |
| Bulgaria | 20.8 | 175.5 | 132.8 | 55.1 | 53.9 | 57.7 | 90.4 | 882.9 | 34.5 | .. | .. |
| Croatia [c] | 453.7 | 40.7 | 466.7 | 1 444.1 | 130.5 | 44.0 | 11.8 | 16.3 | 10.6 | 10.1 | 7.9 |
| Czech Republic | 3.0 | 16.5 | 21.0 | 22.6 | 16.9 | 18.3 | 17.7 | 11.9 | 10.7 | 6.8 | 7.2 |
| Hungary [b] | 27.2 | 33.4 | 24.3 | 24.9 | 23.3 | 19.1 | 21.7 | 21.8 | 16.6 | 13.5 | 15.0 |
| Poland [c] | 365.5 | 64.0 | 41.2 | 37.8 | 39.8 | 30.1 | 25.8 | 19.9 | 14.1 | 35.7 | .. |
| Romania [c] | 9.7 | 125.0 | 173.5 | 210.7 | 139.4 | 55.3 | 56.7 | 99.5 | 49.3 | 40.7 | 48.9 |
| Slovakia [b] | 4.3 | 15.0 | 16.9 | 23.1 | 17.7 | 15.2 | 14.6 | 11.1 | 8.0 | 7.9 | 9.3 |
| Slovenia | 361.5 | 68.4 | 196.7 | 45.1 | 27.1 | 17.1 | 14.0 | 12.1 | 10.7 | 9.3 | 11.6 |
| The former Yugoslav Republic of Macedonia [c] | 433.6 | 79.2 | 1 083.7 | 454.0 | 105.8 | 11.1 | 3.6 | 2.3 | 3.1 | 1.5 | 6.2 |
| Yugoslavia [c] | 400.0 | 100.6 | 4 886.4 | -62.5 | 229.6 | 74.1 | 74.5 | 41.7 | 39.6 | 24.6 | 113.5 |
| Estonia [d] | .. | 122.2 | 570.3 | 93.3 | 71.3 | 34.6 | 23.8 | 19.7 | 13.2 | 0.9 | .. |
| Latvia | .. | 104.5 | 609.8 | 112.0 | 60.0 | 24.1 | 14.9 | 21.6 | 6.5 | 4.0 | 3.2 |
| Lithuania | 15.3 | 183.9 | 632.5 | 246.1 | 70.0 | 43.2 | 34.0 | 21.7 | 12.0 | -0.4 | 2.5 |
| Armenia | 5.9 | 31.8 | 352.3 | 739.1 | 3 640.2 | 210.9 | 62.3 | 41.7 | 20.5 | 15.2 | 25.2 |
| Azerbaijan | 4.4 | 82.9 | 870.4 | 700.8 | 575.7 | 354.8 | 53.0 | 58.2 | 19.8 | 18.1 | 33.3 |
| Belarus | 14.1 | 112.1 | 910.9 | 1 073.8 | 69.6 | 618.7 | 58.5 | 96.8 | 109.4 | 323.9 | 197.1 |
| Georgia | 6.0 | 33.9 | 457.3 | 2 114.7 | 23 327.5 | 87.5 | 145.4 | 12.7 | 26.8 | .. | .. |
| Kazakhstan | 11.0 | 80.5 | 1 053.1 | 993.1 | 1 542.5 | 178.2 | 30.9 | 22.5 | 2.7 | 12.3 | 49.6 |
| Kyrgyzstan | 0.5 | 62.1 | 638.3 | 745.6 | 176.3 | 54.8 | 29.1 | 60.3 | 20.0 | 43.2 | -10.0 |
| Republic of Moldova | 14.2 | 90.1 | 815.0 | 773.0 | 284.4 | 43.8 | 32.2 | 24.9 | 16.6 | 22.5 | 34.3 |
| Russian Federation | 13.0 | 94.9 | 1 065.7 | 798.2 | 260.2 | 131.4 | 64.3 | 21.6 | 14.3 | 52.2 | 59.5 |
| Tajikistan | 8.0 | 76.2 | 550.2 | 917.5 | 142.2 | 145.2 | 425.3 | 71.9 | 96.4 | 26.4 | 31.5 |
| Turkmenistan | .. | 83.8 | 1 010.1 | 1 467.7 | 622.2 | 686.2 | 915.3 | 121.5 | 27.1 | .. | .. |
| Ukraine | 39.0 | 107.9 | 1 419.0 | 2 286.1 | 737.8 | 408.5 | 71.5 | 13.5 | 5.9 | 18.1 | 40.2 |
| Uzbekistan | .. | 82.3 | 706.1 | 1 159.3 | 806.8 | 278.4 | 99.8 | 76.3 | 60.3 | 50.0 | .. |

*Source:* UNECE Common Database, derived from national statistics.

[a] Calculated from reported annual average wages.

[b] Gross wages in total economy. For Hungary for 1990-1992; for Slovakia for 1990-1991.

[c] Net wages in industry. For Poland and Romania for 1990-1992.

[d] Manufacturing for 1991-1993.

## APPENDIX TABLE B.11
### Merchandise exports of eastern Europe, the Baltic states and the CIS, 1980, 1988-2000
*(Billion dollars)*

|  | 1980 | 1988 | 1989 | 1990 | 1991 | 1992 | 1993 | 1994 | 1995 | 1996 | 1997 | 1998 | 1999 | 2000 |
|---|---|---|---|---|---|---|---|---|---|---|---|---|---|---|
| **Eastern Europe** | 56.367 | 65.020 | 63.850 | 61.733 | 57.241 | 59.333 | 62.675 | 72.937 | 94.777 | 100.206 | 107.428 | 119.174 | 117.842 | 132.951 |
| Albania | 0.320 | 0.230 | 0.302 | 0.231 | 0.101 | 0.072 | 0.123 | 0.139 | 0.202 | 0.213 | 0.137 | 0.207 | 0.352 | 0.261 |
| Bulgaria | 7.160 | 7.554 | 6.651 | 5.232 | 3.433 | 3.992 | 3.769 | 3.935 | 5.345 | 4.890 | 4.940 | 4.194 | 4.006 | 4.812 |
| Czechoslovakia | 10.475 | 12.381 | 11.988 | 10.728 | 11.319 | .. | .. | .. | .. | .. | .. | .. | .. | .. |
| Czech Republic | .. | .. | .. | .. | .. | 8.767 | 14.463 | 15.882 | 21.273 | 22.180 | 22.779 | 26.351 | 26.242 | 28.979 |
| Slovakia | .. | .. | .. | .. | .. | 3.500 | 5.458 | 6.714 | 8.585 | 8.822 | 9.640 | 10.775 | 10.277 | 11.908 |
| Hungary | 8.609 | 9.999 | 9.673 | 9.731 | 10.226 | 10.681 | 8.921 | 10.701 | 12.867 | 15.704 | 19.100 | 23.005 | 25.012 | 28.092 |
| Poland | 13.071 | 14.573 | 14.665 | 18.291 | 14.912 | 13.187 | 14.202 | 17.240 | 22.887 | 24.440 | 25.756 | 28.229 | 27.404 | 31.651 |
| Romania | 9.217 | 8.971 | 8.076 | 4.570 | 4.266 | 4.363 | 4.892 | 6.151 | 7.910 | 8.085 | 8.431 | 8.302 | 8.503 | 10.367 |
| Yugoslavia (SFR) | 7.514 | 11.311 | 12.496 | 12.950 | 12.984 | 14.772 | .. | .. | .. | .. | .. | .. | .. | .. |
| Bosnia and Herzegovina | .. | 1.550 | 2.100 | 1.850 | .. | .. | .. | .. | 0.024 | 0.058 | 0.193 | 0.352 | 0.518 | 0.675 |
| Croatia | .. | 2.300 | 2.600 | 4.020 | 3.310 | 4.353 | 3.709 | 4.260 | 4.633 | 4.512 | 4.171 | 4.541 | 4.303 | 4.432 |
| Slovenia | 1.836 | 3.278 | 3.408 | 4.118 | 3.874 | 6.681 | 6.083 | 6.828 | 8.316 | 8.310 | 8.369 | 9.050 | 8.546 | 8.731 |
| The former Yugoslav Republic of Macedonia | .. | .. | 0.654 | 1.113 | 1.095 | 1.199 | 1.055 | 1.086 | 1.204 | 1.147 | 1.237 | 1.311 | 1.186 | 1.320 |
| Yugoslavia | .. | 4.298 | 4.461 | 4.651 | 4.704 | 2.539 | .. | .. | 1.531 | 1.846 | 2.677 | 2.858 | 1.493 | 1.723 |
| **Baltic states** | .. | .. | .. | .. | .. | 2.139 | 4.197 | 4.324 | 5.844 | 6.877 | 8.467 | 8.759 | 7.665 | 8.889 |
| Estonia | .. | .. | .. | .. | .. | 0.444 | 0.802 | 1.305 | 1.838 | 2.079 | 2.934 | 3.236 | 2.938 | 3.226 |
| Latvia | .. | .. | .. | .. | .. | 0.843 | 1.401 | 0.988 | 1.304 | 1.443 | 1.673 | 1.812 | 1.723 | 1.855 |
| Lithuania | .. | .. | .. | .. | .. | 0.852 | 1.994 | 2.031 | 2.705 | 3.355 | 3.860 | 3.711 | 3.004 | 3.809 |
| **CIS: total** | .. | .. | .. | .. | .. | .. | .. | 89.991 | 110.622 | 121.936 | 122.913 | 104.716 | 104.179 | 143.754* |
| CIS: non-CIS | .. | .. | .. | .. | .. | 51.242 | 52.547 | 62.652 | 80.007 | 87.897 | 88.698 | 76.757 | 82.346 | 114.943* |
| Armenia | .. | .. | .. | .. | .. | .. | 0.156 | 0.216 | 0.271 | 0.290 | 0.233 | 0.221 | 0.232 | 0.301 |
| Non-CIS | .. | .. | .. | .. | .. | 0.026 | 0.029 | 0.058 | 0.101 | 0.162 | 0.138 | 0.140 | 0.175 | 0.227 |
| Azerbaijan | .. | .. | .. | .. | .. | 1.484 | 0.725 | 0.653 | 0.637 | 0.631 | 0.781 | 0.606 | 0.929 | 1.745 |
| Non-CIS | .. | .. | .. | .. | .. | 0.754 | 0.351 | 0.378 | 0.352 | 0.341 | 0.403 | 0.374 | 0.718 | 1.510 |
| Belarus | .. | .. | .. | .. | .. | .. | .. | 2.510 | 4.707 | 5.652 | 7.301 | 7.070 | 5.909 | 7.380 |
| Non-CIS | .. | .. | .. | .. | .. | 1.194 | 0.789 | 1.031 | 1.777 | 1.888 | 1.922 | 1.910 | 2.287 | 2.926 |
| Georgia | .. | .. | .. | .. | .. | .. | .. | 0.156 | 0.154 | 0.199 | 0.240 | 0.193 | 0.238 | 0.330 |
| Non-CIS | .. | .. | .. | .. | .. | 0.068 | 0.069 | 0.039 | 0.057 | 0.070 | 0.102 | 0.085 | 0.131 | 0.194 |
| Kazakhstan | .. | .. | .. | .. | .. | .. | .. | 3.231 | 5.250 | 5.911 | 6.497 | 5.436 | 5.592 | 9.140 |
| Non-CIS | .. | .. | .. | .. | .. | 1.398 | 1.501 | 1.357 | 2.366 | 2.732 | 3.515 | 3.266 | 4.100 | 6.749 |
| Kyrgyzstan | .. | .. | .. | .. | .. | 0.317 | 0.396 | 0.340 | 0.409 | 0.505 | 0.604 | 0.514 | 0.454 | 0.505 |
| Non-CIS | .. | .. | .. | .. | .. | 0.077 | 0.112 | 0.117 | 0.140 | 0.112 | 0.285 | 0.283 | 0.271 | 0.297 |
| Republic of Moldova | .. | .. | .. | .. | .. | 0.470 | 0.483 | 0.565 | 0.746 | 0.795 | 0.875 | 0.632 | 0.464 | 0.472 |
| Non-CIS | .. | .. | .. | .. | .. | 0.166 | 0.178 | 0.159 | 0.279 | 0.252 | 0.267 | 0.203 | 0.211 | 0.196 |
| Russian Federation [a] | .. | .. | .. | .. | .. | .. | .. | 66.862 | 79.869 | 86.889 | 86.627 | 73.000 | 73.700 | 102.796* |
| Non-CIS | .. | .. | .. | .. | .. | 42.376 | 44.297 | 53.001 | 65.607 | 70.975 | 69.959 | 58.800 | 62.800 | 89.068* |
| Tajikistan | .. | .. | .. | .. | .. | 0.193 | 0.350 | 0.492 | 0.749 | 0.770 | 0.746 | 0.597 | 0.689 | 0.784 |
| Non-CIS | .. | .. | .. | .. | .. | 0.109 | 0.227 | 0.399 | 0.497 | 0.439 | 0.473 | 0.394 | 0.374 | 0.411 |
| Turkmenistan | .. | .. | .. | .. | .. | .. | .. | 2.145 | 1.881 | 1.682 | 0.751 | 0.594 | 1.190 | 2.500 |
| Non-CIS | .. | .. | .. | .. | .. | 0.908 | 1.049 | 0.494 | 0.951 | 0.610 | 0.300 | 0.442 | 0.700 | 1.200 |
| Ukraine | .. | .. | .. | .. | .. | 7.415 | 7.817 | 10.272 | 13.128 | 14.401 | 14.232 | 12.637 | 11.582 | 14.573 |
| Non-CIS | .. | .. | .. | .. | .. | 3.297 | 3.223 | 4.653 | 6.168 | 6.996 | 8.646 | 8.435 | 8.329 | 10.075 |
| Uzbekistan | .. | .. | .. | .. | .. | .. | .. | 2.549 | 2.821 | 4.211 | 4.026 | 3.218 | 3.200 | 3.230 |
| Non-CIS | .. | .. | .. | .. | .. | 0.869 | 0.721 | 0.966 | 1.712 | 3.321 | 2.689 | 2.425 | 2.250 | 2.090 |
| **Former Soviet Union** | 57.942 | 62.016 | 62.286 | 59.056 | 46.660 | .. | .. | .. | .. | .. | .. | .. | .. | .. |
| **Total** | 114.310 | 127.035 | 126.136 | 120.788 | 103.901 | 112.714 | 119.419 | 167.252 | 211.243 | 229.019 | 238.808 | 232.649 | 229.686 | 285.594* |

*Source:* UNECE secretariat, based on national statistical publications and direct communications from national statistical offices.

*Note:* Trade flows reported include the "new trade" among members of the dissolved federal states: former Czechoslovakia (from 1993), the former SFR of Yugoslavia (from 1992) and the former USSR: for the Baltic states (from 1992) and for the CIS. Data excluding the "new trade" were shown in earlier issues of this publication. Changes in the method of recording trade are reflected from 1993 in data for the Czech Republic (inclusion of OPT transactions, etc.), from 1995 in Latvia (imports registered c.i.f.) and Lithuania (change from special to general system), from 1996 in Hungary (inclusion of trade flows of free trade zones) and from 1997 in Slovakia (inclusion of OPT transactions, etc.).

As from 1991, all trade values are expressed in dollars at prevailing market exchange rates. For earlier years, values reported in national currencies were adjusted by the ECE secretariat to remove distortions stemming from mutually inconsistent national rouble/dollar cross-rates in the valuation of the then important intra-CMEA trade flows. For details on the revaluation, see the note to table 2.1.3 and the discussion in box 2.1.1 in UNECE, *Economic Bulletin for Europe*, Vol. 43, 1991.

[a] Russian Goskomstat data excluding trade by physical persons (shuttle trade), but including trade flows not crossing the Russian borders such as off-board fish sales and natural gas deliveries under debt repayment agreements with former CMEA countries.

## APPENDIX TABLE B.12

**Merchandise imports of eastern Europe, the Baltic states and the CIS, 1980, 1988-2000**
*(Billion dollars)*

|  | 1980 | 1988 | 1989 | 1990 | 1991 | 1992 | 1993 | 1994 | 1995 | 1996 | 1997 | 1998 | 1999 | 2000 |
|---|---|---|---|---|---|---|---|---|---|---|---|---|---|---|
| **Eastern Europe** | 65.443 | 60.158 | 61.185 | 63.408 | 61.610 | 68.388 | 76.285 | 86.128 | 117.026 | 135.887 | 146.195 | 159.491 | 155.447 | 172.641 |
| Albania | 0.320 | 0.280 | 0.385 | 0.381 | 0.409 | 0.524 | 0.421 | 0.549 | 0.650 | 0.913 | 0.620 | 0.795 | 0.903 | 1.070 |
| Bulgaria | 6.321 | 8.131 | 7.325 | 5.584 | 2.700 | 4.530 | 5.120 | 4.272 | 5.638 | 5.074 | 4.932 | 4.957 | 5.515 | 6.494 |
| Czechoslovakia | 10.619 | 12.180 | 11.772 | 11.808 | 10.962 | .. | .. | .. | .. | .. | .. | .. | .. | .. |
| Czech Republic | .. | .. | .. | .. | .. | 10.368 | 14.617 | 17.427 | 25.265 | 27.919 | 27.563 | 28.789 | 28.073 | 32.244 |
| Slovakia | .. | .. | .. | .. | .. | 3.889 | 6.332 | 6.634 | 8.777 | 11.112 | 11.622 | 13.006 | 11.265 | 12.660 |
| Hungary | 9.188 | 9.372 | 8.863 | 8.797 | 11.449 | 11.123 | 12.648 | 14.554 | 15.466 | 18.144 | 21.234 | 25.706 | 28.008 | 32.080 |
| Poland | 14.705 | 12.987 | 12.941 | 12.619 | 15.531 | 16.141 | 18.758 | 21.566 | 29.043 | 37.137 | 42.314 | 47.054 | 45.911 | 48.940 |
| Romania | 11.061 | 5.361 | 5.834 | 6.889 | 5.793 | 6.260 | 6.522 | 7.109 | 10.278 | 11.435 | 11.280 | 11.838 | 10.395 | 13.055 |
| Yugoslavia (SFR) | 13.229 | 11.847 | 14.064 | 17.330 | 14.765 | .. | .. | .. | .. | .. | .. | .. | .. | .. |
| Bosnia and Herzegovina | .. | 1.300 | 1.850 | 1.750 | .. | .. | .. | .. | 0.524 | 1.204 | 1.555 | 2.120 | 2.431 | 2.290 |
| Croatia | .. | 2.900 | 3.750 | 5.133 | 3.811 | 4.346 | 4.166 | 5.229 | 7.510 | 7.788 | 9.104 | 8.383 | 7.799 | 7.923 |
| Slovenia | 2.463 | 2.914 | 3.216 | 4.727 | 4.131 | 6.141 | 6.501 | 7.304 | 9.492 | 9.421 | 9.367 | 10.098 | 10.083 | 10.115 |
| The former Yugoslav Republic of Macedonia | .. | .. | 0.934 | 1.531 | 1.274 | 1.206 | 1.199 | 1.484 | 1.719 | 1.627 | 1.779 | 1.915 | 1.773 | 2.060 |
| Yugoslavia | .. | 4.915 | 5.383 | 6.701 | 5.548 | 3.859 | .. | .. | 2.665 | 4.113 | 4.826 | 4.830 | 3.291 | 3.711 |
| **Baltic states** | .. | .. | .. | .. | .. | 1.802 | 4.101 | 5.251 | 8.006 | 10.110 | 12.809 | 13.768 | 11.888 | 12.954 |
| Estonia | .. | .. | .. | .. | .. | 0.406 | 0.896 | 1.659 | 2.540 | 3.231 | 4.441 | 4.786 | 4.108 | 4.327 |
| Latvia | .. | .. | .. | .. | .. | 0.794 | 0.961 | 1.240 | 1.818 | 2.320 | 2.724 | 3.189 | 2.946 | 3.170 |
| Lithuania | .. | .. | .. | .. | .. | 0.602 | 2.244 | 2.352 | 3.649 | 4.559 | 5.644 | 5.794 | 4.834 | 5.457 |
| **CIS: total** | .. | .. | .. | .. | .. | .. | .. | 63.137 | 79.576 | 86.960 | 94.316 | 81.747 | 62.005 | 70.673* |
| **CIS: non-CIS** | .. | .. | .. | .. | .. | 42.297 | 33.696 | 36.743 | 45.678 | 49.205 | 57.780 | 50.560 | 37.571 | 38.631* |
| Armenia | .. | .. | .. | .. | .. | .. | 0.255 | 0.394 | 0.674 | 0.856 | 0.892 | 0.902 | 0.811 | 0.885 |
| Non-CIS | .. | .. | .. | .. | .. | 0.050 | 0.087 | 0.188 | 0.340 | 0.578 | 0.593 | 0.672 | 0.624 | 0.711 |
| Azerbaijan | .. | .. | .. | .. | .. | 0.940 | 0.629 | 0.778 | 0.668 | 0.961 | 0.794 | 1.077 | 1.036 | 1.172 |
| Non-CIS | .. | .. | .. | .. | .. | 0.333 | 0.241 | 0.292 | 0.440 | 0.621 | 0.443 | 0.673 | 0.711 | 0.797 |
| Belarus | .. | .. | .. | .. | .. | .. | .. | 3.066 | 5.564 | 6.939 | 8.689 | 8.549 | 6.674 | 8.477 |
| Non-CIS | .. | .. | .. | .. | .. | 0.843 | 1.119 | 0.974 | 1.887 | 2.369 | 2.872 | 2.995 | 2.385 | 2.476 |
| Georgia | .. | .. | .. | .. | .. | .. | .. | 0.338 | 0.385 | 0.687 | 0.944 | 0.884 | 0.602 | 0.726 |
| Non-CIS | .. | .. | .. | .. | .. | 0.228 | 0.167 | 0.066 | 0.231 | 0.417 | 0.603 | 0.617 | 0.377 | 0.479 |
| Kazakhstan | .. | .. | .. | .. | .. | .. | .. | 3.561 | 3.807 | 4.241 | 4.301 | 4.350 | 3.687 | 5.052 |
| Non-CIS | .. | .. | .. | .. | .. | 0.469 | 0.494 | 1.384 | 1.154 | 1.295 | 1.969 | 2.290 | 2.089 | 2.295 |
| Kyrgyzstan | .. | .. | .. | .. | .. | 0.421 | 0.448 | 0.317 | 0.522 | 0.838 | 0.709 | 0.842 | 0.600 | 0.554 |
| Non-CIS | .. | .. | .. | .. | .. | 0.071 | 0.112 | 0.107 | 0.168 | 0.351 | 0.273 | 0.401 | 0.341 | 0.256 |
| Republic of Moldova | .. | .. | .. | .. | .. | 0.640 | 0.628 | 0.659 | 0.841 | 1.072 | 1.172 | 1.024 | 0.587 | 0.777 |
| Non-CIS | .. | .. | .. | .. | .. | 0.179 | 0.184 | 0.183 | 0.272 | 0.420 | 0.567 | 0.584 | 0.345 | 0.517 |
| Russian Federation [a] | .. | .. | .. | .. | .. | .. | .. | 38.661 | 46.709 | 47.373 | 53.568 | 44.600 | 31.000 | 33.769* |
| Non-CIS | .. | .. | .. | .. | .. | 36.984 | 26.807 | 28.344 | 33.117 | 32.798 | 39.365 | 32.500 | 22.200 | 22.171* |
| Tajikistan | .. | .. | .. | .. | .. | 0.254 | 0.630 | 0.547 | 0.810 | 0.668 | 0.750 | 0.711 | 0.663 | 0.675 |
| Non-CIS | .. | .. | .. | .. | .. | 0.132 | 0.374 | 0.314 | 0.332 | 0.285 | 0.268 | 0.265 | 0.148 | 0.115 |
| Turkmenistan | .. | .. | .. | .. | .. | .. | .. | 1.468 | 1.364 | 1.011 | 1.183 | 1.008 | 1.500 | 1.780 |
| Non-CIS | .. | .. | .. | .. | .. | 0.030 | 0.501 | 0.782 | 0.619 | 0.450 | 0.531 | 0.530 | 1.000 | 1.100 |
| Ukraine | .. | .. | .. | .. | .. | 6.892 | 9.533 | 10.745 | 15.484 | 17.603 | 17.128 | 14.676 | 11.846 | 13.956 |
| Non-CIS | .. | .. | .. | .. | .. | 2.049 | 2.652 | 2.907 | 5.488 | 6.427 | 7.249 | 6.779 | 5.103 | 5.916 |
| Uzbekistan | .. | .. | .. | .. | .. | .. | .. | 2.603 | 2.748 | 4.712 | 4.186 | 3.125 | 3.000 | 2.850 |
| Non-CIS | .. | .. | .. | .. | .. | 0.929 | 0.958 | 1.202 | 1.630 | 3.195 | 3.047 | 2.256 | 2.250 | 1.800 |
| **Former Soviet Union** | 52.218 | 58.044 | 64.983 | 64.963 | 45.405 | .. | .. | .. | .. | .. | .. | .. | .. | .. |
| **Total** | 117.661 | 118.202 | 126.168 | 128.371 | 106.901 | 112.487 | 114.082 | 154.516 | 204.609 | 232.957 | 253.320 | 255.006 | 229.340 | 256.268* |

*Source:* UNECE secretariat, based on national statistical publications and direct communications from national statistical offices.

*Note:* See appendix table B.11.

[a] Russian Goskomstat data excluding trade by physical persons (shuttle trade), but including trade flows not crossing the Russian borders such as off-board fish sales and natural gas deliveries under debt repayment agreements with former CMEA countries.

## APPENDIX TABLE B.13

### Balance of merchandise trade of eastern Europe, the Baltic states and the CIS, 1980, 1988-2000

*(Billion dollars)*

|  | 1980 | 1988 | 1989 | 1990 | 1991 | 1992 | 1993 | 1994 | 1995 | 1996 | 1997 | 1998 | 1999 | 2000 |
|---|---|---|---|---|---|---|---|---|---|---|---|---|---|---|
| **Eastern Europe** | -9.076 | 4.861 | 2.665 | -1.675 | -4.369 | -9.055 | -13.610 | -13.190 | -22.250 | -35.680 | -38.767 | -40.317 | -37.605 | -39.690 |
| Albania | – | -0.050 | -0.083 | -0.150 | -0.308 | -0.452 | -0.298 | -0.410 | -0.448 | -0.701 | -0.483 | -0.589 | -0.551 | -0.809 |
| Bulgaria | 0.839 | -0.577 | -0.674 | -0.352 | 0.732 | -0.538 | -1.352 | -0.336 | -0.293 | -0.184 | 0.008 | -0.763 | -1.509 | -1.682 |
| Czechoslovakia | -0.144 | 0.201 | 0.216 | -1.080 | 0.356 | .. | .. | .. | .. | .. | .. | .. | .. | .. |
| Czech Republic | .. | .. | .. | .. | .. | -1.601 | -0.154 | -1.545 | -3.992 | -5.739 | -4.784 | -2.438 | -1.831 | -3.265 |
| Slovakia | .. | .. | .. | .. | .. | -0.389 | -0.874 | 0.080 | -0.192 | -2.290 | -1.983 | -2.231 | -0.988 | -0.752 |
| Hungary | -0.579 | 0.627 | 0.810 | 0.934 | -1.223 | -0.442 | -3.727 | -3.853 | -2.599 | -2.440 | -2.134 | -2.701 | -2.996 | -3.988 |
| Poland | -1.634 | 1.586 | 1.724 | 5.672 | -0.619 | -2.955 | -4.555 | -4.326 | -6.156 | -12.697 | -16.558 | -18.825 | -18.507 | -17.289 |
| Romania | -1.844 | 3.610 | 2.242 | -2.320 | -1.528 | -1.897 | -1.630 | -0.958 | -2.368 | -3.351 | -2.849 | -3.536 | -1.892 | -2.688 |
| Yugoslavia (SFR) | -5.715 | -0.536 | -1.568 | -4.380 | -1.780 | 14.772 | .. | .. | .. | .. | .. | .. | .. | .. |
| Bosnia and Herzegovina | .. | 0.250 | 0.250 | 0.100 | .. | .. | .. | .. | -0.500 | -1.146 | -1.362 | -1.768 | -1.913 | -1.615 |
| Croatia | .. | -0.600 | -1.150 | -1.113 | -0.501 | 0.007 | -0.457 | -0.969 | -2.877 | -3.276 | -4.933 | -3.842 | -3.496 | -3.491 |
| Slovenia | -0.626 | 0.365 | 0.192 | -0.609 | -0.257 | 0.540 | -0.418 | -0.476 | -1.176 | -1.111 | -0.998 | -1.048 | -1.537 | -1.384 |
| The former Yugoslav Republic of Macedonia | .. | .. | -0.280 | -0.418 | -0.179 | -0.007 | -0.144 | -0.398 | -0.515 | -0.480 | -0.542 | -0.604 | -0.587 | -0.740 |
| Yugoslavia | .. | -0.617 | -0.922 | -2.050 | -0.844 | -1.320 | .. | .. | -1.134 | -2.267 | -2.149 | -1.972 | -1.798 | -1.988 |
| **Baltic states** | .. | .. | .. | .. | .. | 0.337 | 0.096 | -0.927 | -2.162 | -3.232 | -4.343 | -5.009 | -4.224 | -4.065 |
| Estonia | .. | .. | .. | .. | .. | 0.038 | -0.094 | -0.353 | -0.702 | -1.152 | -1.507 | -1.550 | -1.170 | -1.102 |
| Latvia | .. | .. | .. | .. | .. | 0.049 | 0.440 | -0.252 | -0.514 | -0.877 | -1.051 | -1.377 | -1.223 | -1.316 |
| Lithuania | .. | .. | .. | .. | .. | 0.250 | -0.250 | -0.322 | -0.944 | -1.204 | -1.784 | -2.083 | -1.831 | -1.648 |
| **CIS: total** | .. | .. | .. | .. | .. | .. | .. | 26.854 | 31.046 | 34.975 | 28.597 | 22.970 | 42.174 | 73.081* |
| **CIS: non-CIS** | .. | .. | .. | .. | .. | 8.945 | 18.851 | 25.909 | 34.329 | 38.692 | 30.919 | 26.197 | 44.774 | 76.311* |
| Armenia | .. | .. | .. | .. | .. | .. | -0.099 | -0.178 | -0.403 | -0.566 | -0.660 | -0.682 | -0.580 | -0.584 |
| Non-CIS | .. | .. | .. | .. | .. | -0.024 | -0.058 | -0.130 | -0.239 | -0.416 | -0.455 | -0.532 | -0.449 | -0.484 |
| Azerbaijan | .. | .. | .. | .. | .. | 0.544 | 0.096 | -0.125 | -0.031 | -0.330 | -0.013 | -0.471 | -0.107 | 0.573 |
| Non-CIS | .. | .. | .. | .. | .. | 0.421 | 0.110 | 0.086 | -0.088 | -0.280 | -0.040 | -0.299 | 0.007 | 0.713 |
| Belarus | .. | .. | .. | .. | .. | .. | .. | -0.556 | -0.857 | -1.287 | -1.388 | -1.480 | -0.765 | -1.097 |
| Non-CIS | .. | .. | .. | .. | .. | 0.351 | -0.330 | 0.057 | -0.110 | -0.481 | -0.950 | -1.085 | -0.098 | 0.450 |
| Georgia | .. | .. | .. | .. | .. | .. | .. | -0.182 | -0.231 | -0.488 | -0.704 | -0.692 | -0.364 | -0.396 |
| Non-CIS | .. | .. | .. | .. | .. | -0.160 | -0.098 | -0.027 | -0.174 | -0.347 | -0.501 | -0.532 | -0.246 | -0.285 |
| Kazakhstan | .. | .. | .. | .. | .. | .. | .. | -0.330 | 1.443 | 1.670 | 2.196 | 1.086 | 1.906 | 4.087 |
| Non-CIS | .. | .. | .. | .. | .. | 0.929 | 1.007 | -0.027 | 1.212 | 1.437 | 1.547 | 0.976 | 2.011 | 4.455 |
| Kyrgyzstan | .. | .. | .. | .. | .. | -0.104 | -0.052 | 0.023 | -0.113 | -0.333 | -0.105 | -0.328 | -0.146 | -0.050 |
| Non-CIS | .. | .. | .. | .. | .. | 0.005 | – | 0.010 | -0.028 | -0.239 | 0.012 | -0.118 | -0.070 | 0.041 |
| Republic of Moldova | .. | .. | .. | .. | .. | -0.170 | -0.145 | -0.094 | -0.095 | -0.277 | -0.297 | -0.392 | -0.123 | -0.306 |
| Non-CIS | .. | .. | .. | .. | .. | -0.013 | -0.006 | -0.024 | 0.007 | -0.168 | -0.300 | -0.381 | -0.134 | -0.321 |
| Russian Federation [a] | .. | .. | .. | .. | .. | .. | .. | 28.201 | 33.160 | 39.516 | 33.059 | 28.400 | 42.700 | 69.027* |
| Non-CIS | .. | .. | .. | .. | .. | 5.392 | 17.490 | 24.657 | 32.490 | 38.177 | 30.594 | 26.300 | 40.600 | 66.897* |
| Tajikistan | .. | .. | .. | .. | .. | -0.061 | -0.280 | -0.055 | -0.061 | 0.102 | -0.004 | -0.114 | 0.026 | 0.109 |
| Non-CIS | .. | .. | .. | .. | .. | -0.023 | -0.147 | 0.085 | 0.165 | 0.154 | 0.205 | 0.129 | 0.225 | 0.295 |
| Turkmenistan | .. | .. | .. | .. | .. | .. | .. | 0.677 | 0.517 | 0.670 | -0.432 | -0.414 | -0.310 | 0.720 |
| Non-CIS | .. | .. | .. | .. | .. | 0.878 | 0.548 | -0.288 | 0.332 | 0.160 | -0.231 | -0.088 | -0.300 | 0.100 |
| Ukraine | .. | .. | .. | .. | .. | 0.523 | -1.716 | -0.473 | -2.356 | -3.202 | -2.896 | -2.038 | -0.265 | 0.617 |
| Non-CIS | .. | .. | .. | .. | .. | 1.248 | 0.571 | 1.746 | 0.680 | 0.569 | 1.397 | 1.657 | 3.227 | 4.159 |
| Uzbekistan | .. | .. | .. | .. | .. | .. | .. | -0.054 | 0.073 | -0.501 | -0.159 | 0.093 | 0.200 | 0.380 |
| Non-CIS | .. | .. | .. | .. | .. | -0.060 | -0.237 | -0.236 | 0.082 | 0.126 | -0.358 | 0.169 | – | 0.290 |
| **Former Soviet Union** | 5.724 | 3.972 | -2.697 | -5.907 | 1.255 | .. | .. | .. | .. | .. | .. | .. | .. | .. |
| **Total** | -3.351 | 8.833 | -0.032 | -7.583 | -3.000 | 0.227 | 5.337 | 12.736 | 6.634 | -3.937 | -14.512 | -22.357 | 0.345 | 29.326* |

*Source:* UNECE secretariat, based on national statistical publications and direct communications from national statistical offices.

*Note:* See appendix table B.11.

[a] Russian Goskomstat data excluding trade by physical persons (shuttle trade), but including trade flows not crossing the Russian borders such as off-board fish sales and natural gas deliveries under debt repayment agreements with former CMEA countries.

## APPENDIX TABLE B.14

**Merchandise trade of eastern Europe and the Russian Federation, by direction, 1980, 1988-2000**

*(Shares in total trade, per cent)*

|  | 1980 | 1988 | 1989 | 1990 | 1991 | 1992 | 1993 | 1994 | 1995 | 1996 | 1997 | 1998 | 1999 | 2000 |
|---|---|---|---|---|---|---|---|---|---|---|---|---|---|---|
| **Eastern Europe**, *to and from:* | | | | | | | | | | | | | | |
| *Exports* | | | | | | | | | | | | | | |
| World | 100.0 | 100.0 | 100.0 | 100.0 | 100.0 | 100.0 | 100.0 | 100.0 | 100.0 | 100.0 | 100.0 | 100.0 | 100.0 | 100.0 |
| ECE transition economies | 48.5 | 46.5 | 44.4 | 38.1 | 28.5 | 23.0 | 28.2 | 26.3 | 25.8 | 26.1 | 26.1 | 22.4 | 18.8 | 19.0 |
| Former Soviet Union | 27.1 | 27.2 | 25.5 | 22.3 | 17.9 | 12.4 | 9.8 | 9.0 | 8.9 | 9.4 | 10.3 | 7.4 | 4.9 | 4.9 |
| Eastern Europe | 21.4 | 19.3 | 18.9 | 15.8 | 10.7 | 10.7 | 18.5 | 17.4 | 16.9 | 16.7 | 15.8 | 15.0 | 14.0 | 14.1 |
| Developed market economies | 35.7 | 38.7 | 42.6 | 49.5 | 59.8 | 63.0 | 58.0 | 62.5 | 64.5 | 65.0 | 66.4 | 71.4 | 75.2 | 74.6 |
| Developing economies | 15.8 | 14.8 | 13.0 | 12.4 | 11.7 | 14.0 | 13.8 | 11.2 | 9.7 | 8.9 | 7.5 | 6.2 | 6.0 | 6.4 |
| *Imports* | | | | | | | | | | | | | | |
| World | 100.0 | 100.0 | 100.0 | 100.0 | 100.0 | 100.0 | 100.0 | 100.0 | 100.0 | 100.0 | 100.0 | 100.0 | 100.0 | 100.0 |
| ECE transition economies | 42.0 | 40.5 | 36.4 | 26.6 | 25.5 | 24.7 | 29.3 | 26.1 | 25.3 | 23.8 | 22.1 | 18.3 | 18.6 | 21.9 |
| Former Soviet Union | 26.8 | 26.3 | 23.5 | 18.3 | 20.2 | 17.9 | 16.5 | 14.1 | 13.2 | 12.7 | 11.2 | 8.6 | 8.5 | 11.8 |
| Eastern Europe | 18.8 | 19.6 | 18.7 | 14.3 | 8.1 | 6.8 | 12.8 | 12.0 | 12.1 | 11.1 | 10.9 | 9.6 | 10.1 | 10.2 |
| Developed market economies | 38.7 | 41.1 | 44.0 | 53.3 | 58.3 | 64.4 | 61.5 | 65.0 | 65.8 | 66.6 | 68.1 | 72.4 | 71.1 | 67.6 |
| Developing economies | 19.3 | 18.4 | 19.5 | 20.1 | 16.1 | 10.9 | 9.2 | 9.0 | 8.9 | 9.6 | 9.8 | 9.3 | 10.3 | 10.5 |
| **Former Soviet Union/Russian Federation**, *to and from:* | | | | | | | | | | | | | | |
| *Exports* | | | | | | | | | | | | | | |
| World | 100.0 | 100.0 | 100.0 | 100.0 | 100.0 | 100.0 | 100.0 | 100.0 | 100.0 | 100.0 | 100.0 | 100.0 | 100.0 | 100.0 |
| ECE transition economies | 34.5 | 29.4 | 26.6 | 21.8 | 25.9 | 22.3 | 18.1 | 15.1 | 16.8 | 18.2 | 19.5 | 18.1 | 17.8 | 20.0 |
| Eastern Europe | 34.5 | 29.4 | 26.6 | 21.8 | 25.9 | 20.7 | 16.8 | 11.7 | 13.2 | 14.3 | 14.9 | 14.2 | 13.3 | 14.5 |
| Developed market economies | 42.2 | 38.9 | 41.8 | 49.5 | 56.5 | 57.9 | 59.7 | 66.6 | 60.6 | 58.1 | 58.6 | 60.0 | 58.0 | 55.6 |
| Developing economies | 23.3 | 31.7 | 31.6 | 28.7 | 17.6 | 19.9 | 22.2 | 18.3 | 22.6 | 23.7 | 21.9 | 21.9 | 24.2 | 24.4 |
| *Imports* | | | | | | | | | | | | | | |
| World | 100.0 | 100.0 | 100.0 | 100.0 | 100.0 | 100.0 | 100.0 | 100.0 | 100.0 | 100.0 | 100.0 | 100.0 | 100.0 | 100.0 |
| ECE transition economies | 31.5 | 32.4 | 27.6 | 24.7 | 26.0 | 15.9 | 10.6 | 14.1 | 15.5 | 12.6 | 13.7 | 12.0 | 9.6 | 10.9 |
| Eastern Europe | 31.5 | 32.4 | 27.6 | 24.7 | 26.0 | 15.0 | 10.0 | 11.7 | 12.4 | 10.6 | 11.1 | 9.8 | 8.2 | 9.4 |
| Developed market economies | 46.4 | 46.3 | 50.1 | 52.9 | 58.1 | 62.4 | 60.6 | 70.3 | 69.5 | 67.8 | 68.3 | 68.2 | 68.3 | 69.3 |
| Developing economies | 22.1 | 21.3 | 22.3 | 22.4 | 15.9 | 21.7 | 28.8 | 15.6 | 15.0 | 19.6 | 18.0 | 19.8 | 22.1 | 19.8 |

*Source:* UNECE Common Database, derived from national statistics.

*Note:* Data for 1980-1990 refer to the east European CMEA countries (Bulgaria, Czechoslovakia, German Democratic Republic, Hungary, Poland and Romania) and to the former Soviet Union. Trade data in national currencies were revalued at consistent rouble/dollar cross-rates (see the note to appendix table B.11). As from 1991, eastern Europe covers Bulgaria, former Czechoslovakia (from 1993, Czech Republic and Slovakia including their mutual trade), Hungary, Poland and Romania, and the second panel reflects non-CIS trade of the Russian Federation only.

Partner-country grouping has been recently revised with subsequent revisions back to 1980. Thus, the earlier reported "Transition economies" group is now replaced by "ECE transition economies", which covers the Baltic states, CIS and the east European countries including the successor states of the former SFR of Yugoslavia. The "Eastern Europe" partner-group now covers Albania, Bulgaria, the Czech Republic, Hungary, Poland, Romania, Slovakia and the successor states of the former SFR of Yugoslavia, which earlier were in the "Other socialist countries" subgroup. The rest of subgroup "Other socialist countries", which in previous series covered China, Cuba, Democratic People's Republic of Korea, Mongolia and Viet Nam, is now included in the "Developing countries" group.

## APPENDIX TABLE B.15

**Exchange rates of eastern Europe, the Baltic states and the CIS, 1980, 1988-2000**
*(Annual averages, national currency units per dollar)*

| | Unit[a] | 1980 | 1988 | 1989 | 1990 | 1991 | 1992 | 1993 | 1994 | 1995 | 1996 | 1997 | 1998 | 1999 | 2000 |
|---|---|---|---|---|---|---|---|---|---|---|---|---|---|---|---|
| Albania | lek | .. | .. | .. | 8.90 | 24.20 | 75.03 | 102.06 | 94.62 | 93.14 | 104.33 | 148.93 | 150.63 | 137.69 | 143.71 |
| Bulgaria | lev[b] | 0.86 | 0.83 | 0.84 | 0.79 | 17.45 | 23.42 | 27.85 | 54.13 | 67.08 | 177.88 | 1 681.87 | 1760.37 | 1.8364 | 2.1233 |
| Czechoslovakia | koruna | 5.37 | 14.37 | 15.06 | 18.56 | 29.56 | 28.30 | .. | .. | .. | .. | .. | .. | .. | .. |
| Czech Republic | koruna | .. | .. | .. | .. | .. | .. | 29.15 | 28.79 | 26.54 | 27.14 | 31.70 | 32.29 | 34.57 | 38.60 |
| Slovakia | koruna | .. | .. | .. | .. | .. | .. | 30.80 | 31.93 | 29.71 | 30.68 | 33.62 | 35.23 | 41.36 | 46.20 |
| Hungary | forint | 32.64 | 50.41 | 59.07 | 63.21 | 74.73 | 78.98 | 91.91 | 105.11 | 125.69 | 152.65 | 186.79 | 214.40 | 237.15 | 282.18 |
| Poland | Zloty[c] | 3.05 | 430.64 | 1 439 | 9 500 | 10 576 | 13 627 | 18 136 | 22 723 | 2.42 | 2.70 | 3.28 | 3.49 | 3.96 | 4.35 |
| Romania | leu | 4.47 | 16.00 | 16.00 | 22.43 | 71.84 | 307.98 | 760.12 | 1 654 | 2 033 | 3 085 | 7 183 | 8 876 | 15 333 | 21 709 |
| Yugoslavia (SFR) | dinar[d] | 24.64 | 25 239 | 28 760 | 11.32 | 19.64 | .. | .. | .. | .. | .. | .. | .. | .. | .. |
| Bosnia and Herzegovina | dinar | .. | .. | .. | .. | .. | .. | .. | .. | .. | .. | .. | .. | .. | .. |
| Croatia | kuna[e] | .. | .. | .. | .. | 18.80 | 264.30 | 3 577.63 | 6.00 | 5.23 | 5.43 | 6.10 | 6.36 | 7.11 | 8.28 |
| Slovenia | tolar | .. | .. | .. | .. | 27.57 | 81.29 | 113.24 | 128.81 | 118.52 | 135.37 | 159.69 | 166.13 | 181.77 | 222.68 |
| The former Yugoslav Republic of Macedonia | denar[f] | .. | .. | .. | .. | 19.69 | 508.07 | 23.26 | 43.25 | 38.05 | 39.92 | 50.40 | 54.48 | 56.90 | 65.90 |
| Yugoslavia | dinar[g] | .. | .. | .. | 10.65 | 19.73 | 750.00 | .. | 1.55 | 4.74 | 4.96 | 5.72 | 9.23 | 10.94 | 16.06 |
| Estonia | kroon[h] | .. | .. | .. | .. | .. | 12.11 | 13.22 | 12.98 | 11.46 | 12.03 | 13.88 | 14.07 | 14.68 | 16.97 |
| Latvia | lats[i] | .. | .. | .. | .. | .. | .. | 0.67 | 0.56 | 0.53 | 0.55 | 0.58 | 0.59 | 0.59 | 0.61 |
| Lithuania | litas[j] | .. | .. | .. | .. | .. | .. | 4.37 | 3.98 | 4.00 | 4.00 | 4.00 | 4.00 | 4.00 | 4.00 |
| Armenia | dram | .. | .. | .. | .. | .. | .. | 8.66 | 288.35 | 405.93 | 413.47 | 490.70 | 504.92 | 535.06 | 539.53 |
| Azerbaijan | manat | .. | .. | .. | .. | .. | .. | .. | 1 169 | 4 417 | 4 295 | 3 987 | 3 869 | 4 119 | 4 474 |
| Belarus | rouble[k] | .. | .. | .. | .. | .. | .. | 2 177 | 4 017 | 11 538 | 13 472 | 26 729 | 58 971 | 274 512 | 881 750 |
| Georgia | lari[l] | .. | .. | .. | .. | .. | .. | .. | 1.10 | 1.29 | 1.26 | 1.30 | 1.39 | 2.02 | 1.98 |
| Kazakhstan | tenge | .. | .. | .. | .. | .. | .. | .. | 35.54 | 60.95 | 67.30 | 75.43 | 78.35 | 119.47 | 142.13 |
| Kyrgyzstan | som | .. | .. | .. | .. | .. | .. | .. | 10.86 | 10.83 | 12.81 | 17.37 | 21.37 | 39.73 | 47.79 |
| Republic of Moldova | leu | .. | .. | .. | .. | .. | .. | .. | 4.07 | 4.50 | 4.60 | 4.62 | 5.37 | 10.52 | 12.43 |
| Russian Federation | rouble[m] | 0.65 | 0.61 | 0.63 | 0.59 | 1.74 | 192.75 | 927.46 | 2204 | 4559 | 5121 | 5785 | 9.71 | 24.62 | 28.13 |
| Tajikistan | samoni[n] | .. | .. | .. | .. | .. | .. | .. | .. | 107.59 | 292.89 | 560.64 | .. | 1 235.57 | 1.83 |
| Turkmenistan | manat | .. | .. | .. | .. | .. | .. | .. | 19.50 | 110.42 | 3 509 | 4 143 | .. | 5 200 | 5 200 |
| Ukraine | hryvnia[o] | .. | .. | .. | .. | .. | .. | 4 796 | 31 700 | 147 314 | 1.83 | 1.86 | 2.51 | 4.13 | 5.44 |
| Uzbekistan | sum[p] | .. | .. | .. | .. | .. | .. | 932.15 | 9.96 | 29.81 | 40.15 | 66.43 | .. | 124.64 | 236.58 |
| *Memorandum item:* | | | | | | | | | | | | | | | |
| Former GDR | mark[q] | 3.30 | 8.14 | 8.14 | 8.14 | 1.66 | 1.56 | 1.65 | 1.62 | 1.43 | 1.50 | 1.73 | 1.76 | .. | .. |

*Source:* UNECE Common Database, derived from national, IMF and CIS statistics. Annual averages are unweighted arithmetic averages of monthly values. Change or redenomination of currency is indicated by a vertical bar.

*Note:* Under the central planning system with its state foreign trade monopoly, exchange rates served primarily statistical and accounting purposes (notably the conversion of foreign trade values for statistics expressed in domestic currency), without direct impact on domestic price formation. Market-based exchange rates and a meaningful link to domestic currency values emerged only with the transformations from 1989 onward. The official exchange rates of the earlier period are therefore not suitable for the conversion to dollars of macroeconomic and other data of these countries expressed in domestic currency. These strictures should be kept in mind in the interpretation and use of the data for the 1980s shown above.

[a] Currency unit of the last period shown. For prior periods, see footnotes.
[b] The leva was redenominated at 1:1,000 from 5 July 1999.
[c] The zloty was redenominated at 1:10,000 from 1 January 1995.
[d] The dinar was redenominated at 1:10,000 from 1 January 1990.
[e] The kuna replaced the Croat dinar on 3 May 1994 at 1:1,000; the 1994 average is shown in kuna terms.
[f] The denar (which had replaced the Yugoslav dinar 1:1 on 26 April 1992) was redenominated 1:100 on 1 May 1993; the 1993 average is shown in terms of that unit.
[g] The dinar was further redenominated on 1 July 1992 (1:10), 1 October 1993 (1:1 million), 1 January 1994 (1:1 trillion) and 24 January 1994 (1:13 million). Average annual exchange rates not available for 1993-1994.
[h] The kroon replaced the Soviet rouble in June 1992 with a peg to the deutsche mark (8:1); the average shown for 1992 refers to June-December.
[i] The lats replaced an earlier Latvian rouble at 1:200 on 18 October 1993; the 1993 average is shown in lat terms.
[j] The litas replaced the earlier talonas at 1:100 on 1 June 1993; the 1993 average is shown in litas terms.
[k] The Belarus rouble was redenominated 1:10 on 10 August 1994; the 1994 average here assumes this applied to the entire year. The Belarus rouble was further redenominated at 1:1,000 since January 2000. Annual averages were calculated from end-of-period monthly rates.
[l] The lari replaced the lari-kupon on 25 September 1995; the annual average for 1994 is shown in million lari-kupon, and that for 1995 in lari.
[m] 1980-1991: Soviet rouble/dollar rate used in the conversion of foreign trade data for statistical purposes. The rouble was redenominated at 1:1,000 from 1 January 1998.
[n] A new currency, the samoni, was put into circulation on 30 October 2000. Made up of 100 dirams, the samoni was to be used in parallel to the Tajik rouble for 5 months, with the exchange rate of 1,000 roubles to the samoni.
[o] The hryvnia replaced the former karbovanets on 2 September 1996 at 1:100,000; the average for 1996 is shown in hryvnia terms.
[p] Sum-kupon in 1993.
[q] German Democratic Republic mark through 1990, deutsche mark thereafter.

## APPENDIX TABLE B.16

**Current account balances of eastern Europe, the Baltic states and the CIS, 1990-2000**
*(Million dollars)*

|  | 1990 | 1991 | 1992 | 1993 | 1994 | 1995 | 1996 | 1997 | 1998 | 1999 | 2000 |
|---|---|---|---|---|---|---|---|---|---|---|---|
| **Eastern Europe**[a] | -5 726 | -2 146 | -967 | -7 896 | -2 589 | -1 897 | -13 258 | -14 263 | -17 538 | -20 726 | -17 776 |
| Albania | -118 | -168 | -51 | 14 | -43 | -15 | -62 | -254 | -45 | -133 | -163 |
| Bosnia and Herzegovina | .. | .. | .. | .. | -177 | -193 | -748 | -1 060 | -789 | -971 | -909 |
| Bulgaria | -1 710 | -77 | -361 | -1 098 | -32 | -198 | 164 | 1 046 | -61 | -652 | -702 |
| Croatia[b] | -621 | -589 | 329 | 623 | 854 | -1 442 | -1 091 | -2 325 | -1 531 | -1 390 | -399 |
| Czech Republic | -122 | 1 708 | -456 | 456 | -787 | -1 369 | -4 292 | -3 211 | -1 336 | -1 567 | -2 273 |
| Hungary[c] | 123 | 267 | 325 | -3 455 | -3 911 | -2 480 | -1 678 | -981 | -2 298 | -2 081 | -1 496 |
| Poland[c] | 716 | -1 359 | -269 | -2 868 | 677 | 5 310 | -1 371 | -4 309 | -6 862 | -11 558 | -9 946 |
| Romania | -3 337 | -1 012 | -1 564 | -1 174 | -428 | -1 774 | -2 571 | -2 137 | -2 968 | -1 469 | -1 359 |
| Slovakia | -767 | -786 | 173 | -601 | 665 | 391 | -2 098 | -1 827 | -1 982 | -980 | -713 |
| Slovenia[b] | 518 | 129 | 926 | 192 | 573 | -99 | 31 | 11 | -147 | -783 | -612 |
| The former Yugoslav Republic of Macedonia[b] | -409 | -259 | -19 | 15 | -158 | -222 | -289 | -276 | -308 | -113 | -113 |
| Yugoslavia | .. | .. | .. | .. | -400 | -1 037 | -600 | -1 561 | -660 | -764 | -339 |
| **Baltic states** | .. | .. | 548 | 353 | -59 | -788 | -1 400 | -1 890 | -2 426 | -2 095 | -1 483 |
| Estonia | .. | .. | 36 | 22 | -167 | -158 | -398 | -563 | -478 | -247 | -315 |
| Latvia | .. | .. | 191 | 417 | 201 | -16 | -279 | -345 | -650 | -654 | -493 |
| Lithuania | .. | .. | 321 | -86 | -94 | -614 | -723 | -981 | -1 298 | -1194 | -675 |
| **CIS** | .. | .. | -538 | 10 267 | 5 286 | 4 409 | 6 139 | -4 092 | -6 737 | 23 680 | 47 638 |
| Armenia | .. | .. | -50 | -67 | -104 | -218 | -291 | -307 | -403 | -307 | -278 |
| Azerbaijan | .. | 153 | 488 | -160 | -121 | -401 | -931 | -916 | -1 365 | -600 | -168 |
| Belarus | .. | .. | 131 | -435 | -444 | -458 | -516 | -788 | -866 | -194 | -220 |
| Georgia | .. | .. | -248 | -354 | -277 | -216 | -275 | -375 | -416 | -195 | -262 |
| Kazakhstan | .. | -1 300 | -1 900 | -641 | -905 | -213 | -751 | -799 | -1 236 | -236 | 743 |
| Kyrgyzstan | .. | .. | -61 | -88 | -84 | -235 | -425 | -138 | -364 | -180 | -77 |
| Republic of Moldova | .. | .. | -152 | -155 | -82 | -95 | -192 | -275 | -335 | -47 | -121 |
| Russian Federation[d] | -6 300 | 2 500 | 1 142 | 12 792 | 8 434 | 7 484 | 11 753 | 2 060 | 687 | 24 731 | 46 317 |
| Tajikistan | .. | .. | -53 | -208 | -170 | -89 | -70 | -56 | -108 | -36 | -62 |
| Turkmenistan | -308 | 447 | 926 | 776 | 84 | 23 | – | -580 | -934 | -751* | 100* |
| Ukraine | .. | .. | -526 | -765 | -1 163 | -1 152 | -1 184 | -1 335 | -1 296 | 1 658 | 1 481 |
| Uzbekistan | .. | .. | -236 | -429 | 118 | -21 | -980 | -584 | -102 | -164 | 184 |
| **Total above**[a] | .. | .. | -957 | 2 723 | 2 637 | 1 724 | -8 519 | -20 245 | -26 701 | 859 | 28 379 |
| *Memorandum items:* | | | | | | | | | | | |
| **CETE-5** | 469 | -41 | 698 | -6 276 | -2 783 | 1 753 | -9 408 | -10 317 | -12 624 | -16 969 | -15 040 |
| **SETE-7**[a] | -6 195 | -2 105 | -1 665 | -1 620 | 193 | -3 650 | -3 850 | -3 946 | -4 913 | -3 757 | -2 736 |
| Asian CIS | .. | .. | -1 133 | -1 170 | -1 459 | -1 370 | -3 723 | -3 755 | -4 928 | -2 468 | 181 |
| Three European CIS[e] | .. | .. | -547 | -1 355 | -1 689 | -1 705 | -1 892 | -2 397 | -2 496 | 1 417 | 1 140 |

**Source:** National balance of payments statistics; IMF, *Balance of Payments Statistics* (Washington, D.C.), various issues and *Staff Country Reports* (www.imf.org); ECE secretariat estimates.

[a] Totals excluding Bosnia and Herzegovina and Yugoslavia.

[b] Excludes transactions with the republics of the former SFR of Yugoslavia: Croatia (1990-1992), Slovenia (1990-1991) and The former Yugoslav Republic of Macedonia (1990-1992).

[c] Convertible currencies. Hungary until 1995; Poland until 1992.

[d] 1990-1992 excluding transactions with the Baltic and CIS countries.

[e] Belarus, Republic of Moldova and Ukraine.

## APPENDIX TABLE B.17

**Inflows of foreign direct investment [a] in eastern Europe, the Baltic states and the CIS, 1990-2000**

*(Million dollars)*

|  | 1990 | 1991 | 1992 | 1993 | 1994 | 1995 | 1996 | 1997 | 1998 | 1999 | 2000 |
|---|---|---|---|---|---|---|---|---|---|---|---|
| **Eastern Europe** [b] | 480 | 2 332 | 3 125 | 4 165 | 3 575 | 9 230 | 7 979 | 9 462 | 15 478 | 18 748 | 20 104 |
| Albania [c] | – | – | 20 | 58 | 53 | 70 | 90 | 48 | 45 | 41 | 143 |
| Bosnia and Herzegovina | .. | .. | .. | .. | – | – | – | – | 100 | 90 | 150 |
| Bulgaria [c] | 4 | 56 | 42 | 40 | 105 | 90 | 109 | 505 | 537 | 819 | 1 002 |
| Croatia | – | – | 16 | 120 | 117 | 114 | 511 | 533 | 932 | 1 479 | 926 |
| Czech Republic | 132 | 513 | 1 004 | 654 | 869 | 2 562 | 1 428 | 1 300 | 3 718[d] | 6 324[d] | 4 595 |
| Hungary | 311 | 1 459 | 1 471 | 2 339 | 1 146 | 4 454 | 2 275 | 2 173 | 2 036 | 1 970 | 1 700 |
| Poland (cash basis) [c] | 10 | 117 | 284 | 580 | 542 | 1 132 | 2 768 | 3 077 | 5 129 | 6 471 | 8 294 |
| Romania | – | 40 | 77 | 94 | 341 | 419 | 263 | 1 215 | 2 031 | 1 041 | 1 025 |
| Slovakia | 18 | 82 | 100 | 168 | 250 | 202 | 330 | 220 | 684 | 390 | 2 075 |
| Slovenia | 4 | 65 | 111 | 113 | 128 | 177 | 194 | 375 | 248 | 181 | 176 |
| The former Yugoslav Republic of Macedonia [c] | – | – | – | – | 24 | 9 | 11 | 16 | 118 | 32 | 170 |
| Yugoslavia | .. | .. | .. | .. | .. | .. | – | 740 | 113 | 112 | 25 |
| **Baltic states** | .. | .. | 119 | 238 | 460 | 454 | 685 | 1 142 | 1 863 | 1 139 | 1 173 |
| Estonia | .. | .. | 82 | 162 | 215 | 202 | 151 | 267 | 581 | 305 | 387 |
| Latvia | .. | .. | 29 | 45 | 214 | 180 | 382 | 521 | 357 | 347 | 408 |
| Lithuania | .. | .. | 8 | 30 | 31 | 73 | 152 | 355 | 926 | 486 | 379 |
| **CIS** | .. | .. | 1 777 | 1 875 | 1 770 | 4 065 | 5 288 | 8 846 | 6 726 | 6 737 | 5 329 |
| Armenia [c] | .. | .. | – | 1 | 8 | 25 | 18 | 52 | 221 | 122 | 104 |
| Azerbaijan [c] | .. | .. | – | 60 | 22 | 330 | 627 | 1 115 | 1 023 | 510 | 129 |
| Belarus | .. | .. | 7 | 18 | 11 | 15 | 73 | 200 | 149 | 444 | 88 |
| Georgia | .. | .. | – | – | 8 | 6 | 40 | 203 | 265 | 82 | 131 |
| Kazakhstan [e] | .. | .. | 100 | 228 | 635 | 964 | 1 137 | 1 321 | 1 151 | 1 468 | 1 246 |
| Kyrgyzstan | .. | .. | – | 10 | 38 | 96 | 47 | 83 | 109 | 44 | -2 |
| Republic of Moldova | .. | 25 | 17 | 14 | 12 | 67 | 24 | 79 | 74 | 39 | 128 |
| Russian Federation | – | 100 | 1 454 | 1 211 | 690 | 2 066 | 2 579 | 4 865 | 2 762 | 3 309 | 2 714 |
| Tajikistan [c] | .. | .. | 9 | 9 | 12 | 20 | 25 | 30 | 24 | 21 | 22 |
| Turkmenistan [c] | – | – | 11 | 79 | 103 | 233 | 108 | 108 | 64 | 80 | 100* |
| Ukraine | .. | .. | 170 | 198 | 159 | 267 | 521 | 623 | 743 | 496 | 595 |
| Uzbekistan [c] | .. | .. | 9 | 48 | 73 | -24 | 90 | 167 | 140 | 121 | 75 |
| **Total above** [b] | .. | .. | 5 021 | 6 278 | 5 806 | 13 749 | 13 952 | 19 450 | 24 067 | 26 624 | 26 607 |
| *Memorandum items:* |  |  |  |  |  |  |  |  |  |  |  |
| CETE-5 | 476 | 2 236 | 2 970 | 3 853 | 2 935 | 8 527 | 6 995 | 7 146 | 11 815 | 15 336 | 16 839 |
| SETE-7 [b] | 4 | 96 | 155 | 312 | 640 | 703 | 984 | 2 316 | 3 663 | 3 412 | 3 265 |
| Asian CIS | .. | .. | 129 | 435 | 899 | 1 651 | 2 092 | 3 079 | 2 998 | 2 449 | 1 804 |
| Three European CIS [f] | .. | .. | 194 | 229 | 181 | 349 | 617 | 902 | 966 | 979 | 811 |
| Poland (accrual basis) | 89 | 291 | 678 | 1 715 | 1 875 | 3 659 | 4 498 | 4 908 | 6 365 | 7 270 | .. |

*Source:* National balance of payments statistics; IMF, *Balance of Payments Statistics* (Washington, D.C.), various issues and *Staff Country Reports* (www.imf.org); ECE secretariat estimates.

*Note:* Changes in coverage are available in UNECE, *Economic Survey of Europe, 2001* No. 1, chap. 5, box 5.3.1.

[a] Inflows into the reporting country.

[b] Excluding Bosnia and Herzegovina and Yugoslavia.

[c] Net of residents' investments abroad. Bulgaria, 1990-1994; Poland, 1990-1992.

[d] The Czech data for 1998-1999 have recently been revised to incorporate inter-company loans, increasing FDI inflows from $2,720 million and $5,108 million, respectively. Both figures also reflect reinvested profits which is not the case for the preceding years. Also see the note.

[e] Drawings less repayments.

[f] Belarus, Republic of Moldova and Ukraine.

# Corrections to Economic Survey of Europe, 2001 No. 1

*Page 238, table 6.5.5 should be replaced by the following:*

### TABLE 6.5.5

**Estimation results on service-to-consumer-goods price ratio**

| Variables | Base version | Exchange rate regime effect — With inflation acceleration | Exchange rate regime effect — Without inflation acceleration |
|---|---|---|---|
| Constant | 2.060734*** | 1.108583*** | 2.451394*** |
| Service-to-non-food price ratio lagged | 0.444020*** | 0.446326*** | 0.329490*** |
| Productivity in industry | 0.242327*** | 0.174235*** | 0.247501*** |
| Exchange rate effect [a] | | 0.007960*** | 0.005915*** |
| Productivity in services | -0.184074* | 0.128094* | -0.106654** |
| GDP/capita (PPP) | 0.027596** | 0.006321 | |
| Inflation acceleration | | | |
| Country effect | | | |
| Czech Republic | -0.001539** | -0.002185*** | |
| Hungary | 0.001177** | 0.002089*** | |
| Poland | -0.003233** | -0.003756** | |
| Romania | 0.000553 | 0.000522* | |
| Slovenia | 0.003063*** | 0.003482*** | |
| Estonia | 0.001503** | 0.001395** | |
| Latvia | -0.004271*** | -0.004174*** | |
| Lithuania | -0.000796** | -0.000586*** | |
| Russian Federation | -0.006278** | -0.006452** | |
| Sample | | 1991-1998 | |
| Included observations | | 8 | |
| Number of cross-sections used | | 9 | |
| Total panel (unbalanced) observations | | 56 | |
| Adjusted R-squared | 0.954151 | 0.954108 | 0.442372 |
| Standard error of regression | 0.065048 | 0.065078 | 0.226850 |
| Means of dependent variable | | 4.567562 | |
| Standard deviation of dependent variable | | 0.303785 | |
| Estimation method | | GLS[b] | |
| Czech Republic | | 1994-1998 | |
| Hungary | | 1992-1998 | |
| Poland | | 1992-1998 | |
| Romania | | 1991-1998 | |
| Slovenia | | 1993-1998 | |
| Estonia | | 1993-1998 | |
| Latvia | | 1992-1998 | |
| Lithuania | | 1993-1998 | |
| Russian Federation | | 1995-1998 | |

**Source and note:** As for table 6.1.1 and table 6.5.1, except that inflation acceleration is not in logs. (The consumer goods prices refer to the consumer price index less food and less services.)

[a] Exchange rate regime without any formal commitment (managed or free float).

[b] Cross-section weights.

## OTHER RECENT PUBLICATIONS OF ECONOMIC ANALYSIS FROM THE UNITED NATIONS ECONOMIC COMMISSION FOR EUROPE

- *Economic Survey of Europe, 2000 No. 1*, Sales No. E.00.II.E.14 (April)

    This issue contains the secretariat's annual review of developments in 2000 and the outlook for 2001. Special chapters focus on domestic savings in the transition economies, the relationship between foreign direct investment and economic growth and the impact of economic transformation in the transition economies on their exchange rate.

- *Economic Survey of Europe, 2000 No. 2/3*, Sales No. E.00.II.E.28 (December)

    The papers presented at the Spring Seminar of May 2000 on "The Transition Process After Ten Years", are gathered in this issue, together with the comments of discussants and a summary of the general discussion. The papers cover long-run developments (Ivan Berend), macroeconomic policies and achievements (Stanislaw Gomulka), changes in production structures (Michael Landesmann) and the social costs and consequences (Michael Ellman).

- *Economic Survey of Europe, 2000 No. 1*, Sales No. E.00.II.E.12 (April)

    This issue reviews developments in 1999 and discusses the outlook for 2000. The "new economy", south-east Europe and capital inflows into the transition economies since 1989 are among the topics included in the current analysis, while there are special chapters on "Economic Convergence in Europe" and "The Fertility Decline in the Transition Economies, 1989-1998.

- *Economic Survey of Europe, 1999 No. 3*, Sales No. E.99.II.E.4 (November)

    This issue contains the proceedings of the 1999 Spring Seminar on the Reform of Pension Systems in the ECE Region. Included are papers by John Eatwell, Lawrence Thompson and Maria Augusztinovics, together with discussants' comments, and an introduction and summary of the day's discussion by the secretariat.

- *Economic Survey of Europe, 1999 No. 2*, Sales No. E.99.II.E.3 (July)

    In addition to an updated summary of economic developments in Europe, the CIS and North America to mid-1999, this issue contains the secretariat's analysis of "Postwar Reconstruction and Development in South-East Europe" which has received considerable attention in the course of the year.

- *Economic Survey of Europe, 1999 No. 1*, Sales No. E.99.II.E.2 (April)

    This issue provides a detailed survey of economic developments in Europe, the CIS and North America in 1998 and early 1999. There are special sections dealing with the start of EMU, monetary policy in the transition economies, a review of economic adjustment problems in Romania and a chapter devoted to the decline of fertility in the transition economies between 1982 and 1997.

\* \* \* \* \*

More details about other publications and activities of the United Nations Economic Commission for Europe, which pay special attention to issues concerning the transition economies, can be found at the secretariat's website: http://www.unece.org

\* \* \* \* \*

To obtain copies of publications contact:

| | |
|---|---|
| Publications des Nations Unies | United Nations Publications |
| Section de Vente et Marketing | 2 United Nations Plaza |
| Organisation des Nations Unies | Room DC2-853 |
| CH-1211 Genève 10 | New York, NY 10017 |
| Suisse | USA |
| Tel: (4122) 917 2612 / 917 2606 / 917 2613 | Tel: (1212) 963 8302 / (1800) 253 9646 |
| Fax: (4122) 917 0027 | Fax: (1212) 963 3489 |
| E-mail: unpubli@unog.ch | E-mail: publications@un.org |